INVERTEBRATE PALAEONTOLOGY AND EVOLUTION

To the memory of
P. C. Sylvester-Bradley
1913–78

INVERTEBRATE PALAEONTOLOGY AND EVOLUTION

E. N. K. CLARKSON

Senior Lecturer in Geology,
Edinburgh University

London
GEORGE ALLEN & UNWIN
Boston   Sydney

First published in 1979
Reprinted (twice) 1980

GEORGE ALLEN & UNWIN LTD
40 Museum Street, London WC1A 1LU

© E. N. K. Clarkson, 1979

British Library Cataloguing in Publication Data
Clarkson, E. N. K.
Invertebrate palaeontology and evolution
1. Invertebrates, Fossil
I. Title
562   QE770   78-40638

ISBN 0-04-560007-4
ISBN 0-04-560008-2 Pbk

Cover photograph
An assemblage of Upper Ordovician trilobites *(Cyamops stensioei)*
from Kallhöln, Dalarna, Sweden (×5).

Typeset in 10 on 12 Baskerville by A. Brown and Sons, Hull, North Humberside
and printed in Great Britain
by Biddles Ltd., Guildford, Surrey

Contents

Preface

This book is primarily intended as a basic text for undergraduate students reading Geology or Earth Science. Because of its geological applications invertebrate palaeontology is normally studied as part of a geological curriculum, but since it is concerned with the remains of once-living organisms (including the tracks and trails that they made) it is also in part a zoological subject. Since the biological grounding of students reading Geology is very variable and may be limited, I have tried in the first part of the text to give an adequate basis in zoological and taxonomic concepts and in evolution theory, in order that the principles governing the evolutionary diversification of the various invertebrate groups can be understood. In Part One stratigraphical principles and other more geological aspects of invertebrate palaeontology are also discussed.

In Part Two the main invertebrate phyla are covered in turn with particular reference to their morphology, taxonomy, palaeozoology and stratigraphic use. Here I have had to be very selective, since the amount of available information is immense.

Within the past thirty years the development of invertebrate palaeontology in terms of both information and concepts has been almost explosive. The traditional topics of palaeontography (the description of fossils), taxonomy (their classification) and their inseparable companion biostratigraphy (the use of fossils in producing a relative time scale for geology) have been greatly expanded and refined. The form and structure, including microstructure, of well-preserved fossils is now known in great detail, particularly due to the scanning electron microscope which first came into use in 1965. There has also been increasing attention paid to how the invertebrates of the past actually functioned, behaved and related to their physical environment and to each other. Palaeoecology and functional morphology have therefore become integral to the development of palaeontology, as has the global distribution of former organisms.

The selection of particular topics must inevitably reflect the writer's own interests and knowledge, but it has only been possible to mention a very few of them. There are many vital topics (e.g. chemical evolution and microfossils) that I have had to leave out altogether for reasons of space, but perhaps the very limitations of this text will be enough to encourage some of its readers to go to original works for further information. For a text of this kind no bibliography can possibly hope to be comprehensive. I have therefore confined the bibliography to an annotated reference list for each chapter, noting only major reference works in the English language (e.g. the '*Treatise*'), and mentioning some original research papers which either are classic or will at least provide a starting point for the student who wishes to probe more deeply into literature of the subject.

During the writing of this book I have received great help from many friends and colleagues, who have given much time and energy in bringing it to completion. I received initial encouragement from Professor A. Williams, Professor P. C. Sylvester-Bradley, Professor G. Y. Craig and Dr C. T. Scrutton.

Individual chapters were read and amendments suggested by Dr C. T. Scrutton (Chs 5, 8, 10), Dr C. R. C. Paul (Ch. 9), Dr J. Miller (Chs 1–3, 11), Dr M. G. Bassett (Ch. 7), Professor H. B. Whittington (Chs 11, 12), Dr G. P. Larwood (Ch. 6), Dr R. Goldring (Ch. 4) and Dr N. Eldredge (Ch. 2). Dr N. A. Locket also read Chapter 11 and gave valuable advice on presentation. Dr R. Cowen gave detailed criticisms of early drafts of Chapters 8 and 9 and read the whole book when it was finished. Likewise, Professor P. C. Sylvester-Bradley gave great help with the first drafts of Chapters 8 and 10 and read the final typescript. Dr S. M. Manton and Dr J. Nudds permitted me to quote unpublished work. I should point out, however, that any inaccuracies remaining are entirely my responsibility.

Many of the photographs which illustrate this text have been donated by Professor H. B. Whittington, Dr C. H. C. Brunton, Dr P. Crowther, Dr K. Towe, Professor H. Stuermer, Dr A. B. Smith, Dr P. M. Kier, Dr G. P. Larwood, Mr J. Jameson, Dr A. H. F. Robertson and Dr D. B. Macurda. Others were taken specially by Mr C. Chaplin, to whom I am most grateful, from specimens borrowed from the Royal Scottish Museum through the courtesy of Dr C. D. Waterston and through the kind offices of Mr W. Baird, who combed through the collections for especially photogenic fossils.

All the drawings were done from my pencil originals, or from the illustrations of other authors, by Mrs Sharon Chambers, whose skill and accuracy have contributed much to the final presentation of this book.

The writing of a text such as this makes considerable inroads into one's spare time and calls for singular forbearance amongst members of one's family. My wife Cynthia has not only been immeasurably tolerant during these last few years (as have my sons John, Peter, Thomas and Matthew) but also given constant advice on various matters in the text, especially on the arthropod and graptolite sections, and undertaken much editing.

A number of secretaries have assisted with the work at various stages, and I am particularly grateful to Mrs Janette Brunton, who has coped uncomplainingly with several drafts and whose accurate work, patience and attention to detail have greatly aided in the production of the final typescript. I am also very grateful to Mrs Betty MacCormack for her meticulous editorial work.

Finally, I should like to thank Roger Jones of George Allen & Unwin, who has guided the development of this book since its inception and to whose continued assistance and encouragement I owe very much.

E. N. K. Clarkson
Edinburgh, 1979

Part One
General Palaeontological Concepts

Please note that the letters 'i' and 'o' have been omitted from the labelling of parts of illustrations in order to avoid confusion with numerals

Grammoceras thouarsense
Ammonites from the Upper Lias of Whitby, North Yorkshire, England (× 1.5). (Specimen in Royal Scottish Museum, Edinburgh)

1 Principles of palaeontology

INTRODUCTION

Some 4600 million years ago the Earth came into being, probably forming from a condensing cloud of cosmic dust and gas. It is likely that all the planets in the solar system originated in the same way and that the gas clouds from which they formed began as concentric rings shed in succession by centrifugal force from a rapidly rotating sun.

Of all the nine planets in the solar system only Earth, as far as is known, supports life, but it is a striking fact that life on Earth began very early indeed, within the first 30 per cent of the planet's history. There are remains of simple organisms (bacteria and blue-green algae) in rocks 3300 million years old, so life presumably originated before then. These simple forms of life seem to have dominated the scene for the next 2000 million years, and evolution at that time was very slow. But the blue-green algae and photosynthetic bacteria were instrumental in changing the environment, for they gave off oxygen into an atmosphere that was previously devoid of it, so that animal life eventually became possible.

Only when some of the early living beings of this Earth had reached a high level of physiological and reproductive organisation (and most particularly when sexual reproduction originated) was the rate of evolutionary change accelerated, and with it all manner of new possibilities were opened up to evolving life. But this was not until comparatively late in geological history, and there are no fossil animals known from sediments older than about 700 million years. Needless to say, these are all invertebrate animals lacking backbones. All of them are marine; there is no record of terrestrial animals until much later. In terms of our understanding of the history of life, perhaps the most significant of all events took place about 570–600 million years ago at about the beginning of the Cambrian period, for at this stage there was a sudden proliferation of different kinds of marine invertebrates. During this critical period the principal invertebrate groups were established, and they then diversified and expanded. Some of these organisms acquired hard shells and were capable of being fossilised, and only because of this can there be any chance of understanding the history of invertebrate life.

The stratified sedimentary rocks laid down since the early Cambrian, and built up throughout the whole of Phanerozoic time, are distinguished by a rich heritage of the fossil remains of the invertebrates that evolved through successive historical periods; their study is the domain of invertebrate palaeontology and the subject of this book.

OCCURRENCE OF INVERTEBRATE FOSSILS IN PHANEROZOIC ROCKS

Fossil invertebrates occur in many kinds of sedimentary rock deposited in the seas during the Phanerozoic. They may be very abundant in limestones, shales, siltstones and mudstones but on the whole are not common in sandstones. Sedimentary ironstones may have rich fossil remains. Occasionally they are found in some coarse rocks such as greywackes and even conglomerates. The state of preservation of fossils varies greatly, depending on the structure and composition of the original shell, the nature and grain size of the enclosing sediment, the chemical conditions at the time of sedimentation, and the subsequent processes of diagenesis (chemical and physical changes) taking place in the rock after deposition.

In nearly all cases only those animals with hard shells have been preserved as fossils, and rapid burial was normally a prerequisite for fossilisation also. The soft-bodied elements in the fauna, and those forms with thin organic shells, did not normally survive diagenesis and hence have left little or no evidence of their existence other than records of their activity in the form of trace fossils. What we can see in the rocks is therefore only a narrow band in a whole spectrum of the organisms that were once living; only very rarely have there been found beds containing some or all of the soft-bodied elements in the fauna as well. These are immensely significant for palaeontology.

The oldest such fauna is of late Precambrian age, some 700 million years old, and is our only record of animal life before the Cambrian. Another such 'window' is known in Middle Cambrian rocks from British Columbia, where in addition to the normally expected trilobites and brachiopods there is a great range of soft-bodied and thin-shelled animals – sponges, worms, jellyfish, small shrimp-like creatures and animals of quite unknown affinities – which are the only trace of a diverse fauna which would otherwise be quite unknown (see Chapter 12). There are similar 'windows' at other levels in the geological column including a recently-discovered one in the southern Welsh Precambrian, likewise illuminating.

The fossil record is, as a guide to the evolution of ancient life, unquestionably limited, patchy and incomplete. Usually only the hard-shelled elements in the biota (apart from trace fossils) are preserved, and the fossil assemblages present in the rock may have been transported some distance, abraded, damaged and mixed with elements of other faunas. Even if thick-shelled animals were originally present in a fauna they may not be preserved; in sandy sediments in which the circulating waters are acidic, for instance, calcareous shells may dissolve within a few years before the sediment is compacted into rock. Since the sea floor is not always a region of continuous sediment deposition, many apparently continuous sedimentary sequences contain numerous small-scale breaks (diastems) representing periods of winnowing and erosion. Any shells on the sea floor during these erosion periods would probably be transported or destroyed – another limitation on the adequacy of the fossil record.

On the other hand, some marine invertebrates found in certain rock types have been preserved abundantly and in exquisite detail, so that it is possible to infer much about their biology from their remains. Many of the best-preserved fossils come from limestones or from silty sediments with a high calcareous content. In these (Fig. 1.1) the original calcareous shells may be retained in the fossil state with relatively little alteration, depending upon the chemical conditions within the sediment at the time of deposition and after. A sediment often consists of components derived from various environments, and when all of these, including decaying organisms, dead shells and sedimentary particles, are thrown together the chemical balance is unstable. The sediment will be in chemical equilibrium only after diagenetic physico-chemical alterations have taken place. These may involve recrystallisation and the growth of new minerals (authigenesis) as well as cementation and compaction of the rock (lithification), and during any one of these processes the fossils may be altered or destroyed. Shells that are originally of calcite preserve best; aragonite is a less stable form of calcium carbonate secreted by certain living organisms (e.g. corals) and is often recrystallised to calcite during diagenesis or dissolved away completely.

Calcareous skeletons preserved in more sandy or silty sediments may dissolve after the sediment has hardened or during weathering of the rock long after its induration. There may be left **moulds** (often miscalled casts) of the external and internal surfaces of the fossil, and if the sediment is fine enough the details these show may be very good. Some methods for the study of such moulds are described on p. 111 with reference to brachiopods. If a fossil encloses an originally hollow space, as for instance between the pair of shells of a bivalve or brachiopod, this space may either be left empty or become filled with sediment. In the latter case there is preserved a sediment **core,** which comes out intact when the rock is cracked open. This bears upon its surface an **internal mould** of the fossil shell, whilst **external moulds** are left in the cavity from which it came. In rare circumstances the core or the shell, or both, may be replaced by entirely different mineral, as happens in fossils preserved in ironstones. Cores may sometimes be of pyrite. Graptolites are often preserved like this, anaerobic decay having released hydrogen sulphide, which reacted with ferric (Fe^{2+}) ions in the water to allow an internal pyrite core to form. Sometimes a core of silica is found within an unaltered calcite shell. This has happened with some of the Cretaceous sea urchins of southern England. They lived in or on a sediment of calcareous mud along with many sponges, which secreted spicules of biogenic silica as a skeleton. In alkaline conditions (above pH 9), which may sometimes be generated

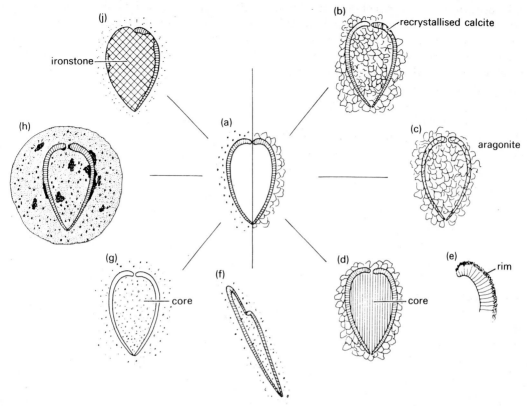

Figure 1.1 Possible processes of fossilisation of a bivalve shell: (a) original shell, buried in mud (left) or carbonate (right); (b) the shell was calcite, was buried in a carbonate sediment and was preserved intact other than as a small recrystallised patch; (c) shell originally of aragonite now recrystallised to calcite which destroys fine structure; (d) original calcite shell retained surrounding a diagenetic core of silica; (e) a silica rim growing on the outside of the shell; (f) tectonic distortion of a shell preserved in mudstone; (g) shell preserved in mud with original shell material leached away, leaving an external and internal mould, surrounding a mudstone core; (h) a calcareous concretion growing round the shell and inside (if the original cavity was empty), with patches of pyrite in places; (j) ironstone replacement of core and part of shell.

during bacterial decay, the solubility of the silica increases markedly, the silica so released will travel through the rock and precipitate wherever the pH is lower. The inside of a sea urchin decaying under different conditions would trap just such an internal microenvironment within which the silica could precipitate as a gel. Such siliceous cores retain excellent features preserving the internal morphology of the shell. On the other hand, silica may replace calcite as a very thin shell over the surface of a fossil as a result of some complex surface reactions. These siliceous crusts may retain a very detailed expression of the surface of the fossil, and, since they can be freed from the rock by dissolving the limestone with hydrochloric acid, individual small fossils preserved in this way can be studied in three dimensions. Some of the most exquisite of all trilobites and brachiopods are known from material such as this.

Fossils are often found in concretions: calcareous or siliceous masses formed around the fossil shortly after its death and burial. Concretions form under certain conditions only, where a delicate chemical balance exists in the water and sediment, by processes as yet not fully understood.

DIVISIONS OF INVERTEBRATE PALAEONTOLOGY

Invertebrate palaeontology is normally studied as a subdivision of geology, as it is within Earth science that its greatest applications lie. It can also be seen as

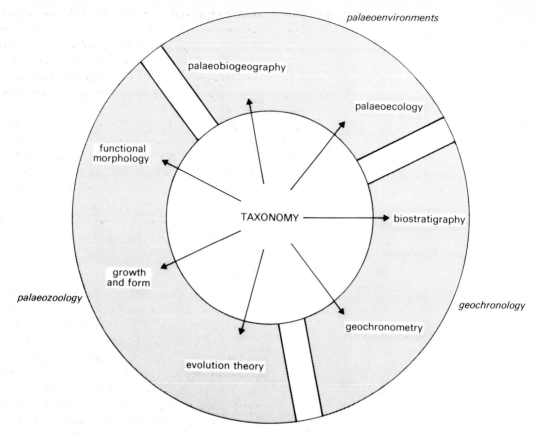

Figure 1.2 The various subdisciplines of palaeontology.

a biological subject, but one that has the unique perspective of geological time. Within the domain of invertebrate palaeontology there are a number of inter-related topics (Fig. 1.2), all of which have a bearing on the others and which also link up with other sciences.

At the heart of invertebrate palaeontology stands **taxonomy:** the classification of fossil and modern animals into ordered and natural groupings. These groupings, known as **taxa,** must be named and arranged in a hierarchical system in which their relationships are made clear, and as far as possible must be seen in evolutionary perspective.

Evolution theory is compounded of various disciplines – pure biology, comparative anatomy, embryology, genetics and population biology – but it is only the palaeontological aspect that allows the predictions of evolutionary science to be tested against the background of geological time, permits the tracing of evolving lineages, and illustrates some of

the patterns of evolution that actually have occurred.

The rates at which animals have evolved have varied through time, but most animal types (**species**) have had a geological life of only a few million years. Some of these evolved rapidly, such as the ammonites, others very slowly (Chapter 2). A rock succession of marine sediments built up over many millions of years may therefore have several fossil species occurring in a particular sequence, each species confined to one part of the succession only and representing the time when that species was living. Herein lies the oldest and most general application of invertebrate palaeontology: **biostratigraphy.** Using the sequence of fossil faunas, the geological column has been divided up into a series of major geological time units (periods), each of which is further divided into a hierarchy of small units. The whole basis for this historical chronology is the documented sequence of fossils in the rocks. But different kinds of fossils have different stratigraphical values, and certain parts of the geological record are more closely subdivided than others. The 'relative' time scale based on fossils is generally very good for those parts of the world where correlations have been accurately drawn. Some 'absolute' ages based on radiometric dating have been fixed at particular points to this relative scale, and these provide a framework for understanding the geological record in terms of real time (i.e. known periods of millions of years) rather than as just a purely relative scale. This is only possible at certain horizons, however, and for practical purposes the geological time scale based on biostratigraphy is unsurpassed for Phanerozoic sediments.

Whilst stratigraphy is the basis of the primary discipline of geochronology, a small facet of palaeontological study has a bearing on what may be termed 'geochronometry'. By counting daily growth rings in extinct corals and bivalves, information has been observed bearing upon the number of days in the lunar month and year in ancient times. This has helped to confirm geophysical estimates on the slowing of the Earth's rotation (Ch. 5).

Since stratigraphical applications of palaeontology have always been so important, the more biological aspects of palaeontology were relatively neglected until comparatively recently. **Palaeoecology**, which has developed particularly since the early 1950s, is concerned with the relationships of fossil animals to their environment, both as individuals (**autecology**) and in the faunal communities in which they occur naturally; the latter is sometimes known as **synecology**.

Since the soft parts of fossil animals are not normally preserved, but only their hard shells, there are relatively few ways in which their biology and life habits can be understood. Studies in **functional morphology**, however, which deal with the interpretation of the biology of fossilised skeletons or structures in terms of their original function, have been successfully attempted with many kinds of fossils, restricted in scope though these endeavours may necessarily be. **Ichnology** is the study of trace fossils: the tracks, surface trails, burrows and borings made by once-living animals and preserved in sediments. This topic has proved valuable both in understanding the behaviour of the animals that lived when the sediment was being deposited and in interpreting the contemporaneous environment. Finally, it is only by the integration of taxonomic data

on local faunas that the global distribution of marine invertebrates through time can be elucidated. Such studies of **palaeobiogeography** (or **palaeozoogeography** in the case of animals) can be used in conjunction with geophysical data in understanding the former relative positions and movements of continental masses.

All these aspects of palaeontology are inter-related, and an advance in one may have a bearing upon any other. Thus a particular study in functional morphology may give information on palaeoecology and possibly some feedback to taxonomy as well. Likewise, recent refinements in taxonomic practice have enabled the development of a much more precise stratigraphy.

Taxonomy

It has often been said that to identify a fossil correctly is the first step, and indeed the key, to finding out further information about it. This illustrates clearly that sound classification and nomenclature lie at the root of all biological and palaeontological work; without them no coherent and ordered system of data storage and retrieval is possible. Taxonomy, or **systematics** as biological classification is sometimes known, is the oldest of all the biological disciplines, and the principles outlined by Carl Gustav Linnaeus (1707–78) in his pioneer *Systema Naturae* are still in use today, though greatly modified and extended.

Species concept

The fundamental unit of taxonomy is the species. Animal species (e.g. Sylvester-Bradley 1956) are groups of individuals that generally look like each other and can interbreed together to produce offspring of the same kind. They cannot interbreed with other species. Since it is reproductive isolation alone that defines species it is only really possible to distinguish closely related species if their breeding habits are known. Of all the described 'species' of living animals, however, only about a sixth are 'good', or properly defined, species. Information upon the reproductive preferences of the other five-sixths of all naturally occurring animal populations is just not documented.

The differentiation of most living and all fossil species therefore has to be based upon other and technically less valid criteria.

Of these by far the most important, especially in palaeontology, is **morphology,** the science of form, for most natural species tend to be composed of individuals of similar enough external appearance to be identifiable as of the same kind. Distinguishing species of living animals by morphological criteria alone is not without hazards, especially where the species in question are similar and closely related. Supplementary information such as the analysis of species-specific proteins may be of help in some cases where there is good reason for doing so (e.g. for disease-carrying insects). For the rest some degree of subjectivity in taxonomy has to be accepted, though this can be minimised as far as possible if enough morphological criteria are used.

Nomenclature and identification of fossil species

In the formal nomenclature of any species, living or fossil, taxonomists follow the biological system of Linnaeus where each species is defined by two names: the **generic** and **specific** (or **trivial**) **names.** For example, all cats, large and small, are related, and one particular group has been placed in the **genus** *Felis*. Of the various species of *Felis* the specific names *F. domesticus*, *F. leo* and *F. tigris* formally refer to the domestic cat, lion and tiger respectively. In full taxonomic nomenclature the author's name and the date of publication are given after the species, e.g. *Felis domesticus* Linnaeus 1778 (see below for further discussion).

In palaeontology it can never be known for certain whether a population with a particular morphology was reproductively isolated or not. Hence the definition of species in palaeontology, as in most living specimens, must be based almost entirely on morphological criteria. Moreover, only the hard parts of the fossil animal are preserved, and much useful data has vanished. A careful examination and documentation of all the anatomical features of the fossil has to be the main guide in establishing that one species is different from a related species. In rare cases this can be supplemented by a comparison of the chemistry of the shell, as has proved especially useful in the erection of higher taxonomic categories. Within any interbreeding population there is usually quite a spread of morphological variation. On a broader scale there may be both geographical and stratigraphic variation, and all these must be care-fully documented if the species is to be ideally established. Such studies may be very significant in evolutionary palaeontology.

When a palaeontologist is attempting to distinguish the species in a newly discovered fauna, say of fossil brachiopods, he has to separate the individual fossils out into groups of morphologically similar individuals. There may, to take an example, perhaps be eight such groups, each distinguished by a particular set of characters. Some of these groups may be clearly distinct from one another; in others the distinction may be considerably less, increasing the risk of greater subjectivity. These groups are provisionally considered as species, which must then be identified. This is done by consulting palaeontological monographs or papers containing detailed technical descriptions and illustrations of previously described brachiopod faunas of similar age, and comparing the species point by point. Some of the species may prove to be identical with already described species, or show only minor variation of a kind that would be expected in a local variant within the same species. Other species in the fauna may be new, and if so a full technical description with illustrations must be prepared for each new species, which should be published in a palaeontological journal or monograph. This description is based upon **type specimens,** which are always thereafter kept in a museum or research institute. Usually one of these, the **holotype,** is selected as the reference specimen and fully illustrated; comparative detail may be added from other specimens called the **paratypes.** There are various other kinds of type specimens; for example, a **neotype** may be erected when a holotype has been lost, or when a species is being redescribed in fuller and more up-to-date terms when no type specimen has previously been designated.

The new species must be named and allocated to an existing genus, or if there is no described genus to which it pertains then a new genus must be erected also. To show the method, consider a paper published by Walmsley (1965). In this the species *Isorthis orbicularis* (J. de C. Sowerby) is fully redescribed. It was first described by Sowerby in 1839 as *Orthis orbicularis* at a time when brachiopods were still poorly known and relatively few genera had been erected. As the knowledge of brachiopods grew, so the number of genera increased, and the genus *Isorthis* was erected in 1929 by R. Kozlowski to include all relatives of *Orthis* with a particular set of characters. The type species, i.e. species of *Isorthis* from which this new genus was first described, was *I. szjanochai* Kozlowski 1929. Walmsley recognised the species *orbicularis* as closely related to *I. sjanochai*, and the correct name is now *I. orbicularis* (J. de C. Sowerby 1839). The parentheses around the author's name in such a case refer to the species having been described originally under a different generic name. There are several new species of *Isorthis* in the same paper, e.g. *I. amplificata* Walmsley 1965. In one case a particular group defined at the species level, *I. scuteformis*, was divided into two **subspecies,** *I. scuteformis scuteformis* and *I. scuteformis uskensis,* which are of similar ages but occur in different localities.

All taxonomic work such as this must follow a particular set of rules, which have been worked out by a series of International Commissions and are documented in full in the opening pages of each volume of the *Treatise.*

Taxonomic hierarchy

Whilst all taxonomic categories above the species level are to some extent artificial and subjective, ideally they should as far as possible reflect evolutionary relationships.

Similar species are grouped in **genera** (singular **genus**), genera in **families,** families in **orders,** orders in **classes** and classes in the largest division of the animal kingdom: **phyla** (singular **phylum**). There may be various subdivisions of these categories, e.g. superfamilies, suborders, etc., and in certain groups there is even a case for erecting 'superphyla'. There are only about twenty phyla in the animal kingdom, and only about a dozen of these, e.g. Mollusca, Brachiopoda, etc., leave any fossil remains.

In taxonomy higher taxa are usually distinguished by their suffix (i.e. -ea, -a, etc.). As an example, here is documented the classification of the Silurian brachiopod *Dicoelosia biloba* according to a taxonomic scheme in which the author of the taxon and the year of publication are quoted.

Phylum Brachiopoda Dumeril 1806
Class Articulata Huxley 1869
Order Orthida Schuchert & Cooper 1932
Suborder Orthidina Schuchert & Cooper 1932

Superfamily Enteletacea Waagen 1884
Family Dicoelosiidae Cloud 1948
Genus *Dicoelosia* King 1850
Species *biloba* (Linnaeus 1778)

Use of statistical methods

Inevitably palaeontological taxonomy carries a certain element of subjectivity since the information coded in fossilised shells does not give a complete record of the structure and life of the animals that bore them. There are particular complications that cause trouble. For instance, palaeonotological taxonomy can do little to distinguish **sibling species,** which look alike and live in the same area but cannot interbreed. **Polymorphic species,** in which many forms are present within one biological species, may likewise be hard to speciate correctly. In particular, where **sexual dimorphism** is strong the males and females of the one species may be so dissimilar in appearance that they have sometimes been described as different species, and the true situation may be hard to disentangle (p. 177).

When it comes to the distinction of closely related species, however, there are a number of statistical tests that may help to give a higher degree of objectivity. One simple bivariate test in common use, for example, can be used when a series of growth stages are found together. If a collection of brachiopods is made from locality A, the length/width ratios or some other appropriate parameters may be plotted on a graph as a scatter diagram. A line of best fit (e.g. a reduced major axis) may then be drawn through the scatter. This gives a simple $y = ax + b$ graph, where a is the gradient and b the intercept on the y axis. A similar scatter from a population collected from locality B may be plotted on the same graph, and the reduced major axis drawn from this too. The relative slopes and separations of the two axes may then be compared statistically. If these lie within a certain threshold the populations can then be regarded as one of the same species; if outside it then the species are different.

This is only one of a whole series of possible tests, and more elaborate techniques of multivariate analysis are becoming increasingly important in taxonomic evolutionary studies.

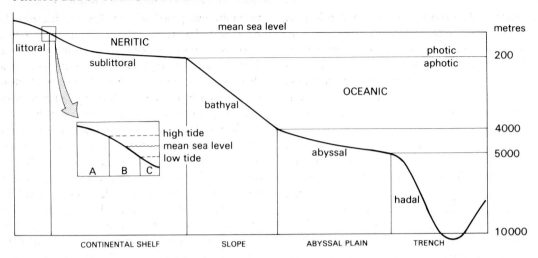

Figure 1.3 Modern marine environments. A, B and C in the inset refer to supralittoral, littoral and sublittoral environments. (Based on Laporte, L. (1968) in *Ancient Environments* (Englewood Cliffs, N.J.: Prentice Hall).)

Palaeobiology

Various categories may be included in palaeobiology: palaeoecology (here discussed with palaeobiogeography), functional morphology and ichnology, each of which requires some discussion.

Palaeoecology

Since ecology is the study of animals in relation to their environment, palaeoecology is the study of ancient organisms in their environmental context. All animals are adapted to their environments in all of its physical, chemical and biological aspects. Each species is precisely adapted to a particular ecological niche in which it feeds and breeds. It is the task of palaeoecology to find out about the nature of these adaptations and about the relationships of the animals with each other and their environments; it involves the exploration of both present and past **ecosystems** (Schäfer 1972).

Palaeoecology must always remain a partial and incomplete science for so much of the information available for the study of modern ecosystems is simply not preserved in ancient ones. The animals themselves are all dead and their soft parts have gone; the original physics and chemistry of the environment is not directly observable and can only be inferred from such secondary evidence as is available; the shells may have been transported away from their original

environments by currents, and the fossil assemblages that are found may well be mixed or incomplete; post-depositional diagenetic processes may have altered the evidence still further. Yet palaeoecology is still a valid science provided that both its strengths and limitations are fully understood. Much is now known about the post-mortem history of organic remains (**taphonomy**), which includes burial processes (**biostratinomy**). This helps to disentangle the various factors responsible for deposition of a particular fossil assemblage, so that assemblages preserved *in situ*, which can yield valuable palaeoecological information, can be distinguished from assemblages that have been transported.

Modern environments and vertical distribution of animals. Figure 1.3 shows the main environments within the Earth's oceans at the present day and the nomenclature for the distribution of marine animals within the oceans.

Modern marine environments are graded accordingly to depth. The **littoral** environments of the shore grade into the subtidal **shelf,** and at the edge of the shelf the continental **slope** goes down to depth; this is the **bathyal** zone. Below this lie the flat **abyssal plains** and the **hadal** zones of the deep ocean **trenches.** There is often a pronounced zonation of life forms in depth zones more or less parallel with the shore. In addition there is a general decrease in **abundance** (number of individuals) but not necessarily **diversity** (number of species) on descent into deeper water from the edge of the shelf. The faunas of the abyssal and hadal regions were originally derived from those of shallow waters but are highly adapted for catching the limited food available at great depths. These regions are, however, impoverished relative to the shallow water regions.

Only the shelf and slope environments are normally preserved in the geological record, the trench sediments rarely so. The abyssal plains are underlain by basaltic rock, formed at the mid-oceanic ridges and slowly moving away from them to become finally consumed at the subduction zones lying below the oceanic trenches; it moves as rigid plates. The ocean basins are very young geologically, the oldest sediments known therein being of Triassic age. These are now approaching a subduction zone and are soon to be consumed without trace. Hence there are very few indications of abyssal sediment now uplifted and on the continents.

What is preserved in the geological record is therefore only a fraction, albeit the most populous part, of the biotic realm of ancient times. The sediments of the continental shelf include those of the littoral, lagoonal, shallow subtidal, median and outer shelf realms. Generally sediments become finer towards the edges of the shelf, the muddier regions lying offshore. There may be reefs close to the shore or where there is a pronounced break in slope.

Horizontal distribution of marine sediments:

Distribution. The main controls affecting the distribution of recent and fossil animals are temperature, the nature of the substrate, salinity and water turbulence. The large-scale distribution of animals in the oceans is largely a function of temperature whilst the other factors generally operate on a more local scale. Tropical shelf regions carry the most diverse faunas, and in these the species are very numerous but the number of individuals of any one species is relatively few. In temperate through boreal regions the species diversity is less, though numbers of individuals per species can be very large.

Salinity in the sea is of the order of 35 parts per thousand. Most marine animals are **stenohaline,** i.e. confined to waters of near-normal salinity. A few are **euryhaline,** i.e. very tolerant of reduced salinity. The brackish water environment is physiologically 'difficult', and faunas living in brackish waters are normally composed of very few species, especially bivalves and gastropods belonging to specialised and often long ranged genera. These same genera can be found in sedimentary rocks as old as the Jurassic, and their occurrence in particular sediments which lack normal marine fossils is a valuable pointer to reduced salinity in the environment in which they lived.

Water turbulence may exercise a substantial control over distribution, and the characters of faunas in high and low energy environments are often very disparate. Robust, thick-shelled and rounded species are normally adapted for high energy conditions, whilst thin-shelled and fragile forms point to a much quieter water environment, and it may be possible to infer much about relative turbulence in a fossil environment merely from the type of shells which occur.

Modern and ancient communities. In shallow cold-temperature seas marine invertebrates are normally found in recurrent ecological communities or associations, which are usually substrate-related. In these a particular set of species are usually found together since they have the same habitat preferences. Within these communities the animals either do not compete directly, being adapted to microniches within the same habitat, or have a stable predator–prey relationship.

Community structure is normally well defined in cold temperature areas, but in warmer seas where diversity is higher it is generally less clear.

Petersen (1918), working on the faunas of the Kattegat, first studied and defined some of these naturally occurring communities. He also recognised two categories of bottom-dwelling animals: **infaunal** (buried and living within the sediment) and **epifaunal** (living on the sea floor or on rocks or seaweed). It was soon found that parallel communities occur, with the same genera but not the same species, on the opposite sides of the Atlantic. Since this pioneer work a whole science of community ecology has grown up, having its counterpart in palaeoecology. Much effort has been expended in trying to understand the composition of fossil communities, the habits of the animals composing them, community evolution through time and, as far as possible, the controls acting upon them (e.g. Thorson 1957, 1971). This is perhaps the most active field of palaeoecology at present, as a host of recent original works testifies (e.g. Craig 1954; Ziegler *et al.* 1968; Bretsky 1969; Boucot 1975; Duff 1975; Tipper 1975; Scott & West 1976).

As Fürsich (1977) makes it clear, however, most fossil assemblages 'lack soft-bodied animals. Where possible trace fossils can be used to compensate for this but they are no real substitute. The term **'community'** cannot therefore be applied to these samples and the term **'association'** meaning a non-random recurrent assemblage usually representing the remnant of a community is preferred.' Using biostratinomic and sedimentological data the 'degree of distortion' from the original community can in some cases be estimated.

Feeding relationships and communities. All modern animals feed on either plants, other animals, organic detritus or degradation products. The tiny plants of the plankton are the **primary producers (autotrophs)** as are seaweeds. Small planktonic animals are the **primary consumers** (herbivores and detritus eaters); there are **secondary** (carnivores) and **tertiary consumers** (top carnivores) in turn. Each animal species is therefore part of a **food web** of **trophic** (i.e. feeding) relationships, wherein are a number of **trophic levels.** In palaeoecology it is rarely possible to draw up a realistic food web (though this is one of the more important aspects of modern ecology), but most fossil animals can be usually assigned to their correct feeding type and so the trophic level may be reasonably estimated.

Of primary consumers the following types are important:

(a) Filterers or suspension feeders, which are infaunal or epifaunal animals sucking in suspended organic material from the water;
(b) epifaunal 'collectors' or detritus feeders, which sweep up organic material from the sea floor; some infaunal bivalves and worms are also collectors;
(c) swallowers or deposit feeders, which are infaunal animals unselectively scooping up mud rich in organic material.

Secondary and tertiary consumers, the carnivores, may prey on any of these, but it is usually the communities of the primary trophic level that are normally most commonly preserved because of their sheer number of individuals.

In many living communities most of the **'biomass'** is actually contributed by very few taxa, not more than five usually (the **trophic nucleus**), but there may be representatives of a number of other species in small numbers. In this system competition between the species concerned seems to be minimised. It is thus mutually beneficial since the different taxa are exploiting different resources within the environment. Living communities are therefore generally well balanced, the number of species and individuals of particular species being controlled by the nature and availability of food resources. Fossil assemblages may be tested according to this concept (Tipper 1975, Fürsich 1977). If they are 'unbalanced' then either (a) there may have been soft-bodied unpreservable organisms which originally completed the balance, or (b) the assemblage has been mixed through transportation and thus does not reflect the true original community.

Faunal provinces. Marine zoogeography (Ekman 1953, Briggs 1974) is primarily concerned with the global distribution of marine faunas and with the definition of **faunal provinces.** These are large geographical regions of the sea (and most particularly the continental shelves) within which the faunas at the specific, generic, and sometimes familiar level have a distinct identity. In faunal provinces many of the animals are **endemic,** i.e. not found outside a particular province. Such provinces are often separated from neighbouring provinces by fairly sharp boundaries, though in other cases the boundaries may be more gradational.

Figure 1.4 shows the main zoogeographical regions of the present continental shelves as defined by Briggs (1974). There can be distinguished tropical shelf, warm temperate, cold temperate and polar regions, whose limits are controlled by latitude but also by the spread of warmer or colder water through major marine currents. Tropical shelf faunas occur in four separate provinces. Of these the large Indo-West Pacific province, extending from southern Africa to eastern Australia, is the richest and most diverse and has been a major centre of dispersal throughout the Tertiary. Smaller and generally poorer provinces of tropical faunas are found in the East Pacific, West Atlantic and (least diverse of all) East Atlantic. These are separated both by land barriers (e.g. the Panama Isthmus) and by regions of cooler water (e.g. where

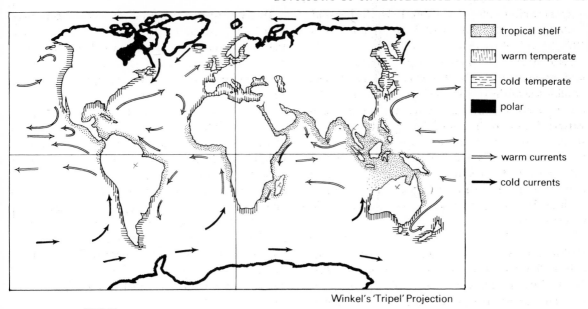

tropical shelf
warm temperate
cold temperate
polar

warm currents
cold currents

Winkel's 'Tripel' Projection

Figure 1.4 Distribution of modern marine shelf-living animals in faunal provinces. (Based on Briggs, 1974)

the cold Humboldt Current sweeping up the western side of South America restricts the tropical fauna to within a few degrees south of the equator).

The shelf faunas of cooler regions of the world are likewise restricted by temperature, and again warm or cold currents exercise a strong control of their distribution. Some zoogeographic 'islands' can be isolated by regions of warmer or cooler water. For example, the tip of Florida carries a tropical shelf fauna, isolated to the north-east and north-west by cooler water areas, but though this is part of the West Atlantic tropical shelf province it has been isolated for some time and therefore its fauna has diverged somewhat from that of the Caribbean shelf. Likewise some oceanic islands (e.g. the Galapagos) may be considered as part of a general zoogeographic region or province, but as their shelves have been initially colonised by chance migrants they may have very many endemic elements which have evolved in isolation. How many of these endemics there are may depend largely on how long the islands have been isolated.

The development of today's faunal provinces has been charted throughout most of the Tertiary. At present Mesozoic and Palaeozoic provinces are likewise being documented (Middlemiss *et al.* 1971; Hallam 1973; Hughes 1975). When used in conjunction with palaeocontinental maps it may be possible to see how the distribution pattern of ancient faunas related to the position of ancient continental masses and their shelves. Sometimes palaeozoogeographical and geological evidence may have a bearing upon palaeotemperatures and even allow some inference to be made regarding ancient ocean current systems.

Modern and ancient reefs Throughout geological time animals have not only become adapted to particular environments but also themselves created new habitats and environments. Within these there has been scope for almost unlimited ecological differentiation.

Perhaps the most striking examples of such **biogenic** environments are the reefs of the past and present. Reefs (Fig. 1.5) are massive accumulations of limestone built up by lime-secreting algae and by various kinds of invertebrates. Through the activities of these frame builders great mounds may be built up to sea level, with caves and channels within them providing a residence for innumerable kinds of animals, all ecologically differentiated for their particular niches.

In the barrier and patch reefs of the tropical seas of today, which grow up to the surface, the warm oxygen-rich turbulent waters allow rapid calcium metabolism and hence continuous growth. The

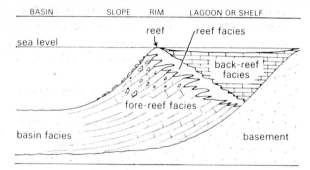

Figure 1.5   Generalised section through a reef (algal or coral).

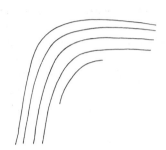

Figure 1.6   Crest of Permian algal reef.

principal frame builders are algae and corals, but there are many other kinds of invertebrates in the reef community: sponges, bryozoans and molluscs amongst others. Some of these add in minor ways to the reef framework; others break it down by boring and grazing. The growth of the reef to sea level continually keeps pace with subsidence, but it is also being continually eroded. In a typical coral reef complex the reef itself is a hard core of cemented algal and coral skeletons facing seawards, and as the reef subsides it grows outwards over a fore-reef slope of tumbled boulders broken from the reef front. Behind it is a lagoon with a coral sand sediment and tidal flats along the shore colonised by blue-green algae. Green algae are commonest in the back-reef facies; red algae are the main lime secreters of the reef itself.

Large reefs such as the above are known as **bioherms;** they form discrete mounds rising from the sea floor. **Biostromes,** on the other hand, are flat laminar communities of reef-type animals and barely rise above the sea floor.

Throughout geological history there have been various kinds of reef communities, which have arisen, flourished and become extinct. In all of these the frame builders have included algae, but the invertebrate frame builders have been of different kinds; the present corals are only the last in a series of reef-building animals (Newell 1972).

The oldest known reefs are over 2000 million years old, made up entirely of sediment-trapping and possibly lime-secreting blue-green algae: the prokaryotic stromatolites. Some of these reefs reached considerable dimensions. One is reported from the Great Slave Lake region in Canada as being over 450 m thick, and separating a shallow-water carbonate platform from a deep-water turbidite-filled basin. But there are no preserved metazoans in these reefs, and their ecological structure may have been very simple.

With the rise of frame-building metazoans in the Lower Cambrian a new kind of reef community made its appearance. Stromatolitic reefs were invaded by the sponge-like archaeocyathids, growing in clumps and thickets on the reef surface. When these earliest of reef invertebrates became extinct in late Middle Cambrian time there were no more reef animals until some 60 million years later; the only reefs were stromatolitic. In Middle Ordovician time these algae were joined by corals and stromatoporoids (lime-secreting sponge relatives) as well as by red algae, which together formed a reef environment attracting a host of other invertebrates including brachiopods and trilobites. For unknown reasons this type of reef complex did not continue beyond late Devonian time. The reefs that arose in the Carboniferous were mainly algal, stromatoporoids and corals no longer playing such an important part in their construction.

The same kind of reef continued into the Permian, and many of these fringed the shrinking inland seas of that time. They rose at the edge of deeper basins in which the water periodically became more saline as it evaporated; there was likewise evaporation in giant salt pans behind the reef as water drawn through the reef dried out in the lagoons behind. In the Permian reefs of Texas and northern England, which are very large, the reef front rose as a vertical wall of laminated algal sheets, turning over at the reef crest where it reached sea level (Fig. 1.6). The upper surface was intensely colonised by stromatolites, which died out towards the lagoonal back-reef facies. This kind of reef in general morphology therefore does not closely approximate the standard pattern of Figure 1.5. When the Permian shelf seas had dried out completely the Permian reef-complex type vanished, and there were no more reefs until the slow beginnings of the coral–algal reef system that arose in the late Triassic. Coral reefs have expanded and flourished since then, other than during a catastrophic period in the early Cretaceous from which there are no reefs known (though corals must have been living somewhere at that time). When the corals recovered they were joined in many places by the peculiar rudistid bivalves, also reef formers, which at one period almost supplanted corals as the dominant reef frame-builders. Yet these too died out in the late Cretaceous leaving corals the undisputed and dominant reef-building invertebrates.

There has been some decline in the spread of coral reefs and in the number of coral genera since the beginning of the Tertiary. They are now confined to the Indo-Pacific region and on a smaller scale to the West Atlantic. This decline may still be continuing, though the reefs were not significantly affected by the Pleistocene glaciation. The future of world reef communities, the most complex of all marine ecosystems, remains to be seen.

Functional morphology, growth and form

The functions of particular organs in fossils cannot be established by many of the methods available to zoologists. But it is still possible to go some way towards explaining how particular organs worked when the animal that bore them was alive. Such functional interpretation is, however, limited, and in many ways it is hard to go beyond a certain point, even when the function of a particular organ is known (p. 269).

Palaeontologists are often presented either with organs whose function is not clear and which have no real counterparts in living animals, or with fossils of bizarre appearance which are so modified from the normal type for the taxon that they testify to extreme adaptations. Some attempts can be made towards interpreting these morphologies in terms of adaptation and mode of life, which in turn may lead to a clearer understanding of evolutionary processes. If these problems are to be tackled then a coherent methodological scheme is needed. One particular system of approach, the paradigm method of Rudwick (1961) (see Bibliography following Ch. 7), has been much discussed, but it lies largely outside the scope of this work so only some examples of its application are mentioned.

Two related aspects of palaeozoology which can be deduced from fossilised remains alone are growth and form. Following the classic work of Thompson (1917) it has been understood that the conformation of the parts of any organism is the result of interacting forces, dictated by physico-mathematical laws, which have operated through growth.

Different marine invertebrate skeletons may be functionally convergent in the way they grow, and the same kind of growth patterns turn up frequently in representatives of many phyla. This is because there are only relatively few ways in which an animal can grow and yet produce a hard covering. Invertebrate skeletons, by contrast with those of vertebrates, are generally external, and this narrows down the spectrum of growth possibilities still further. Some of these types of skeletons are:

(a) External shells growing at the edges only by accretion of new material along a particular marginal zone of growth. Very often such growth results in a logarithmic spiral shell as in brachiopods, cephalopods and in coralla of certain simple corals.

(b) External skeletons of plates – disjunct, contiguous or overlapping – normally secreted along a single zone which may be but is not always marginal. A good example is the echinoderm skeleton in which the plates once formed are permanently locked into place, though they may thereafter grow individually by accretion of material in concentric zones.

(c) External skeletons all formed at the one time. The arthropod exoskeleton is most typical. Growth here is difficult for the skeleton has to be periodically moulted, a process known as **ecdysis.** When the old exoskeleton is cast the arthropod takes up water or air and swells to the next larger size, and the new cuticle which underlies the old one then hardens. Growth is thus rapid and episodic and is only possible during moulting. The disadvantage of this system is the vulnerability to damage and predation during moulting.

Growth systems and their geometrical consequences are thus concomitant both on the nature of the skeletal material and on the spatial constraints operating during growth.

Ichnology

Ichnology is the study of the behaviour of once living animals by examination of the tracks, trails, borings and markings that they made when alive. These are called **trace fossils** or **ichnofossils.** They are usually preserved only at the interface of two types of sediment, such as for instance where the markings made on a mud surface by a crawling trilobite were filled in by fine sand, which later hardened. Ichnology is by no means a recent development in geological science – indeed it can be traced back to the 1820s – but it is only within the last two decades that it has come to be a really powerful tool for understanding certain past sedimentary environments and one of our major clues to the behaviour of the animals that lived within these environments. The different kinds of traces made in shallow water sediments have been fully documented in the vast mudflats of the North Sea by German workers as summarised by Schäfer (1972), and this knowledge has been applied to the geological record. Since very good summaries of trace fossil work exist elsewhere (Crimes & Harper 1970, Frey 1976) only a few points will be made here to show something of the scope and methods of ichnology.

Classification of trace fossils

Trace fossils are all sedimentary structures made by the activities of once living animals, mainly invertebrates (Fig. 1.7). But similar traces can be made by quite different kinds of organisms. There are various ways of studying trace fossils and hence different systems of classifying them (Simpson 1976). Three of these in particular are important: morphological and preservational; behavioural; and phylogenetic.

Morphological and preservational classification. Trace fossils are all given 'form-generic names', that is, the names are used only for the distinction of various types and do not attempt to identify or suggest their maker. A simplified system of grouping them with interpretation is:

(a) Tracks or trails on a bedding plane originally made upon the sediment-water interface, e.g. *Cruziana* (trilobite trails), *Nereites* (worm trails).
(b) Radially symmetrical horizontal markings, e.g. *Asteriactites* (resting marks of starfish).
(c) Tunnels and shafts within the sediment, e.g. *Skolithos* (vertical worm tubes), *Chondrites* (branching galleries probably made by a probing worm).
(d) Traces with a **spreite** (a web-like structure usually with a series of concentric markings) as often found joining two branches of a U-tube and representing former positions of the tube within the sediment, e.g. *Rhizocorallium* (horizontal U-tubes), *Diplocraterion* (vertical U-tubes), *Zoophycos* (inclined spirals) – all of which could have been made by different kinds of animals.
(e) Pouch-shaped markings, e.g. *Pelecypodichnus* (bivalve burrows), *Rusophycus* (trilobite resting traces).
(f) Others, e.g. *Palaeodictyon* (net-like structure of uncertain origin).

Whilst this descriptive classification has its usefulness the characters selected for description must be arbitrary and therefore far from objective. A classification based on preservational features alone is equally hard to apply.

Behavioural classification. Seilacher (1964) designed a very useful scheme based upon the behaviour of the organism that made the traces. In modified form this classification has now six categories, but since they overlap to some extent this scheme also is not a perfect system, though it is useful. The categories are:

(a) Crawling traces (Repichnia): tracks of moving surface dwellers, e.g. *Cruziana*.
(b) Resting traces (Cubichnia), e.g. *Rusophycus*, *Asteriactites*.
(c) Grazing or surface feeder traces (Pascichnia). Surface feeders make characteristic patterned markings on the sediment; the same sort of patterns may be seen made by Recent snails grazing algae on stones in rock pools.

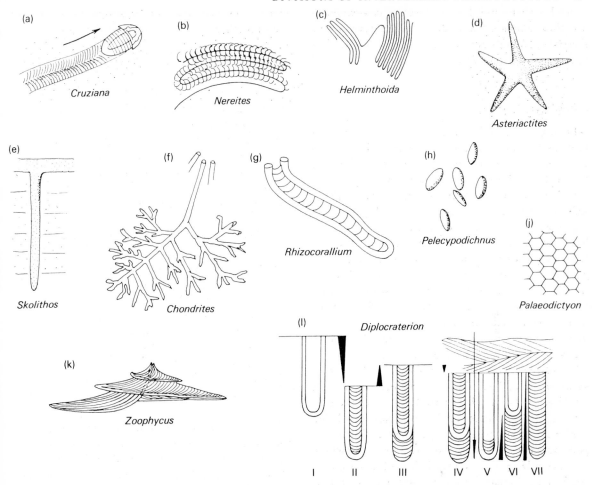

Figure 1.7 Trace fossils: (a) *Cruziana;* (b) *Nereites;* (c) *Helminthoida;* (d) *Asteriactites;* (e) *Skolithos;* (f) *Chondrites;* (g) *Rhizocorallium* (with spreite); (h) *Pelecypodichnus;* (j) *Palaeodictyon;* (k) *Zoophycus;* (l) *Diplocraterion* – from left to right (I) normal U-tube; (II) U-tube descending as sediment is removed; (III) U-tube ascending as sediment is rapidly deposited; (IV–VII) U-tubes initiated at various times showing movement in relation to erosion or deposition of sediments on top. Arrows show directions of movement before final plane erosion and deposition of overlying current bedded sediment. ((a)–(k) redrawn from Simpson, 1976, (l) redrawn from Goldring, 1962.)

Helminthoida, for instance, is a trace appearing as a series of parallel lines joined by tight loops, each line being equidistant from its neighbours. This is made by a surface feeder economically exploiting a food supply, such as bacteria, upon the sediment surface, covering as much ground as possible but with the minimum of effort and never recrossing its own trail or feeding along previously grazed surfaces.

(d) Feeding traces (Fodinichnia): excavations made by deposit feeders, i.e. animals living on the surface but actively mining within it for their food. They include some U-shaped tubes with internal spreites, which show that the U-burrow has been progressively deepened. There are also various kinds of radial structures, including the many-branched *Chondrites*, which was probably made by a worm exploiting a rich deposit successively, again avoiding any area of previously ingested sediment.

(e) Dwelling structures (Domichnia): permanent burrows and borings of suspension feeders, i.e. animals living within the sediment but straining off particles from the water above. They may be subcylindrical tubes (e.g. *Skolithos*), but there are also some U-shaped burrows, though normally without spreites.

(f) Escape structures (Fugichnia). Various kinds of vertical tubes suggest that animals moved upwards to escape from being buried by an influx of sediment, or downwards when the sediment above was eroded. The original tubes are usually therefore modified Domichnia. *Diplocraterion,* for example, is a vertical U-tube found in high energy environments, with spreite structures either above or below it or both. It is interpreted as the permanent dwelling burrow of an animal that lived below the sediment. When sediment was removed by rapid erosion the animal burrowed down more deeply in order to remain concealed at its habitual depth, leaving a spreite above. Rapid sedimentation caused the organism to move upwards, producing a spreite below of somewhat different morphology since it was no longer linking parallel arms. Such a structure gives information about the environment, suggesting very rapid erosion and deposition, and the related up-and-down movement of the organism that made it led Goldring (1962) to give it the appropriately descriptive specific name *D. yoyo.*

Phylogenetic classification.   Generally there are few indications as to the identity of a particular trace maker, and often even its phylum can hardly be established since animals of many phyla can make the same kind of traces. In the case of some arthropod ichnofossils, however, the trace maker is known with certainty; trilobites, for instance, have actually been found within *Rusophycus* burrows. But since this is rare indeed there is no point in trying to erect classification based upon the identity of the trace maker, and form-generic names have to be given.

No means of classification is therefore entirely satisfactory or complete, and the one that is followed in any particular case depends rather upon the purpose of the study.

Uses of ichnology

Sedimentary environment.   By contrast with body fossils, trace fossils are always found in place. If the enclosing sediment had moved they would have been destroyed. They often occur in sedimentary suites where there are no body fossils present and especially in clastic sequences. Rather than being destroyed by diagenesis they are commonly improved by it, since lithological contrast between the fossil and the enclosing sediment may thereby be enhanced. Furthermore they tend to be restricted to a narrow facies range. As guides to the nature of the sedimentary environment in certain types of sequence they have proved invaluable.

If trace fossils are present in a clastic sequence then the sediment must have been well aerated. Conversely, a sequence lacking in traces may well have been anoxic. Thus black shales with abundant traces, even if deep water in origin, cannot possibly have been anoxic. Graptolitic shales, and their Mesozoic equivalents which contain only thin-shelled planktonic bivalves, are generally lacking in trace fossils, which suggests a euxinic origin.

Rates of sedimentation can often be inferred from the relative abundance of trace fossils, as with *Diplocraterion,* and the presence of definite borings in 'hardgrounds' shows clearly that the sediment must have been indurated when the borings were formed, as may sometimes be confirmed by encrusters upon its surface. Inevitably a temporary break in deposition must be recognised to have allowed such induration. Minor variations in facies can often be inferred from the presence of particular trace fossil assemblages, even if the nature of their makers is not known.

There are also rather generalised trace fossil assemblages which give information of a grosser quality, especially as pertains to original depth of deposition (Seilacher 1967). Since physical protection is very important to shallow water dwellers, partly due to turbulence and partly because the light allows other animals to see them, a high proportion of invertebrates are burrowing suspension feeders (filterers) and the traces are mainly Cubichnia and Domichnia. Further away in quieter waters deposit feeders (swallowers) are abundant, and hence Fodinichnia become increasingly common, likewise Repichnia. In the lightless environment of the deep sea there is less need for protection, and because of the rich supply of surface-dwelling bacteria there are very many Fodinichnia and especially Pascichnia. The latter make complex spiral or meandering patterns for systematically grazing from food-rich surface layers. On this basis there have been erected three main facies: a shallow water *Cruziana* facies of burrows, arthropod tracks and resting traces; an intermediate *Zoophycos* facies, with mainly sediment miners; and finally a deepwater *Nereites* facies, with grazing trails of complex form.

Stratigraphy. Generally trace fossils are of very little stratigraphic use. The very characters that make them so valuable environmentally – i.e. long range, strict facies control, etc. – are just the opposite of what is needed for good zone fossils. Yet some short-ranged types are useful in local stratigraphy, and most particularly trace fossils have been of immense value in understanding the explosion of life at the Precambrian–Cambrian boundary (Ch. 3), and particularly of metazoan life, for many trace fossils can only be made by an animal with a gut. Other examples have been given by Crimes and Harper (1970).

Fossil behaviour. In a sense, all ichnology relates to the study of fossil behaviour. The complex grazing trails that have already been mentioned testify to a degree of behavioural complexity in their ancient makers like that of many present organisms. Trace fossils have been used directly in analysing the life habits and locomotion of trilobites and other arthropods (Ch. 11).

Stratigraphy

Sedimentary rocks have been built up by layer upon layer of sediment, which sometimes has been much the same for long periods of time but at other times has changed its character rapidly. The individual layers within sedimentary rocks are separated by bedding planes. These bedding planes are time horizons, and the history of a rock sequence is reflected in its layering. Each layer in the rock sequence must have been laid down on a pre-existing layer, so that the oldest rocks are at the bottom of an exposed sequence, the youngest at the top, unless the succession has been tectonically inverted. This is the Principle of Superposition, recognised as long ago as 1669 by the Italian scientist Nicolaus Steno.

Stratigraphy is concerned with the study of stratified rocks, their classification into ordered units and their historical interpretation. It bears not only upon past geological events but also upon the history of life and is perhaps the most basic part of all geology.

Much of stratigraphy is concerned with chronology; the geological record has to be divided up into time periods, standardised, as far as possible, all over the world. One of the primary aims of stratigraphy has been to produce an accurate chronology in which not only the order of events but also their dates are known. Stratigraphical classification is basic to all of this.

There are three principal categories of stratigraphical classification, **lithostratigraphy, biostratigraphy** and **chronostratigraphy,** all of which are ways of ordering rock strata into meaningful units.

Lithostratigraphy

Lithostratigraphy is concerned with the erection of units based upon the characters of the rocks and differentiated on types of rock, e.g. siltstone, limestone, clay, etc. It is useful in local areas and essential in geological mapping, but there is always the danger that even in a small area rock units cut across time planes. For instance, if a shoreline has been advancing in one direction a particular suite of sediments, probably of the same general kind, will be left in its wake. Though this bed will appear in the rock record as a single uniform layer, it will not all have been deposited at the one time; since it cuts across time planes it is said to be **diachronous.** Such diachronism is common in the geological record. Furthermore many suites of sediments are laterally impersistent; different sedimentary facies may have existed at the same time within a small space – a sandstone layer, for instance, passing into a shale some distance away. Lithostratigraphy is thus only of real value within a relatively small region.

The divisions erected in lithostratigraphy are arranged in a hierarchial system: **group, formation, member** and **bed.** A bed is a distinct layer in a rock sequence. A member is a group of beds united by certain common characters. A formation is a group of members, again united by characters with features in common. It is the primary unit of lithostratigraphy and is most useful in geological mapping. Hence it is formations that are normally represented by different colours on geological maps and cross-sections, and a formation is normally defined for its mapping applications. Finally a group ranks above a formation; it is composed of two or more formations and is often used for simplifying stratigraphy on a small scale map.

Biostratigraphy

In biostratigraphy the fossil contents of the beds are used in interpreting the historical sequence. It is based upon the principle of the irreversibility of evolution. This means that at any one moment in the Earth's history there was living a unique and special assemblage of animals, characteristic of that period and of no other. As time went on these were replaced by others; each successive fossil assemblage is a pale reflection of the life at the time the enclosing sediments were deposited. Thus during the early Palaeozoic trilobites and brachiopods were the most common fossils; by the Mesozoic the most abundant preservable invertebrates were the ammonites; they too became extinct, and snails and bivalves are the commonest relics of Caenozoic time. This is how it appears on a broad scale. But when the time ranges of individual fossil species are examined it is evident that some of these lasted for only a fraction of geological time, characterising very precisely a particular brief historical period.

In any local area, once the sequence of fossil faunas has been precisely established through assiduous collection and documentation from exposed sections, this known succession can be used for correlation with other areas. Certain fossil species have been found to be particularly good stratigraphical markers. They characterise short sections of the geological succession known as **zones.** To take an example, ammonites are particularly good zone fossils for Mesozoic stratigraphy. The Jurassic period lasted some 55 million years, and in the standard British succession there are over sixty ammonite zones with which it is subdivided, so the zones are defined historical periods which have an average duration of less than a million years each.

The practical problems in biostratigraphy are, however, very complex, and some parts of the geological succession are much more closely zoned than others. The main problems are:

(a) Many kinds of fossils, especially those of bottom-dwelling invertebrates, are facies-controlled.

They lived in particular environments only, e.g. lime-mud sea floor, reef, sand or silty sea floor. They were often highly adapted for particular conditions of temperature, salinity or substrate and are not found preserved outside this environment. This means that they can only be used for correlating particular environments and thus are not universally applicable.

(b) Some kinds of fossils are very long-ranged. Their rates of evolutionary change were very slow. They can only be used in a broad and general sense for long period correlation and are of very little use for establishing close subdivisions.

(c) Such good zone fossils as the graptolites are delicate and only preserved in quiet environments, being destroyed by more turbulent conditions.

(d) Since fossil species could migrate following their own environment through time, there is always a possibility of diachronous faunas. The zone as defined in one area may not therefore be exactly time-equivalent to that in another region.

In the example of a graptolite, therefore, for the reasons outlined in (c) and (d), the total range or **biozone** of a species is not likely to be preserved in any one area, and it is therefore hard to draw ideal isochronous boundaries or time lines.

Ideally, zone fossils should have a particular combination of characters to make them fully suitable for biostratigraphy. These would be:

(a) a wide horizontal distribution, preferably intercontinental;

(b) a short vertical range so that they could be used to define a very precise part of the geological column:

(c) enough morphological characters to enable them to be identified and distinguished easily;

(d) strong hard shells to enable them to be commonly preserved;

(e) independence of facies, as would be expected from a free-swimming animal.

All of these conditions are seldom fulfilled in fossils used for zonation; perhaps the neritic ammonites came closest to it, and it is not surprising that the principles of really precise stratigraphical correlations were first worked out fully with these fossils, notably by the German palaeontologist A. Oppel in the 1850s.

It was Oppel too who first recognised that there are various ways of using fossils in stratigraphy, which partially circumvent the difficulties mentioned, and hence different types of biozones. There are four main kinds of biozones generally used (Hedberg 1976). **Assemblage zones** are beds or groups of beds with a natural assemblage of fossils. They may be based on all the fossils preserved therein or on only certain kinds. They are usually very much environmentally controlled and therefore of use only in local correlation. **Range zones** are perhaps of more general application. A range zone usually represents the total range of a particularly useful selected element in the fauna. One may therefore refer to the *Psiloceras planorbis* zone, based upon the eponymous ammonite that defines the lowest zone of the European Jurassic, above which is the *Schlotheimia angulata* zone. Each range zone is always named after a particular species which occurs within it. Where there are a number of zonally useful species, or where the ranges of individual species are long, a more precise time definition may be given by the use of overlapping stratigraphical ranges. Such zones are therefore called **concurrent range zones. Acme** or **peak zones** are useful locally. An acme zone is a body of strata in which the maximum abundance of a particular species is found, though not its total range. Such acme zones may be narrow but are often useful as marker horizons in geological mapping. Finally an **interval zone** is an interval between two distinct biostratigraphical horizons. It may not have any distinctive fossils, or indeed any fossils at all, being simply a convenient way of referring to a group of strata bracketed between two named biostratigraphically defined zones.

Biostratigraphical units, unlike litho- and chronostratigraphical units, are not hierarchially arranged, apart from in the case of **subzones** which are local divisions where a zone can be divided more finely in a particular region than elsewhere.

Chronostratigraphy

Chronostratigraphy is more far reaching than either bio- or lithostratigraphy but has its roots in both of them. Its purpose is to organise the sequence of rocks on a global scale into chronostratigraphical units, so that all local as well as worldwide events can be related to a single standard scale. Hence it is concerned with the age of strata and their time relations. To do this a hierarchical classification of time-equivalent units must be employed. The conventional hierarchial system used is:

Chronostratigraphical units	Geochronological units
Eonothem	Aeon
Erathem	Era*
System*	Period*
Series*	Epoch*
Stage*	Age
Chronozone	Chron

*These terms are in most common use.

Chronostratigraphical units relate quite simply to geochronological units; thus the rocks of the Cambrian System were all deposited during the Cambrian period. Most of these terms are self-explanatory, but it should be recognised that they are all, at least in theory, worldwide in extent.

The *Psiloceras planorbis* chronozone is a time unit equivalent to the time in which the said ammonite was in existence, even if it was confined to certain parts of the world only. It is hard indeed, however, to be able to delimit chronozones accurately, since most fossils were confined to certain geographical regions or provinces as are most of the animals living today. There are relatively few well-established chronozones or 'world instants' as they have been called, and so 'chronozone', though it has a real meaning, is not a term applicable to most practical stratigraphy. A **stage,** on the other hand, is a group of successive zones having great practical use, especially since it is normally the basic working time unit of chronostratigraphy, the narrowest that can actually be used on a regional scale.

It is usually at the stage level that rocks of widely different facies can be correlated. As an example there are some difficulties in making precise zonal correlations between Ordovician trilobite–brachiopod faunas and time-equivalent faunas with graptolites. Graptolites are rarely preserved in the siltstones and limestones favoured by the shelly fossils, and the latter being benthic could not inhabit the stagnant muds in which the graptolites were best preserved. In some areas, of course, the faunas do alternate in vertical sequence since the sites of deposition of these two facies fluctuated with oscillating shorelines, but though precise zone-to-zone correlations are possible at some levels it is found in practice that Ordovician graptolite zones correlate best with stages defined on shelly fossils.

Fossils give a relative chronology which can be used as the primary basis of chronostratigraphy. But it is often hard to correlate precisely beds of equivalent age in widely separated areas. The fossil sequences, though well documented with any one area, may contain very few elements in common, if indeed any at all, for they belong to different faunal provinces which are hard to correlate. Sometimes, however, the boundaries of such provinces may have oscillated to-and-fro. There may thus appear elements of adjacent faunal provinces in vertical succession; thus facili-

tating stratigraphical correlation. And at most stratigraphical horizons there are usually some ubiquitous worldwide fossils, so intercontinental correlation is not impossible.

In chronostratigraphy the relative sequence given by the fossils is supplemented and enhanced by absolute dates which can be affixed at certain points wherever appropriate rocks occur. These are usually lavas bracketed between fossiliferous sediments, and their occurrence is not too common. It is most unlikely, therefore, that radiometric dating will supersede palaeontological correlation; the two are entirely complementary, and the great success of chronistratigraphy, in spite of its limitations, owes much to both.

BIBLIOGRAPHY

Books, treatises, symposia

Ager, D. V. 1963. *Principles of palaeoecology*. New York: McGraw-Hill. (Useful basic text)

Barrington, E. J. W. 1967. *Invertebrate structure and function*. London: Nelson. (Invaluable zoology text)

Beerbower, J. R. 1968. *Search for the past: an introduction to palaeontology*. Englewood Cliffs, NJ: Prentice-Hall. (Palaeontology textbook)

Boucot, A. 1975. *Evolution and extinction rate controls*. Amsterdam: Elsevier. (Advanced text, mainly dealing with brachiopod distribution)

Brasier, M. D. 1979. *Microfossils*. London: George Allen & Unwin.

Briggs, J. C. 1974. *Marine zoogeography*. New York: McGraw-Hill. (Delimitation of faunal provinces)

Ekman, S. 1953. *Zoogeography of the sea*. London: Sidgwick & Jackson. (An older but still useful text on marine life zones and faunal provinces)

Frey, R. W. (ed.) 1976. *The study of trace fossils*. Berlin: Springer-Verlag. (Invaluable compilation of original papers and introductory chapters)

Hallam, A. 1973. *Atlas of palaeobiogeography*. Amsterdam: Elsevier. (48 original papers on distribution of fossil faunas)

Harland W. B. (ed.) 1967. *The fossil record – a symposium with documentation*. London: Geological Society of London. (Contains time range diagrams of all fossil orders and charts showing diversity fluctuations through time)

Hedberg, H. D. 1976. *International stratigraphic guide*. New York: Wiley. (Official guide to stratigraphical procedure)

Hedgpeth, J. W. and J. S. Ladd 1957. *Treatise on marine ecology and palaeoecology*. (Vol. 1, ed. Hedgpeth; Vol. 2, ed. Ladd) Mem. Geol. Soc. America no. 67. Lawrence, Kansas. (Standard text, with papers and annotated

bibliography on ecology of all living and fossil phyla)

Hughes, N. F. (ed.) 1975. *Organisms and continents through time*, 1–340. Spec. pap. in palaeont., no. 17. (23 original papers on distribution of faunas; many charted on global maps)

Middlemiss, F. A., P. F. Rawson and G. Newall (eds) 1971. *Faunal provinces in space and time*. Liverpool. Liverpool Geol Soc., Seel House Press. (13 original papers on faunal distribution)

Moore, R. C. and C. Teichert (eds) 1953. *Treatise on invertebrate paleontology*. Geol. Soc. America and Univ. Kansas Press. (The standard reference work on invertebrate fossils – each phylum separately treated. Referred to as '*Treatise*' in chapter bibliographies)

Moore, R. C. (ed.) 1962. '*Treatise*' Part (W). Miscellanea with supplement 1. (Ed. C. Teichert – Trace Fossils 1975) (Description of all known trace fossil genera)

Piveteau J. (ed.) 1952–66. *Traité de paléontologie*. Paris: Masson & Cie. (The French 'Treatise', slightly older than the American *Treatise*, but of very high quality)

Raup, D. M. and S. M. Stanley 1978. *Principles of paleontology* 2nd edn. San Francisco: Freeman. (Excellent text-book emphasising approaches and concepts, but not morphological or stratigraphical details)

Schäfer, W. 1972. *Ecology and palaeoecology of marine environments* (translated edn.), G. Y. Craig (ed.). Edinburgh: Oliver & Boyd. (Standard work on Recent North Sea environments, with applications for palaeoecology)

Scott, R. H. and R. R. West (eds) 1976. *Structure and classification of paleocommunities*. Pennsylvania: Dowden, Hutchinson & Ross. (11 original papers)

Sylvester-Bradley, P. C. (ed.) 1956. *The species concept in palaeontology*. London: Systematics Assoc. Publ. (Basic work of several papers on palaeontological taxonomy)

Tasch, P. 1973. *Paleobiology of the invertebrates*. New York: Wiley. (Palaeontology text)

Thompson, D'Arcy W. 1917. *On growth and form*. (Abridged edn, J. T. Bonner (ed.) 1961). Cambridge: CUP. (Individualistic classic work on physical laws determining growth)

Thorson, G. 1971. *Life in the sea*. London: World University Library. (Simple well-illustrated text; deals with community structure)

Valentine, J. W. 1973. *Evolutionary palaeoecology of the marine biosphere*. Englewood Cliffs, NJ: Prentice-Hall. (Valuable text on evolution and ecology)

Individual references

Bretsky, P. W. 1969. Central Appalachian late Ordovician communities. *Bull. Geol.. Soc. Am.* **80,** 193–212. (Community structure)

Craig, G. Y. 1954. The palaeoecology of the Top Hosie Shale (Lower Carboniferous) at a locality near Kilsyth. *Quart. J. Geol. Soc. Lond.* **110,** 103–19. (Classic palaeoecological study)

Craig, G. Y. and N. S. Jones 1966. Marine benthos, substrate, and palaeoecology. *Palaeontology* **9,** 30–8. (Determinants of distribution of benthos)

Crimes, T. P. and J. C. Harper 1970. *Trace fossils*. Liverpool: Liverpool Geol. Soc., Seel House Press. (Proceedings of an international conference; contains 35 papers)

Duff, K. L. 1975. Palaeoecology of a bituminous shale: the Lower Oxford Clay of central England. *Palaeontology* **18,** 443–82. (Palaeoecological study)

Fürsich, F. T. 1977. Corallian (Upper Jurassic) marine benthic associations from England and Normandy. *Palaeontology* **20,** 337–85. (Palaeoecology, mainly of bivalve communities)

Goldring, R. 1962. Trace fossils and the sedimentary surface in shallow water and marine sediments. In *Deltaic and shallow marine deposits*. L. J. M. U. Van Straaten (ed.), Amsterdam: Elsevier. (Description of *Diplocraterion* and other trace fossils)

Johnson, J. C. 1974. Extinctions of perched faunas. *Geology* **2,** 479–82. (Faunas colonising flooded shelves are most vulnerable to extinction when the sea retreats)

Newell, N. D. 1972. The evolution of reefs. *Scient. Am.* **226,** 54–65. (Informative short paper)

Petersen, C. C. J. 1918. *The sea bottom and its production of fishfood*. Copenhagen: Rep. Danish Biol. Stat. **25,** 1–62. (The first definitive summary of benthic communities)

Schopf, T. J. M. (ed.) 1972. *Models in paleobiology*. San Francisco: Freeman, Cooper & Co. (10 original papers, many significant)

Seilacher, A. 1964. Biogenic sedimentary structures. In *Approaches to palaeoecology*, J. Imbrie and N. D. Newell (eds). New York: Wiley. (Classification of trace fossils)

Seilacher, A. 1967. Bathymetry of trace fossils. *Mar. Geol.* **5,** 413–28. (Depth zonation of modern and ancient traces)

Simpson, S. 1976. Classification of trace fossils. In *The study of trace fossils*, R. W. Frey (ed.), 39–54. Berlin: Springer-Verlag. (Different methods of classification)

Tipper, J. C. 1975. Lower Silurian marine communities: three case histories. *Lethaia* **8,** 287–99. (Palaeoecological study)

Thorson, G. 1957. Bottom communities. In *Treatise on Marine Ecology and Palaeoecology 1*, J. W. Hedgpeth (ed.). Mem. Geol. Soc. Amer. no. 67, 461–534. (Standard work, invaluable, well-illustrated)

Walmsley, V. G. 1965. *Isorthis* and *Salopina* (Brachiopoda) in the Ludlovian of the Welsh Borderland. *Palaeontology* **8,** 454–77. (Pure taxonomy)

Ziegler, A. M., L. R. M. Cocks and R. K. Bambach 1968. The composition and structure of Lower Silurian marine communities. *Lethaia* **1,** 1–27. (The first major work on Lower Palaeozoic communities; very well illustrated)

2 Theory of evolution

'The marriage of palaeontology with the mainstream of evolutionary biology has been something of a disappointment. Though there was never any question of divorce, the lack of much effective interaction between palaeontology and the study of the evolutionary process points to a sort of *de facto* estrangement' (Eldredge 1974a).

The development of evolution theory has been closely bound up with palaeontological evidence from the very beginning, for amongst all the sciences concerned with organic evolution only palaeontology has the unique perspective of geological time. Hence the special role of palaeontology in the understanding of evolution has been largely in documenting the course of evolutionary change as revealed in the rock record.

The means whereby these changes have come about and by which animals and plants have diversified cannot, however, be inferred directly from fossils. It is only since the rise of genetics, cytology, molecular biology and population dynamics that real progress in understanding the mechanism of evolutionary change has been possible. Largely because of the great developments in these fields there has been a tendency for palaeontology, and especially the invertebrate fossil record, to be relegated to a rather secondary role in evolutionary science, and, as Eldredge has maintained, the interplay between palaeontology and the 'mainstream of evolutionary biology' has been less than these various sciences deserve. In recent years, nevertheless, whilst the traditional role of evolutionary palaeontology in documenting the fossil record has continued there have been substantial new developments which illuminate the spectrum of evolutionary science in different ways. In order that the fossil record can be understood in an evolutionary context the following text presents a simplified outline of classic evolution theory (Neo-Darwinism). Fuller treatments, emphasising different aspects of evolution, are found in

such major works as Simpson (1953), Drake (1968), Eaton (1970), Maynard-Smith (1975), Valentine (1974) and Hallam (1977). Whilst Neo-Darwinian theory is generally accepted as being a reasonably satisfactory paradigm for the majority of observable evolutionary phenomena, some modern critics (e.g. von Bertalanffy 1969, Koestler 1971) do not believe it to be comprehensive. Indeed it has been argued by some authorities that Neo-Darwinism can really explain only peripheral phenomena. Whatever the outcome of the present debate may be, however, Neo-Darwinism is at least an essential starting point.

DARWIN, THE SPECIES AND NATURAL SELECTION

Whilst Darwin (1859) is generally regarded as the father of evolution theory, there were many pre-Darwinian scientists who postulated that animals and plants had changed over long periods of time and that new types or 'species' had arisen from pre-existing ones. Most prominent of these was French naturalist Jean-Baptiste Lamarck (1774–1829), who proposed long before Darwin that all living organisms had originated from primitive ancestors and that in the slow process of such changes had become adapted for living in particular environments. The concept of adaptation originated with Lamarck and is very important in understanding evolution. Lamarck, however, linked this with some concepts no longer believed to be tenable. He believed that adaptation had come about through some kind of internal driving force, a 'vital spark' which made animals become more complex. He felt that new organs must arise from new needs and that these 'acquired characters' were inherited, as in his classic postulate that the neck of the giraffe had become longer in response to a 'need' to reach leaves higher up on the tree. Since 'Lamarckism' has largely been tested and rejected scientifically few scientists today accept Lamarck's theory of the inheritance of acquired characters, but all recognise his genius in pinpointing the critical concept of adaptation.

The special point about Darwin is that he provided a logical and testable theory for evolutionary change: one that has stood the test of time and provided a starting point for later developments.

The full title of Darwin's major work of 1859 was *On the Origin of Species by Means of Natural Selection, or the Preservation of Favoured Races in the Struggle for Life.* The main points of the theory are straightforward.

(a) Animal species reproduce more rapidly than is needed to maintain their numbers. Animal populations, however, though fluctuating, tend to remain stable. (Here he was influenced by the Englishman Malthus who had written on this subject some years earlier.)

(b) There must therefore be competition within and between species in the 'struggle for existence', for food, for living space and (within members of the same species) for mates, if the characters that individuals bear are to be transmitted to the next generation.

(c) Within species all animals vary, and this variation is inherited.

(d) In the struggle for life those individuals best fitted to survive in a particular environment are the ones to live and to reproduce. The others are weeded out in the intense competition. The favourable characters that make such survival possible are inherited by future generations, and the accumulation of different favourable characters leads to the separation of species well adapted to particular environments. This is what Darwin called 'natural selection'.

All this seems logical enough, though Darwin's early critics, Mivart, for instance, argued that Darwin had not really shown how favourable characters were actually accumulated, only how those animals less fitted to their environment failed to survive. And in this they were not unsound, for the most serious weakness of the theory as presented by Darwin was that the nature of variation and heredity were largely unknown, so that his views on this were speculative and insufficient for the theory to be seen to work.

The pioneer work of the Austrian monk Gregor Mendel in 1865, and of the later school of T. H. Morgan which began in 1910, laid the foundation for genetic experiment and theory. It was this that supplied the necessary understanding of heredity essential to the amplification of Darwin's theory.

Inheritance and the source of variation

In the cells of all eukaryotes (all animals and most plants) there is a **nucleus** (Fig. 2.1) containing elongated thread-like bodies, the **chromosomes,** which are made of protein and DNA (deoxyribose-nucleic acid). The chromosomes are usually paired (Fig. 2.2a) and consist of strings of **genes,** of which there may be several hundred per chromosome. These genes are the primary units of heredity; they are particles of DNA which carry a kind of programme, or **genetic code** as it is known, in the order of the nucleotide bases which they bear. This genetic code directs the development and functioning of the whole organism. Some genes are 'structural' and control the synthesis of the proteins and other compounds of which the living body is made. Other genes are concerned with organising and regulating these compounds and building them up into the body. Exactly how the genes fulfil these tasks is far from well understood, and one of the principal tasks of modern genetic and molecular biology is to shed light upon this. It has been established, however, that the genes release chemical products which start a whole host of complex reactions. In some kinds of develop-

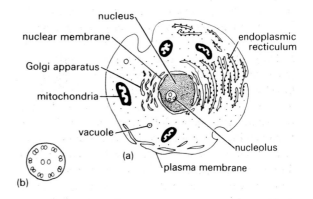

Figure 2.1   (a) Eukaryote cell structure, with organelles; (b) cross-section of flagellum showing paired 9 + 2 structure of micro-fibrils. (Based on Hurry, S. W. (1965) in *The microstructure of cells* (London: John Murray).)

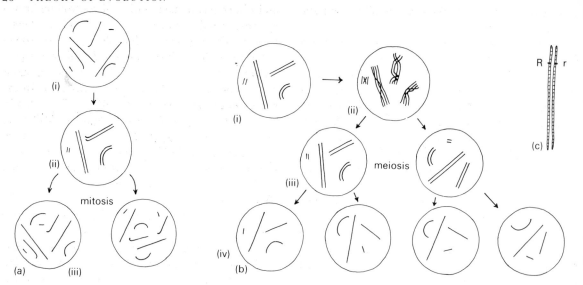

Figure 2.2 (a) Mitosis: (i) pre-mitotic somatic cell with unpaired chromosomes; (ii) homologous chromosomes have paired up prior to longitudinal fission; (iii) chromosomes have divided longitudinally, and moved apart to opposite ends of the cell, which has then split into two cells with identical sets of chromosomes. (b) Meiosis: (i) chromosomes have paired up as in mitosis; (ii) chiasma formation after longitudinal fission results in exchange of sections of homologous chromosome; (iii) first meiotic division results in two daughter cells with paired chromosomes; (iv) second meiotic division resembles mitotic division and results in gamete production. (c) Homologous chromosomes, paired up with dominant (R) and recessive (r) genes at matching loci.

ment the genes are switched on and off in particular sequence, releasing products which react together in synthesising complex molecules. Theories now being developed (e.g. that of Britten & Davidson 1973) suggest that structural genes can be activated and deactivated when needed; evidently the initial stimulus to the switching on of structural genes is given when a sensor gene receives an appropriate stimulus.

Genetically identical organisms reared in different environments will not develop to exactly the same form, as is witnessed by the different appearance of vegetables grown in rich and poor soils respectively. The inherited genetic material in any organism, known as the **genotype,** is reacted upon by the environment (probably through sensor genes) to create a developed individual. This individual, the product of both heredity and environment, is the **phenotype.**

Sometimes single genes control single characters. More often characters are **polygenic;** that is, many genes, each of small effect, contribute to the biochemical pathways that result in the formation of a particular character. Again, some genes are pleiotropic; that is, they affect several characters since the same gene products may be used in different biochemical pathways.

Where does variation come from?

Within the nucleus of any cell in the body of most living organisms there is a specific number of chromosomes with their genes. The chromosome number is always constant for the species. In man there are twenty-three pairs of chromosomes; in the fruit fly/*Drosophila* there are four pairs. The members of each pair are all **homologous,** i.e. similar in appearance and length, except for one pair, the sex chromosomes, upon which the genes regulating sexual characters are located.

When chromosomes are paired like this the organisation is said to be **diploid,** the chromosomes number being conventionally defined as $2N$.

When the body (**somatic**) cells of an animal divide as the organism grows, the chromosomes divide by longitudinal fission, and each **'daughter'** cell inherits an exact copy of the chromosomes and genes in the **parent cell.** This process, known as **mitosis**

(Fig. 2.2a), is effectively the same for all the somatic cells in any one body.

On the other hand, the formation of eggs or sperm (**gametes**) in the testes and ovaries respectively involves a very different process: **meiosis** (Fig. 2.2b). There are two cell divisions. The first of these is a reduction division in which each daughter cell inherits only one chromosome from a homologous pair. As the chromosomes separate during the first meiotic division they are sometimes tangled at certain points known as **chiasmata,** and as they pull apart they break and re-form, each exchanging parts of the same length with its homologous chromosome. The result of such '**crossing over**' is that the unpaired chromosomes passed down to the daughter cell are not identical with those of the parent cell, as they are composed of bits of each of the homologous paired chromosomes. The second meiotic division is like a somatic cell mitosis, so that the end product of the two divisions is four gametes from a single parent cell. Each gamete has half the number of parent chromosomes; the chromosomes are unpaired, and none are identical to any of the parent chromosomes since genetic material was exchanged during crossing over.

These gametes containing unpaired chromosomes are said to be **haploid,** and the chromosome number is N.

When eggs are fertilised by sperm the homologous chromosomes from different parents come together and pair up. In this process of **recombination** a new genotype results from already existing genes, deriving from different sources. The effect of both crossing over and recombination is that variation within the products of reproduction is high. It is mainly upon this variation that natural selection operates.

Significance of alleles

Homologous chromosomes lie side by side matching gene by gene. The homologous genes, however, are not necessarily identical, for they may exist in a number of 'expressions', known as **alleles,** each of which is slightly different chemically and will lead to the development of differences in the characters it controls. Such differences may range from the minor to the very substantial.

Sometimes, as in the genes controlling human eye colour, there may be only two alleles, designated R and r in conventional notation, R controlling brown eye colour and r controlling blue eye colour. Alleles for eye colour may exist on homologous chromosomes in four possible combinations: RR, Rr, rR or rr. In the **homozygous** condition alleles on homologous chromosomes are of the same kind: RR or rr. In the **heterozygous** condition they are of different kinds: Rr and rR. An individual possessing RR will have brown eyes, whilst the possessor of rr will be blue-eyed. But the combinations Rr or rR will also lead to brown eyes, for the allele R is **dominant** over r and masks its effect. In this case r is said to be the **recessive** gene.

Sometimes alleles are multiple; some genes are known to exist in up to eleven alleles. Human blood types, for example, are controlled only by a single gene which occurs as three alleles designated p, q and r. Alleles p and q have equal dominance whilst r is recessive. Of the four human blood groups O has the recessive homozygote rr; group A may carry pp or pr and group B qq or qr, whilst the group AB has the dominant heterozygote pq.

The above examples deal with genes affecting single characters, but there are more complex patterns of variation when two or more genes control single characters. Thus plumage colours in parakeets are under the control of two genes which in different combinations may produce blue, green, yellow or white feathers – a wider combination of colours than would be possible with only a single pair of alleles.

These sources of variation – crossing over, recombination, homo- and heterozygosity, multiple alleles, etc. – are all consequences of diploidy. In addition, many plants increase their potential variability by **polyploidy,** i.e. duplication of chromosomes, allowing a greater number of loci for alleles. Polyploidy, however, is rarely found in animals.

Recessive genes might be expected to get lost during evolution, but in fact the proportions of alleles within populations do not normally change, unless there is a special survival advantage in possessing one particular type of allele. If there is no such advantage a genetic equilibrium is maintained in terms of the Hardy–Weinberg equation:

$$p^2 + 2pq + q^2 = \text{equilibrium}$$

where p^2 and q^2 represent the relative proportions of dominant homozygotes *(RR)* and recessive heterozygotes *(rr)* respectively. The equilibrium so defined will only be shifted when there is a decided selective advantage in possessing a particular allele. Thus if a particular phenotype is preserved by natural selection the balance will alter and the proportion of the allele within the population will increase.

Evolution can thus be considered as based upon changes in gene frequencies through natural selection.

Mutation

Natural selection operates upon the great stock of variation inherent in the gene pool of a population. The genes are continually reshuffled by the mechanism discussed above in new combinations or genotypes. But if evolving organisms are ever to give rise to anything radically new there must be a source of new variation. Where does this variation come from?

It sometimes happens that a gene will suddenly and spontaneously change to a new allele. This kind of random change is knows as a **gene mutation.** It is the result of a chemical change in the DNA helix: a substitution or rearrangement of nucleotide bases, or a duplication or loss of a small part of the sequence.

The new allele thus formed is heritable and will produce changes in one or more characters in the organism. These changes are usually small scale and of limited effect; if they are larger they are very often lethal. Most mutations are neutral, some are disadvantageous, but others may lead to new or somewhat different characters which may carry some selective advantage. An advantageous mutation, which may only have a marginally beneficial effect, will only be able to spread if:

(a) it is recurrent within the population so that new recruits are continually added to the mutant stock;
(b) it carries some selective advantage so that the Hardy–Weinberg equilibrium is shifted in its favour;
(c) the population is small enough for it to be carried and spread relatively quickly.

All alleles have resulted from mutation; multiple alleles have arisen at different times. Some mutant alleles may spontaneously mutate back again to the original gene expression from which they arose. The

rates of mutation vary with the gene. Some are much higher than others, but they are generally low, one mutation in 10 000 for a particular gene being considered as high. The rate of mutation can be increased by radiation, including exposure to ultraviolet light, or by chemical means.

There are not only gene mutations but also **chromosomal mutations,** which may involve the loss, duplication or transfer of a set of genes within the chromosome. These tend to have larger-scale phenotype effects than do gene mutations and are more often disadvantageous.

Spread of mutations through populations

Mutation is the primary source of variation, but it is only through the effects of diploidy in higher organisms that the mutations can spread. As mentioned, it is not likely that mutations will spread very fast in a large population, even if they are dominant and mutation rates are high. But most natural populations are not evenly distributed throughout their species range. They tend to be subdivided into relatively small units of population, known as **gamodemes.** Examples of such natural population groupings are rabbit warrens, schools of fishes, pond communities, the faunas of enclosed lagoons and, in the human sphere, the inhabitants of towns and villages. They tend to be partially isolated from other such gamodemes but not completely so, for though each gamodeme carries its own gene pool there is some possibility of exchange and hence gene flow between the pools.

It is within such small natural populations that favourable mutations, especially those allowing better adaptations, are able to spread. A mutant condition established even in one family may, if it possesses selective advantage, soon spread through the population in the gamodeme within a relatively few generations. It may then spread to other gamodemes in the vicinity, increasing the effective range of the useful mutant condition. Such a mutant condition, especially if occurring in a **peripheral isolate** (i.e. a single gamodeme, or small cluster of gamodemes on the fringe of a larger widespread population), will be the first step towards the formation of a new and distinct species. The accumulation of a few more mutations may then isolate it completely, as in the classic Darwinian mode. The origin of species from such peripheral isolates is known as **allopatric speciation** (p. 31). When the new type is sterile with the parent stock, having diverged enough genetically, the 'parent' and 'daughter' species are then entirely distinct.

Recessive mutations spread comparatively slowly, and only in small populations with close inbreeding is there any likelihood of more rapid spreading. However, recessive mutations can become dominant under certain circumstances. Experiments have shown that changes in temperature alone may be enough to modify the dominance of a gene. And in natural populations there are some genes existing only as modifiers, changes in which may alter dominance to recessiveness in a gene. If a recessive allele is favourable it will be selected for dominance.

Isolation and species formation

It should now be clear that some kind of isolation is normally essential for species formation. It may be a geographical isolation, i.e. some part of a population being cut off from the rest by a barrier, as noted above. But there are other kinds of isolation which may be equally effective in certain cases. There is ecological isolation, in which species become adapted for microhabitats in the same general area. Physiological or biological factors may also isolate populations so that they speciate, e.g. changes in the timing of the breeding cycle or merely size differences precluding copulation. Again, there may be a simple genetic incompatibility; fertilisation may take place, but because of different chromosome lengths or locations of alleles recombination is impossible and the hybrids abort in embryo or are sterile.

Genetic drift

If a population is very small it will contain only a limited and random sample of the total genetic variability within the whole species, i.e. only a fraction of the overall 'gene pool' of that species. The smaller the population the fewer will be the alleles handed down to future generations.

At each meiosis only one chromosome of a pair, with its random sample of alleles, is passed down to the gametes; any alleles in a tiny population that are not recombined in fertilisation are therefore lost, and a special evolutionary process becomes important in these small populations, known as **genetic drift.** This is simply an evolutionary change in gene frequencies through random chance assortment, without selection being involved. Hence certain localised populations descendant from an initial pair or few pairs may become quite distinctive or may even differentiate as new species merely because the founders of the population happened to carry a particular set of alleles.

In any gene pool the effects of all the factors considered so far are continually interacting. Gene frequencies within the gene pool are related at the chromosome level to mutation, crossing over and recombination. They may be 'stabilised' in a particular kind of selection that actually prevents evolution in a large population, for if the species is successful, widespread and well adapted to its environment, the undesirables are weeded out and the only operational selection is **stabilising selection.** This is analogous to what Mivart proposed in his critique of Darwin. On the other hand, gene frequencies begin to alter when the environment changes to become easier or harsher or when a population migrates into a new area and, finding little competition, begins to differentiate ecologically. The intensity of selection under these latter circumstances is much greater. One would therefore expect to find in the fossil record relics of long periods of species stability, separated by breaks representing times when speciation was very rapid, and then again stable periods. And this is exactly what is found, as will be explained at some length in the next section.

In any evolving system the more microenvironments there are the greater will be the possibility of differentiation. Species thus can be said to arise in response to the environments actually available.

Natural selection today

It is well known that natural selection is operating at the present time. Evidence for this comes from various sources. Topical examples could include the rise of myxomatosis-resistant strains of rabbits or antibiotic-resistant viruses. The case of 'industrial melanism' in moths has been extensively studied and quoted. Before the Industrial Revolution in the north of England, the moths inhabiting the lichen-covered trunks of trees had a pale speckled colour harmonising with the environment and rendering them inconspicuous. When records were first kept in the early 1850s these moths made up 99 per cent of the population; the other 1 per cent were dark-coloured types which were more easily picked out by predators. But by the end of the century the woodlands near the cities had become polluted and the lichens, which are very sensitive to unclean air, had been killed, exposing the blackened tree trunks below. The vast majority of the moths in these areas were then the dark-coloured forms, the pale varieties accounting for no more than 1 per cent. Since it had become advantageous to be dark-coloured so that predators would less easily see the moths on the dark trunks, natural selection had been intense in the changing environment. Some seventy species of moths are known to have been affected by such industrial melanism within the very short time span of fifty years. But only a few genes are here affected; it requires more changes to initiate speciation, involving a greater variety of genes controlling different characters.

In the example quoted above the selective advantage of the gene is clear, but in many cases it is less easy to understand the selective advantages of particular characters.

FOSSIL RECORD AND EVOLUTION

To what extent does the special perspective of geological time give us an insight into the nature of evolutionary processes? Most particularly through palaeontology it is possible to trace phylogenies, i.e. the descent of particular evolving lineages. This has in the past been one of the major tasks for palaeontology, and nowadays not only phyletic evolution but also community evolution is receiving much attention. On a more general scale, a unique contribution of palaeontology is an understanding of how evolutionary rates have varied over extended time periods, and from this has emerged a conception of modes or patterns of evolution. A further possibility is that the predictions of genetical theory can be tested, as has already been explained with allopatric evolution. The effects of both inter- and intraspecific competition can be seen in fossil as well as in modern taxa; clearly, the better the fossil record is known in terms of the distribution of past invertebrates, the

greater the possibilities for this kind of work. Some of these topics are here selected for special discussion.

Rates of evolution

Evolutionary rates within evolving groups are far from constant. Most importantly there is a general pattern which has been recognised in many taxa; in this the rate is at first very rapid, later declining and in the survivors becoming almost static. Initially a successful group, expanding its range through migration, will evolve very quickly with early stocks colonising and becoming adapted to different environments. Such adaptive radiations are normally quick; the rapid rate of evolution here is sometimes known as **tachytelic** evolution. It raises some special problems which will be discussed later.

Once the initial burst is over and the main groups are differentiated, the rate of evolution generally slows to a more normal or **horotelic** speed in which the origin of new types more or less equates with the extinction of old ones. If conditions are right the lineage remains vigorous and continually speciating, but if a lineage declines, as it may, the extinction rate exceeds the production of new species. A graph showing 'survivorship' of a particular lineage is thus strongly skewed. In a declining lineage the survivors may often continue for a very long time, evolving very slowly or not at all. This is **bradytelic** evolution. Bradytelic taxa include all 'living fossils': relics of ancient stocks which have continued long after the parent groups which gave rise to them have vanished. Such genera as the brachiopod *Lingula* may be considered here. This genus was already in existence in the Upper Cambrian and since then has changed relatively little. It is able to survive because early on in history it colonised a 'difficult' but persistent environment, that of mudflats and shoals exposed at low tide (though specimens are known from deeper waters). There was relatively little competition for this since the physiological problems imposed by fluctuating salinity and periodic exposure are severe. But because *Lingula* was able to cross this adaptive threshold, which other invertebrates have rarely been able to do, there is no reason why it should not go on living in such an environment indefinitely. It is unlikely to be displaced by competitors.

Some 'bradytelic' animals now live in the quiet and stable waters of the deep sea. Hexactinellid sponges and the monoplacophoran mollusc *Neopilina* (whose most recent fossil relative is Middle Devonian) are amongst these. But the deep sea, though a stable enough environment and a haven for relict stocks, is not an ideal home because of the relatively limited food supply. Those animals which can cope with such unpopular environments have a fair chance of becoming bradytelic, under the influence of stabilising selection. Relics of once important groups living thus are now restricted though not necessarily declining, for competition in the environments in which they live is not particularly strong.

Modes of evolution

Intimately connected with rates of evolution are the various patterns in which animals arise and evolve. It is here that the fossil record is of particular value, for though the nature of these patterns might be predicted it would be unlikely to be confirmed from observations on modern animals alone.

Three main grades or modes of evolution have already been referred to in passing and will be summarised here: **micro-, macro-** and **megaevolution.**

Microevolution

The origins of new species from existing ones largely through changing interactions between the animal and its environment are generally referred to as microevolution. Whilst the genetic and isolating processes by which species differentiate are well understood, it is only comparatively recently that full confirmation of how they have actually operated through time has been clearly seen. Simpson (1953), working mainly with vertebrates, linked genetic theory and palaeontological data in a timely manner, as did Carter (1954). But for a long time most palaeontologists tended to have firmly entrenched concepts of evolutionary change based on somewhat *a priori* arguments. There was a general tenet that species had arisen by what came to be termed **phyletic gradualism.** This implies an even slow methodical change of the whole gene pool and hence of the characters of the entire population. This transformation was thought to have involved most or all of the population and to have occurred over much of the geographical range of the parent species. If this

were true one would expect the fossil record (ideally at least) to consist of a continuous series of graded intermediate forms. Now there are a few instances where this may have happened; for example, the evolution of the Cretaceous sea urchin *Micraster* may in part be based upon such processes, though this is disputed by some. But in fact such gradual transformations are not commonly encountered in the geological record; indeed the situation is quite the reverse. In most continuous rock sequences what is normally seen is a sequence of stable fossil populations separated by abrupt morphological breaks. It is true that in many sedimentary sequences there are innumerable undetected breaks in the sequence (diastems) which represent periods of erosion sandwiched in between depositional interludes. But do diastems, unconformities, other imperfections of the geological record, and the migrations of faunas into the area represented by the sequence under consideration account for all the apparent imperfections of the fossil record? It would have been so accepted some years ago; such arguments had become part of the standard habits of thought or paradigm of most palaeontologists.

In 1972 Eldredge and Gould, having independently documented precise patterns of microevolutionary change in Devonian trilobites and in Pleistocene land snails, concluded that the old arguments were largely untenable and were in any case rather out of touch with genetic theory. Instead they made clear that evolution by allopatric speciation was a more viable model for palaeontology and could sometimes be tested by a close examination of fossil populations. The central concept of allopatric speciation is that new species 'can only arise when a small local population (one or more gamodemes) becomes isolated, at the margin of the geographic range of the parent species'. Eldredge and Gould termed these small marginal populations 'peripheral isolates'. It is in these, as we have seen in previous sections, that gene flow is at its most rapid, whether under the influence of selection or of drift. Should the parent species then become extinct one of the peripheral isolates may migrate into the original range of the ancestral species and then expand rapidly. In the rock sequence this replacement would be represented by an abrupt faunal break. This happened more than once with Eldredge's trilobites, and also with Gould's Bermudan land snails; indeed the relative degree of isolation of the islands they lived on at any particular time could be inferred from the geological record.

Most fossil species tend to exist in 'punctuated equilibria', i.e. as relatively stable populations showing little change during their life span and separated from each other by abrupt faunal breaks suggestive of extinction (at least for that area) and migration of new stock into the region. Only in relatively rare examples, of course, where the successor populations are widespread and preserved through great stratigraphical thicknesses can allopatric speciation directly be seen to have operated. The existence of punctuated equilibria by itself is consistent with almost any model of speciation, and very often little can be inferred from a particular vertical sequence as to how species actively arose. Even so, allopatric speciation seems to be a more valid model for palaeontological evolution than the old concept of phyletic gradualism, or 'stately unfolding' as it came (perhaps pejoratively) to be known. And as such it has found great favour in recent years.

This is not to say that evolution by phyletic gradualism does not happen. It probably does, though is unlikely to be an important means of speciation. And it can still be retained as part of the conceptual armoury of palaeontology, provided it is seen in its true perspective.

A final point in microevolution theory: in what kinds of environments, assuming allopatric speciation to be the norm, would we expect to find the highest rates of speciation? There has been some controversy about this, and there is as yet no positive certainty. Bretsky and Lorenz (1971) have pointed out that:

(a) In unstable environments (usually onshore) faunal diversities are low, but there are large numbers of morphologically variable individuals living in stable kinds of communities. Bretsky and Lorenz have therefore postulated that these communities were stable and persistent and not especially subject to the adverse effects of sudden environmental change. Hence their rates of evolution would be expected to be low.

(b) Stable environments, usually offshore, tend to show much higher faunal diversities, but there are relatively small populations of each species with limited variation. Bretsky and Lorenz have predicted that in these the effects of sudden

environmental changes would be keenly felt, extinctions would be common and hence the rate of evolution would be high. Other authorities claim, however, that morphological diversity is inversely correlated with genetic diversity.

Eldredge (1974b), furthermore, whilst accepting that there are relatively more species in the offshore environment, has noted that onshore areas (a) tend to persist continually through longer periods of time, and (b) are spatially more heterogeneous or 'patchy' in any case. Hence onshore areas are just as likely to be the major sites of speciation if not more so than offshore regions. Studying the geographical range of the Middle Devonian trilobite *Phacops cristata* over most of North America, he found that the highest intraspecific variability occurred in the onshore environments as predicted; such variability decreased away from the onshore regions. Furthermore, comparison of ancestral and descendant forms suggests that some *Phacops* species in later Devonian rocks had actually originated from definitive 'onshore ancestors'. Eldredge has therefore claimed that higher diversities in the offshore regions do not necessarily result from higher rates of speciation therein but exist purely because of normal biogeographical processes, such as marine transgression and regressions. Much more evidence is needed, however, before these issues can be resolved.

Macroevolution

This is broadly synonymous with **adaptive radiation:** the diversification of an initial stock into smaller stocks, each becoming adapted to its own environment. Whilst adaptive radiation can occur at almost any taxonomic level, it is most strikingly illustrated amongst invertebrates in groups showing rapid diversification above the species level. Thus the Jurassic 'evolutionary explosion' of sea urchins, for instance (Fig. 9.8), illustrates a rather fine (if somewhat slow) example of an ancestral stock diversifying into many highly successful orders. The radiation of trilobites in the Ordovician, following the later Cambrian extinctions (Fig. 11.17), is another striking example. There are many such good examples in the invertebrate record, though these have been rather neglected in texts on evolution in favour of those of vertebrates, which perhaps show more clearly how one new adaptive step or key character may initiate a radiation, paving the way for later radiations as new characters are produced. This is generally less clear in invertebrates, though perhaps the echinoids again are especially good examples.

Certain processes particularly appear to have been operative in macroevolution. One is the well-understood principle of change in relative growth of different parts of the body, giving the derivative stock different physical proportions. Such **allometric** changes as these could be relatively gradual through time and may account for much adaptive macroevolutionary change. But not all changes are gradual; some indeed are quite abrupt and hence more difficult to understand. Most particularly, can we explain the formation of new and functional characters or organs? **Aristogenesis** is a term sometimes used for the formation of new characters or structures. Osborn (1934), who developed this concept, imagined 'new adaptive units' as 'originating in the germ plasm' (i.e. presumably through gene mutation) and 'slowly evolving into important functional service'. New characters are one thing, however, and integrated functional organs are another, and though they undoubtedly evolved from small beginnings by the integration of the function and structure of several characters not much at present can be said about such processes.

Abrupt and major changes in whole organs or even in groups of organs likewise demand an explanation. There are two related concepts which may be important here, though they are certainly not the full explanation for such changes and indeed only really help us to understand modifications of existing organs. These processes, **neoteny** and **paedomorphosis,** involve changes in the rate or process of development of functional structures.

Neoteny refers to the onset of sexual maturity in the juvenile animal, thus eliminating the characters of the adult stage. If the life history of the organism is complex and the juveniles look very different from the adults, then the loss of adult morphology will allow a very abrupt morphological change to take place very quickly. Neoteny has been invoked with great success in understanding, for instance, the origin of early fish-like ancestors of vertebrates from the tadpole-like free-swimming larvae of sea-squirts. Likewise the small planktonic copepods may be neotenous pro-

ducts of the onset of maturity in planktonic larvae of benthic crustaceans.

Paedomorphosis is a closely related concept meaning the retention of 'juvenile' characters into the adult. Premature sexual maturity of the whole animal does not need to take place to allow this; it is rather that the genes controlling the development of particular characters are modified so that the full development of the character is not completed. This process can substantially change the appearance of a whole organ or complex of characters and seems to be very important in the fossil record, much more so than has been generally appreciated. Paedomorphosis can account for a number of sudden and abrupt changes which could not be fully understood otherwise, including structural and functional changes in whole organs. An example is given by the eyes of trilobites, discussed in Chapter 11.

There are many such examples in the geological record, suggesting that paedomorphosis and neoteny have been important in macroevolution and indeed in megaevolution as well.

Paedomorphosis may be seen as a special case of **pre-adaptation.** The concept of pre-adaptation, in its simplest form, means only that animals must already possess characters capable of being modified if they are going to be able to adapt to new environments. Sometimes these characters may be in the form of organs that have lost their original function; they are therefore 'spare' and can readily become adapted without inhibiting an existing function. More often, however, they are already functional, but by becoming modified they may then be more flexible in the variety of functions they perform. There are many good examples drawn from the vertebrates, such as the use of a gut diverticulum as an air-breathing lung in Devonian and later lungfish living in rivers and ponds that are liable to dry up; the same structure became the swim-bladder in ray-finned fish.

Numerous examples could be cited in the invertebrates also, the remarkable adaptation of already existing spines, cilia and tube-feet in echinoids for deep burrowing being an ideal case.

However, not only may structures be pre-adapted before modification; they must also become **post-adapted** whilst their function and probably form as well are in the process of being transformed. Pre-adaptation has become recognised as a very critical concept, indeed Carter (1954) has written: 'No adaptation is possible without some grade of pre-adaptation in the structure to be adapted.'

An earlier concept relating to ontogeny as a tool for understanding evolutionary relationships was that of Haeckel, who believed that the ontogeny of any animal was a more-or-less direct recapitulation of its phylogeny. In other words, the young stages of development of any animal are very similar to what the ancestors of that species were like. In embryology and in palaeontology there is very little evidence for this (though one or two possible cases are discussed with reference to ammonites in Chapter 8), and since the widespread acceptance of paedomorphosis Haeckel's 'Biogenetic Law', as it was called, has been very largely discarded.

Megaevolution

Megaevolution has been defined as the origin of new systems of animal organisation. It grades into macroevolution and involves the same basic principles. Yet it is less well understood, for the links between higher taxa – orders, classes and most particularly phyla – are very obscure and rarely represented in the fossil record. Thus the diversification of phyla in the early Cambrian is perhaps the least clear of all the phenomena of palaeontological evolution, and the evidence of relationships given by geology is often very limited, more so than the evidence from comparative anatomy, embryology and to some extent chemical taxonomy, which is somewhat better if still hazy.

As with macroevolution, the early stages in the evolution of a new stock under intense selection pressure were tachytelic and geographically very localised. Usually only very small populations were involved, which may have been quickly superseded by more highly adapted types; hence the likelihood of preservation of the transitional linking forms is remote. It may be wondered whether there are any fundamental differences in the evolutionary process operating at the levels of micro-, macro- and megaevolution. This is uncertain and has been much debated, but in the opinion of the present author there is no reason to believe that radically different processes operate at any of these levels. After all, the origin of a successful new species in a small isolated population need not have differed greatly from the origin of the germ of a new higher taxon in similar circumstances, given that paedomorphosis and pre-adaptation may have played a part in launching the new type.

COMPETITION AND ITS EFFECTS

Darwin emphasised the role of competition as an important factor in natural selection. It is clear that competition for food, living space and mates is a primary control within a single species or population. Such competition may be direct, in terms of aggression and the establishment of a hierarchical order within certain kinds of higher animal societies. On the other hand it may be indirect, in terms of too many individuals passively exploiting the available food supply. But what are the effects of competition between species, especially if one invades the territory of another or if the geographical ranges of two species overlap? Undoubtedly interspecific competition is highly important, as is witnessed by the catastrophic reduction of the native British population of red squirrels after the introduction of the American grey squirrel in the nineteenth century. There are now very few red squirrels left in Britain, after only about 100 years.

An invading species may be expected either to fail to displace an established species, or to take over part of its range and set up a territorial balance, or to eliminate the native species entirely. How often the latter happens is not clear, for in the fossil record abrupt faunal breaks may signify the migration of an allopatrically derived species into an area after the native stock has become extinct through environmental or other causes.

Competitive elimination between **sympatric** species (i.e. species living in the same area) may sometimes be avoided by a variety of adaptive strategies. One way is spatial segregation; the animals come to inhabit and feed only within strictly localised geographically or vertically (depth) restricted habitats. Hence the two sympatric species are not directly competing for food or living space.

Another strategy is character displacement, either in sympatric species or where the ranges of neighbouring species overlap geographically. In the latter case some rather curious reactions between the two species with their similar ecological requirements are et up. Either individuals in the range of overlap may come to look very similar to those of the competing species, or alternatively they may diverge in characters to a much greater degree than normal. Evolutionary convergence may be interpreted as a strategy for coexistence which would minimise the effects of selection, so that it would operate upon all the individuals in the range of overlap as if they were just one species and not two competing ones. Divergence of characters, on the other hand, suggests selection within the two species for slightly different habitats, so that competition is avoided by habitat proliferation.

Cases illustrating these are known from the fossil record. In Cambrian agnostid trilobites there is evidence of depth segregation, as well as geographical separation in closely related species, and at the same time size displacement seems to operate in several species. It has been predicted that, in order to be an effective strategy for avoiding competitive elimination, size displacement would need to be in the ratio of about 1:1·28 for sympatric species. The fossils investigated, including these agnostids, seem to come remarkably close to this theoretical figure.

Eldredge (1974a) has shown how during Devonian times in North America there lived two species of the large trilobite *Phacops*. *P. iowenis* inhabited the western area, whilst successive allopatrically derived subspecies of *P. rana* (probably originating as a European invader) occupied the terrain to the east. The species range of the two did not normally overlap – a case of non-competitive geographical separation – but where the two species are found together *P. rana* and *P. iowenis* mutually reacted; *P. iowenis* diverged in morphology from *P. rana* whilst *P. rana* converged, diverged or stayed 'neutral' in different characters, exhibiting a 'mixed' reaction. On the basis of these studies Eldredge has suggested that character displacement may be no more than a 'magnified microcosm of the general pattern of interaction between two species even when allopatric', provided that allopatry is sustained through competitive exclusion.

SUMMARY OF PALAEONTOLOGICAL EVOLUTION THEORY

Evolution can be defined as a change in gene frequencies through time. In any gene pool most variation results from recombination of dominant and recessive alleles, but new variation is introduced by mutation and under certain circumstances mutant alleles can become dominant. In very small populations much evolutionary change is non-adaptive because of the effects of genetic drift, but in larger populations selection acting upon the variation present in the gene pool and leading to adaptation is the primary force in evolutionary change.

A minority of new species may arise by the 'classic' pathway of slow continuous change in a large population, but allopatric speciation is now believed to be much more important. Most new species arise as peripheral isolates on the fringes of the parent population, migrating wherever there is a vacant niche. Populations once established over a wide area tend to remain static, being stabilised by selection unless there is a marked environmental change which increases the selection pressure. Such stasis is the norm for populations, and the origin of new species is a rather special and relatively rare event. Competition between species in a new environment may be direct in the initial stages before a balance is achieved, but in a more balanced community of longer standing a stable predator/prey ratio is maintained and competition is usually avoided, character displacement between similar types being one of the mechanisms. Some types of biotic association are remarkably persistent over long periods of time.

Evolution can be considered at three gradational levels: micro-, macro- and megaevolution, referring to speciation and adaptation, adaptive radiation and the origin of higher taxa respectively. The same kind of processes, however, may be operating at all of these levels of evolution.

BIBLIOGRAPHY

Books, treatises, symposia

Carter, G. S. 1954. *Animal evolution: a study of recent views of its causes*. London: Sidgwick & Jackson. (Classic textbook)

Darwin, C. 1859. *On the origin of species by means of natural selection, or the preservation of favoured races in the struggle for life*. London: John Murray.

Dobzhansky. T., F. J. Ayala, G. L. Stebbins and J. W. Valentine. 1977. *Evolution*. San Francisco: Freeman. (Comprehensive advanced text)

Drake, E. T. (ed.) 1968. *Evolution and environment*. New Haven and London: Yale Univ. Press. (16 original papers)

Eaton, T. E. 1970. *Evolution*. London: Nelson. (Valuable zoological text)

Gould, S. J. 1977. *Ontogeny and phylogeny*. Cambridge, Mass.: Belknap (Harvard). (Excellent historical and philosophical treatment of ideas on recapitulation, neoteny and paedomorphosis)

Jepsen, G. L., E. Mayr and G. G. Simpson 1949. *Genetics, paleontology and evolution*. Princeton, NJ: Princeton Univ. Press. (Many individual papers; classic, if a little dated)

Koestler, A. 1971. *The case of the midwife toad*. London: Picador. (A penetrating analysis of the Lamarckian–Neo-Darwinian controversy of the 1920s)

Maynard-Smith, T. 1975. *The theory of evolution*. London: Penguin. (Highly informative text)

Savage, J. M. 1963. *Evolution*. New York: Holt, Rinehart & Winston. (Short simple introduction)

Simpson, G. C. 1953. *The major features of evolution*. New York: Columbia Univ. Press. (Classic work combining many aspects of evolution theory)

Valentine, J. W. 1973. *Evolutionary paleoecology of the marine biosphere*. Englewood Cliffs, NJ: Prentice-Hall. (See Ch. 1)

Individual references

Bertalanffy, L. von. 1969. Chance of law? In *Beyond reductionism*, A. Koestler and J. Smythies (eds). London: Hutchinson. (Strongly opposes Neo-Darwinism as a complete explanation).

Bretsky, P. W. and D. M. Lorenz 1971. Adaptive response to environmental stability: a unifying concept in paleoecology. *Proc. North Amer. Paleont. Convention, Chicago, 1969* (E), 522–50. (Advanced)

Britten, R. J. and E. H. Davidson 1973. Repetitive and non-repetitive DNA sequences and a speculation on the origins of evolutionary novelty. *Quart. Rev. Biol.* **46**, 111–33. (Advanced)

Eldredge, N. 1971. The allopatric model and phylogeny in Palaeozoic invertebrates. *Evolution* **25**, 156–67. (Very important work)

Eldredge, N. 1947a. Character displacement in evolutionary time. *Amer. Zool.* **14**, 1083–97. (Advanced)

Eldredge, N. 1974b. Stability, diversity and speciation in Palaeozoic epeiric seas. *J. Paleont.* **48**, 540–8. (Important work stressing allopatric evolution)

Eldredge, N. and S. J. Gould 1972. Punctuated equilibria: an alternative to phyletic gradualism. In *Models in paleobiology*, T. J. M. Schopf (ed.), 82–115. San Francisco: Freeman, Cooper & Co. (Allopatric evolution)

Gould, S. J. 1969. An evolutionary microcosm: Pleistocene and Recent history of the land snail *P. (Poecilizonites)* in Bermuda. *Bull. Mus. Comp. Zool.* **138**, 407–532. (Allopatric evolution)

Osborn, H. F. 1934. Aristogenesis: the creative principle in the origin of species. *Amer. Nat.* **68**, 193–235. (First use of this term)

3 Origin and early diversification of metazoans

The Earth's early atmosphere was probably anoxic, consisting of gases such as methane, hydrogen and ammonia. From these gases there condensed simple organic molecules, especially the amino acids which are the building blocks of protein. By about 3500 million years ago chemical evolution, starting with such building blocks, had progressed far enough to allow the origin of the earliest long-chain self-replicating molecules which are the basis of life, for in a chemical sense life is a function of the mutual relations of nucleic acids and amino acids. The nucleic acids act as templates which join the amino acids up in particular sequences to make proteins; these proteins either form the main structural components of cells or are enzymes, catalysing organic reactions.

Some of the early protein-like substances probably polymerised in spherical bodies, such as have been produced experimentally in conditions simulating those of the early Earth. These may be regarded as cell precursors. From these, by processes still poorly understood, came the earliest true organisms. Since the present state of knowledge of the origins of life and the processes of chemical evolution have been ably documented by such writers as Calvin (1969) and Rutten (1971) it will not be followed further here, nor will the later stages of pre-biotic and early biotic evolution which Sylvester-Bradley (1975), Cloud (1976) and others have treated.

The earliest living organisms were undoubtedly **prokaryotes:** cells devoid of nuclei and with their DNA not arranged in chromosomes. These include the modern blue-green algae and bacteria. **Eukaryotic** cells, in which the chromosomes are borne in membrane-bound nuclei and which contain 'organelles' specialised for various functions, are of a much higher grade of organisation. All animals and plants other than prokaryotes are eukaryotic. There are no intermediates between these and their prokaryotic precursors, and no eukaryotes older than

1000 million years are currently known. According to Margulis (1970), eukaryotic cells may have arisen by the symbiotic association of several kinds of prokaryotes.

Prokaryotic blue-green algae were abundant throughout the later Precambrian, forming stromatolites, which are often columnar or hemispherical structures built up layer by layer through the sediment-binding and carbonate-secreting activities of the algae. The massive stromatolites (Walter 1976) were the first true reef-forming organisms (Ch. 1) and still survive today in hypersaline lagoons in Shark Bay, Western Australia. Together with microscopic floating algae they were the main source of free atmospheric oxygen, which allowed the beginnings of animal life.

The first possibly eukaryotic marine algae are known from the 1000-million-year-old Bitter Springs Formation in Central Australia (Schopf 1968, Knoll & Barghoorn 1974, Schopf & Oehler 1976). If these are eukaryotes they are the earliest-known sexually reproducing organisms and bear the potential for rapid evolution, since their potential for variation, upon which natural selection could act, is so much higher than that of prokaryotes.

EARLIEST METAZOANS

The earliest animals probably arose from Precambrian eukaryotic algae which lost their chlorophyll and began to ingest other organisms. **Protozoans** are animals that remained small and evolved, often to elaborate form, within the confines of a single cell. They are represented in the fossil record by various kinds of microfossils, such as the Foraminiferida, Radiolaria (Fig. 3.1) and Chitinozoa. **Metazoans,** or multicellular animals, may have begun as clustered aggregates of protozoans which remained together after cell division and

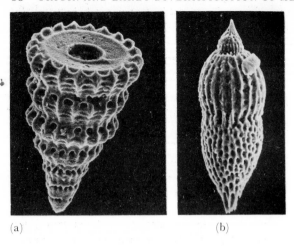

(a)   (b)

Figure 3.1 Radiolarians: (a) *Dictyomitra*; (b) a nassellarian theroperid; from late Cretaceous radiolarites overlying Troodos Massif, Cyprus. (Photographs by courtesy of Dr A. H. F. Robertson)

Figure 3.2 Elements of the Precambrian Ediacara fauna of Australia, with cross-section of *Rangea*. (Not to scale; redrawn from Glaessner and Wade, 1966)

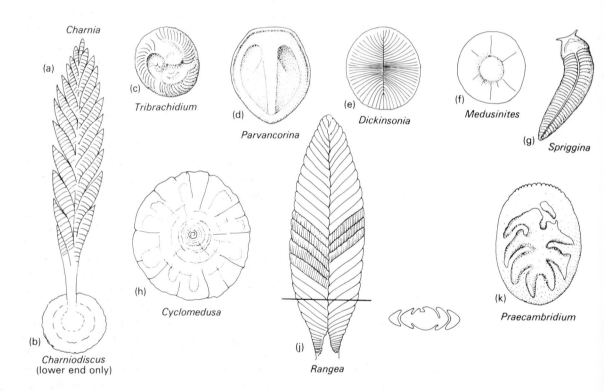

perhaps resembled simple sponges (Ch. 4). In more advanced metazoans different parts of the body became specialised for different functions and tissues developed (i.e. muscle, gut, nerve tissue, etc.). There is no fossil evidence of the earliest stages of metazoan evolution, and the first appearance of fossil metazoans is very abrupt. They are already differentiated into various phyla, and it is not easy to understand the evolutionary relationships of the various kinds on fossil evidence alone. Hence we need to study: the fossil record of Precambrian metazoans, such as it is; trace fossils of the late Precambrian; the palaeontology of the Precambrian/Cambrian boundary and the early Cambrian geological record; and the biological evidence of the nature of metazoan relationships.

Precambrian body fossils

Before 1947 almost nothing was known of the nature of Precambrian metazoan life. The term 'Proterozoic', coined for the later Precambrian (2500–570 million years), was based on the understanding that there were organisms (e.g. stromatolites) living then, but whether or not there was animal life was then unknown.

The discovery in that year of a rich and well-preserved fauna of soft-bodied animals in the Ediacara Hills of South Australia (Fig. 3.2) was the first acceptable faunal record from the Precambrian (Glaessner 1961, 1962). The fossils came from a shallow marine to littoral sequence with flaggy and cross-bedded sandstones; the occasional emergence of shoals is shown by polygonal mudcracks at certain horizons. The fossils are preserved along the interface of fine argillaceous laminae and hard sandstones. Apparently the dead animals came to be stranded upon the tidal mudflats or in the bottom of tidal pools and were preserved when covered by sand. The sand is of unusual texture, rather like a foundry sand, which is a factor in this uncommon mode of preservation.

The date of the Ediacara fauna is estimated radiometrically as 680–700 million years, antedating the beginning of the Cambrian by over 100 million years. Similar faunal types of equivalent age have since turned up in England, south-western Africa and various parts of Russia.

The Ediacara fossils, of which over 1400 specimens are now known belonging to about thirty species, fall into four main categories (Glaessner & Wade 1966, Wade 1972). The majority are jellyfishes (medusoids) and soft corals (pennatulaceans) belonging to Phylum Cnidaria; there are also segmented worms (annelids) and some peculiar organisms of unknown affinities. Sometimes one surface of the fossil alone is preserved, but there are also composite moulds where both surfaces are found compressed together.

Medusoids

Fifteen species of jellyfish have been described. *Mawsonia* is up to 125 mm across, is dome-shaped and has a strongly sculptured surface with large radially arranged lobes. *Cyclomedusa* is much smaller and has a central cone with concentric rings and a wide flat disc. *Kimberella* probably had a much longer bell-shaped dome with distinct longitudinal zones. The tentacles of *Eoporpita* are preserved, but the sculpture of the other nine jellyfish genera is less clear.

Pennatulaceans

Rangea, *Pteridinium* and *Arborea* are elongated feather-shaped bodies. In *Rangea* there is a central shaft about 150 mm long from which spring primary branches at about 45° to the shaft, and some twenty or so secondary branches are produced from each of these. The other genera are somewhat similar, but *Arborea* is up to 600 mm long. *Charnia* is a related genus, not known in Australia but described from beds of similar age in Leicestershire, England. The affinities of these have been much debated, but the morphological comparisons they bear with sea-pens, i.e. alcyonarian corals (Ch. 5) with polyps ranged along the secondary branches, are so close that they are generally believed to pertain to this group.

Annelids

Three species of broad flat bilaterally symmetrical 'worms' are known, referred to as *Dickinsonia*. They have a thin central axis and narrow segments normal to it arranged radially around the front and rear. Some specimens bear gut lobes, but no mouth or eyes have been preserved. A close resemblance to the modern segmented worm *Spinther*, which lives on and eats sponges, has been noted, so *Dickinsonia* is believed to be an annelid. It differs from *Spinther* only in that the distal 'claws' (**parapodia**) situated on the ends of the segments and used for crawling are lacking.

The elegant *Spriggina* has a horseshoe-shaped head with long projecting setae. Some forty segments follow, each terminating in a short parapodium. In many ways it is similar to the modern planktonic swimming annelid *Tomopteris*, but the point has also been made that the ancestors of trilobites may have looked very like *Spriggina*. Its morphology certainly seems to suggest a high degree of organisation.

Fossils of unknown affinities

Praecambridium is an ovoid discoidal form some 5 mm long. The thirteen known specimens show three or four pairs of raised lobes, seemingly arranged in segments, behind a horseshoe-shaped anterior part. Some authorities (Glaessner & Wade 1971) have interpreted this as an arthropod with a soft non-mineralised integument. It has some resemblance to a trilobite protaspis or larval stage, but the analogy is not a close one. A similar though larger form, *Vendia*, has been found in Siberia.

Parvancorina is a shield-shaped animal up to 26 mm long, with a broad anchor-shaped ridge on the upper surface and a narrow incised rim within the margin. Fine striae extend obliquely backwards from the main ridge. Some authors have suggested arthropodan affinities in view of a supposed resemblance to the 'shell' of the little crustacean *Triops*, but this is speculative.

Tribrachidium is discoidal with a central raised platform. Upon this three 'arms' radiate from the centre, and as they near the edge of the platform they are sharply bent, each giving rise to a series of marginal striae. In many ways *Tribrachidium* has a superficial similarity to edrioasteroids (Ch. 9), especially to a few genera which have triradial symmetry. Some specimens have a Y-shaped central mouth and bristle-like appendages in the surface, conceivably precursors of tube-feet. The possibility that this ancient fossil was a kind of 'proto-echinoderm' is therefore not to be disregarded.

Precambrian trace fossils

The earliest appearance of trace fossils may help to date the origins and initial radiation of the metazoans, indeed more readily than the soft-bodied organisms themselves with their limited prospects of preservation.

The earliest of all known trace fossils is approximately 1000 million years old, seemingly a burrowing system of a worm-like organism. There are a very few scattered records in rocks of over 750 million years, but trace fossils do not become abundant or diverse until near the Precambrian/Cambrian boundary.

Most of the early trace fossils probably resulted from the burrowing activity of soft-bodied infaunal worms, but the number of types represented increased greatly in the uppermost Precambrian; horizontal sinuous feeding burrows, surface trails, resting marks and vertical pipes are not uncommon some little distance below the boundary. It is not always easy, however, to date these in absolute terms; nor is it clear what kinds of animals made them.

But in many sequences traces of a kind normally attributable to trilobites ploughing along the surface (*Cruziana*) appear some considerable stratigraphical distance below the trilobites themselves. They are mainly simple forms with paired chevron-like markings, evidently the scratch marks of appendages. In some respects they differ from later *Cruziana*, and they have been interpreted as surface movement trails made by late Precambrian trilobite precursors with thin non-mineralised shells, propelled by 'a simple musculature of low efficiency' (Crimes 1976).

In general, trace fossils are more common than body fossils in the later Precambrian, and whilst they go further back in time they do not take the presumed origin of motile metazoans very far back into the Precambrian, certainly no more than 1000 million years before the present time and probably less. The first good trace and body fossils post-date the widespread late Precambrian glaciation, and it is possible that the intense selection pressures set up by the harsh environmental conditions of the time may have been one of the causes of increased variability thereafter.

Palaeontology of the Precambrian/Cambrian boundary

Large body fossils

The most familiar representatives of the early Cambrian fauna are trilobites, archaeocyathids, certain primitive molluscs, small brachiopods (chiefly with phosphatic shells) and the earliest echinoderms (edrioasteroids). Some of these are found in a few isolated localities alone and did not spread until later; others are widespread from the start. There are also various microfossils, some of which cannot be assigned to known phyla.

All of these appear very abruptly and are fully organised and differentiated on their first appearance, which may seem at first sight difficult to understand.

There are two factors by which the very abrupt appearance of these early fossils can be explained. One is that evolution at this time was tachytelic, perhaps due in part to intense selection pressures operating at that time. The second point is that the acquisition of hard shells as opposed to merely an unmineralised integument must have brought the bearers thereof to a 'fossilisation threshold', a potential not available even to their immediate ancestors. Though the mineralised shell which made fossilisation possible must have been acquired in all the above groups at approximately the same time or at least within a few million years, it was spread out in other taxa throughout the rest of the Cambrian and the early Ordovician. But how did a mineralised external covering originate in the first instance? One suggestion (Glaessner 1962) is that the calcareous or phosphatic material of which it is composed originated as an excretory product, possibly accumulating over the skin and hardening it. Once it was formed quite accidentally, the possession of a shell which could be used as a base for muscle attachment (giving enhanced locomotory prospects), and as a protection would give a tremendous selective advantage to the animal possessing it. Any invertebrates with hard coverings formed in such a way would then have great potential for evolutionary development. Moreover, once hard mouthparts had originated the selection pressure on other phyla to develop hard coverings for protection would be significantly increased.

It is not necessary to invoke drastic changes of environment as an important factor in the rapid diversification in early Cambrian life. But it does seem that just before the Cambrian a definite level of physiological complexity had been reached by many animal groups, which would enable them to cope first with the excretion and later with the secretion of the various materials used in the skeleton.

Something of the vast diversity that life had reached at least by Middle Cambrian times is shown by the Burgess Shale 'window' (Ch. 12). So diverse and 'advanced' is this fauna that it would seem to signify either (a) extremely rapid evolution of most types of animals in the early Cambrian, or (b) an origin for **coelomates** (animals with an internal cavity in which the gut is suspended) well before the Precambrian–Cambrian boundary. It is highly probable that both factors were involved.

Microfossils

In recent years rich and varied faunas (Fig. 3.3) have been documented from the base of the Cambrian and

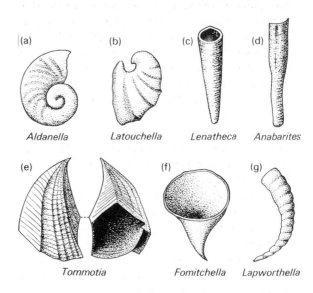

Figure 3.3 Latest Precambrian *(Anabarites)* and earliest Cambrian (others) small shelly fossils from the Siberian Platform. (Drawn from Matthews and Missarzhevsky, 1975.) (a) x 20, (b–e, g) x 15 (approx.), (f) x 40.)

just into the Precambrian (e.g. Matthews & Missarzhevsky 1975). The best-known sequences come from the Siberian Platform, where late Precambrian calcareous beds with dolomitic, oolite and stromatolitic horizons pass without unconformity or major lithological change up into the Cambrian.

The lowermost Cambrian stage (Tommotian) does not contain trilobites but has abundant small fossils, many with phosphatic shells. Some of these (e.g. *Aldanella*) are helically coiled gastropods; there are also hyolithids (an extinct molluscan group) and common sponge spicules. Two other fossil types are of interest: the hyohelminthids (laminated phosphatic tubes open at both ends) and tommotiids (phosphatic conoidal shells usually occurring in symmetrical pairs). These may well be plates of barnacles (cirripede crustaceans), though they are phosphatic rather than calcareous. If so it would suggest a very early origin for these crustaceans.

Similar fossils of Tommotian age are known from Australia, England and Scandinavia. Many of the species continue well up into the Lower Cambrian, where they are found with trilobites. By contrast, there are relatively few microfossils in the latest Precambrian stage: the Yudomian. There are only a few small tubular organisms (e.g. *Anabarites*,

Palaeolina) built of agglutinated particles like the tubes of modern sabellid worms; there is also a possible conodontophorid.

Thus the microfossils of the boundary, like the large fossils, all point to an explosive development of life at this most critical point in life's history.

Biological evidence on metazoan relationships

There are several levels or grades or organisation in multicellular animals living today, the comparative study of which sheds light upon the probable course of their evolution (e.g. Valentine & Campbell 1973).

Undoubtedly the earliest multicellular animals were aggregates of protozoans, loosely integrated but with cells all of the one kind. Some of today's simplest sponges would serve as models for this grade. Sponges (q.v.), however, developed along their own evolutionary pathway and are not really metazoans.

All other multicellular organisms have their cells organised into tissues, i.e. cell layers or masses specialised for different functions, and it is this that defines the metazoans proper.

In the simplest metazoan grade, represented by Phylum Cnidaria, the organisation is **diploblastic;** that is, the body wall consists of two layers: an **ectoderm** and an **endoderm.** These together form the bag-like body which encloses a single cavity, serving as a gut, with a single opening which acts as both mouth and anus.

The next level, which includes all higher types of animals, is the **triploblastic** grade. In this condition there is a third layer, the **mesoderm,** sandwiched between the endoderm and the ectoderm. In its most rudimentary form, as represented by the unsegmented flatworms (the Platyhelminthes), the body is bilaterally symmetrical and the mesoderm forms a more-or-less solid mass of tissue between the endoderm and the ectoderm. The endoderm forms the gut, the ectoderm the outer skin. In all other invertebrates, as well as in the vertebrates, the mesoderm forms a lining to the ectoderm and overlies the endodermal gut also, enclosing a fluid-filled cavity known as the **coelom.** The gut lies freely within the coelom, and since it may be looped or coiled it does not have to follow the contours of the ectoderm. The coelom may have functioned primitively as a hydrostatic skeleton, and this function is retained in the annelid worms. These move by waves of contractions running forwards from the rear. Each contraction forces the incompressible fluid forwards in the direction of movement, and it is against the fluid itself that the muscles contract. Many other uses have been found for the coelom in higher animals, especially as a space for storage of internal organs and materials. The development of the various organs in animals can be followed in embryology. Organs of ectodermal origin include the skin and nervous system, most of the gut and associated organs are endodermal, whilst respiratory and excretory organs and all muscles are mesodermal.

From embryology too, as well as from comparative anatomy, it has been possible to understand the relationships of the main coelomate phyla. There are a number of broad natural divisions in the animal kingdom into groups of phyla, implying that the original triploblastic flatworm-like stock must have differentiated into several ancestral groups, each of which was the foundation of a distinct 'superphylum'. These are:

(a) A 'lophophorate' superphylum including the Brachiopoda and Bryozoa and the small 'worms' known as Sipunculida and Phoronida. These all have the same kind of food-gathering apparatus, known as the **lophophore.** They have, moreover, certain embryological resemblances and for these two reasons are believed to be related. None of these phyla are segmented, but they could have been derived from an animal with a very few segments.

(b) A large group, possibly derived from a sipunculid-like animal, including the Echinodermata and Chordata (vertebrates and their allies). These are united chiefly on embryological grounds, especially the type of larva and the way in which the coelom develops in the fertilised egg. Chordates are fully segmented, and segmentation is likewise apparent in echinoderms in the jointing of the stem, as in the flexible stalk of fossil sea-lilies (crinoids).

(c) The segmented worms (Annelida) and the Arthropoda. These have always been understood to be closely related and may have had a common ancestor. Furthermore, the swimming 'trochophore' larvae of marine annelids are very similar to those of some marine Mollusca, and this latter phylum probably split off early from the main line of descent of the segmented animals. Molluscs are generally unsegmented, though there is an indication of segmentation in the most primitive molluscan group, the Monoplacophora, in which some of the organs are serially arranged. It is therefore quite possible that molluscs derived initially from a segmented coelomate ancestor but lost their segmentation as they became distinct.

The times of origin of all these groups are unknown. Many of them, especially those with calcareous skeletons, did not appear until the Ordovician but could have had ancestors with non-mineralised skeletons long before then. The presence of segmented worms and possible arthropods in the Precambrian Ediacara fauna suggests an origin for the coelom further back in time than 700 million years.

MAJOR FEATURES OF THE PHANEROZOIC RECORD

The diversification of invertebrate life

Once the major groups of invertebrates were established they were able to expand and differentiate, though their relative dominance has greatly fluctuated through time. Nearly all phyla have continued to flourish since they began, and only the archaeocyathids (if they are indeed a separate phylum) have become extinct. Taxa of lesser rank have generally shorter time ranges. For most groups it has been possible to work out some of the details of their phylogeny, depending on how complete their fossil record is. To some extent also, through functional studies it is possible to understand something of the biological quality of the fossil organisms even if these are of great antiquity, and it is very clear that even from the beginning of the fossil record invertebrate life was highly organised and well adapted, even if some of the later representatives of a stock apparently 'improved' on the earlier ones.

Trilobites, for example, which have been extinct for some 250 million years, were undoubtedly highly functional invertebrates, as is witnessed by the complex morphology of their eyes and other environmental sensors. In a sense they could be described as 'primitive', since their appendages are undifferentiated and they lack some of the elaborations of other arthropod groups. The term 'primitive', however, should not be construed as meaning ill adapted, for though in an evolutionary sense they were unable to escape from the limitations of being trilobites they were able to adapt to many specialised niches and were amongst the most successful life forms of their time.

In trilobites, as in all of the invertebrate fossil groups studied here, once the thematic pattern of organisation of each taxon was established it tended to remain very conservative. Thus within any phylum each hierarchial level of organisation can be related to an archetypal pattern of construction – a heritage which though highly functional was at the same time confining and restrictive. Every archetypal pattern, however much evolutionary plasticity was possible within it, remained an archetype for all the phylum. At taxonomic levels lower than the phylum new kinds of animals could sometimes originate, breaking away from the functional system of their ancestors but immediately setting up new archetypes, which were likewise confining but allowed within certain limits all manner of new evolutionary possibilities.

The molluscs, one of the most diverse of all invertebrate phyla, show very well how the potential inherent in a single archetypal plan was realised. A hypothetical 'archimollusc' (Fig. 8.1) had a single shell covering the mass of viscera, a rasping mouth and a posterior cavity in which the gills were located. From this at least six new and independent groups emerged. Some, such as the monoplacophorans and amphineurans, remained quite close to the ancestral type; others – the bivalves, gastropopods, scaphods and most of all the cephalopods – diverged away in various degrees and became adapted to various habitats and modes of life. Each of these classes had an archetypal 'master plan' in which all the features of the archimollusc were present, though modified, added to and altered in a particular functional combination, allowing in each case a focus for new evolutionary potential.

In the molluscs and in most other invertebrate groups the initial differentiation took place very late in the Precambrian or in the earliest Phanerozoic. The earliest evolutionary stages are not preserved and can only be reconstructed by inference and with subjectivity. Likewise, though the formation of new invertebrate archetypes went on at lower taxonomic levels throughout the whole of the Phanerozoic, there is normally little record of the types that carried the potential for further development.

This is probably because most of them lived in very small and rapidly evolving populations confined to very localised areas. The chances of their preservation and discovery are therefore very small, and such critically important intermediates are rarely found.

Extinctions

Throughout the 570 million years of Phanerozoic evolution there has been a continual relay of successive invertebrate faunas. The record is punctuated by a series of extinction periods of greater or lesser effect. Most species and genera normally have limited geological ranges, and since extinction seems to be the eventual lot of most organisms so there have been innumerable minor extinction episodes. Some of the extinction periods involved the more-or-less simultaneous decimation of some or all of the major taxa, and their effects were very far reaching.

There was a significant extinction episode terminating each of the geological periods, so the systems themselves are defined on the basis of the faunas they contain. Those episodes with the greatest effect on evolving faunas, however, were at the end of Cambrian, Ordovician, Devonian, Permian, Triassic and Cretaceous time, the late Permian extinction being the most far-reaching of all of these. At the end of the Cambrian, for instance, the trilobite faunas that had dominated the scene since the Middle Cambrian underwent a worldwide change, the majority of trilobite families becoming extinct and being replaced by entirely new stocks (Whittington 1966). Whilst this abrupt change is a convenient stratigraphical marker, there must have been reasons for it and indeed for all such times of crisis. It is hard to escape the conclusion that some kind of catastrophe took place, though not necessarily a sudden one.

The commonest elements in the Cambrian faunas are trilobites and thin-shelled phosphatic brachiopods, though many other phyla are also represented. Towards the upper limit of the Cambrian the first abundant calcareous-shelled animals come in for the first time, as a precursor to the record of great adaptive radiations of lime-secreting invertebrates which diversified immensely in the early Ordovician. After the extinction period of the late Cambrian the

'explosion' of echinoderms, brachiopods, molluscs, corals and new trilobite families was spectacular, and at the same time the graptolites arose from the late Cambrian dendroids (both of these had organic skeletons).

This Ordovician radiation established the pattern of marine invertebrate life for the rest of the Palaeozoic, and though there were sharp changes in the late Ordovician and Devonian these apparently were less in effect than those of the latest Permian. At that time the last trilobites became extinct, graptolites had long since gone, and all the large productid brachiopods of the late Palaeozoic disappeared, along with most echinoids, Palaeozoic corals and many kinds of molluscs. It was nearly 20 million years before faunal diversity was re-established on a similar scale to that before, and there were many changes, e.g. the brachiopods losing their dominance to bivalves and the complete replacement of coral and echinoderm faunas.

Whereas the rise of the ammonoids in the Devonian was slow, by the Triassic they had become amongst the most common and characteristic of all preservable invertebrates and very diverse. Yet they too nearly became extinct again at the end of the Triassic, having already suffered a late Permian setback. All the ammonites of the Jurassic and Cretaceous were derived from a single small group of late Triassic ammonoids which alone survived the rigours of the end of the Triassic.

The Mesozoic faunas are dominated by ammonites, with bivalves, gastropods, echinoderms and corals as other important faunal elements. Brachiopods, though far less diverse than those of the Palaeozoic, are still very common. In late Cretaceous time the ammonites declined. Over a period of some 10 million years they became reduced in numbers and diversity and confined to only certain parts of the world. When they finally died out at the end of the Cretaceous many other animal types became extinct at about the same time. The belemnites persisted just into the Eocene, but the brachiopods were once more pruned. These changes took place more or less contemporaneously with the final collapse of the dinosaurs, though a direct connection between these events is not necessarily envisaged.

When the earliest Tertiary faunas were established they were not unlike those of today. The hard-shelled faunal assemblages in the Tertiary consist of bivalves, gastropods, echinoderms, bryozoans, scleractinian corals and other forms, with brachiopods becoming a very minor component. Indeed shell debris on an early Tertiary shoreline would not, in terms of groups represented, be very different from that on a modern beach.

The faunal changes in the major taxa are traced in Part Two, but it will be understood that evolution has taken place not only at the level of species, genera, families, etc. but also in natural communities and on a global scale in faunal provinces. When these aspects have been documented as fully as have been the changes in taxa, then our conception of organic evolution in the Phanerozoic will be all the more complete for it will be based on ecological as well as morphological changes.

Possible causes of mass extinctions

It is generally believed (Newell 1963) that most of the main extinction periods can be accounted for by nothing more drastic than fluctuating sea levels. These fluctuations have taken place hundreds of times. Some have been no more than minor localised incursions or regressions of the sea; others have been on a much more widespread scale, probably the result of large scale processes to be understood in terms of plate tectonics.

A flooded continental shelf or epicontinental sea provides a great area of living space and very many habitats. There will be especially many habitats if the temperature gradient across the sea is high. If the shelf seas are contracted then the living space will disappear, many of the habitats will vanish and the ecological disturbance will be profound. A fall in sea level of only a few metres could do untold damage. What have been termed 'perched faunas' (Johnson 1974) (see Bibliography after Ch. 1), i.e. faunas that have evolved rapidly and colonised a shelf or continental sea during a time of maximum flooding, are especially vulnerable when the sea retreats.

Furthermore, if any 'key species', plant or animal, occupying a critical position in a food web becomes extinct, the whole network of feeding relationships dependant upon it will be immediately disrupted, and those animals at higher trophic levels which depend on it for food may well become extinct too.

A shrinking habitat due to natural fluctuations of sea level seems therefore to be a probable and major reason for most extinctions. The Permian extinctions, which were probably the greatest crisis through which life passed during the Phanerozoic, may have involved other factors. Not only were shelf seas greatly reduced, but also much salt was taken out of the sea and deposited as immense quantities of evaporite.

In the late Permian there were a number of inland seas connected to the ocean by narrow channels and bordered by enormous algal reefs. Of these the best known are the seas of western Texas and north-western Europe. Evaporites were deposited both behind the reefs as salt water washed over them and in the basins themselves as they periodically dried out. The low lying regions formerly occupied by these seas were sometimes flooded, only to dry out again; in the Zechstein sea of north-western Europe four such desiccation cycles have been recognised. Whilst the hypersaline waters of these seas were generally hostile to life, except for the algal reefs themselves and a few specialised invertebrates, the amount of salt taken out of the upper waters of the sea and permanently locked away in evaporite sediments must have been considerable; it could have seriously affected marine salinity and hence the life of all stenohaline marine organisms. These became extinct in great numbers. Further evolution and replacement were only possible after the normal salinity balance had been restored after several million years.

BIBLIOGRAPHY

Books, treatises, symposia

Calvin, M. 1969. *Chemical evolution*. New York: Oxford University Press. (Invaluable reference)

Hallam, A. 1977. *Patterns of evolution as illustrated by the fossil record*. Amsterdam: Elsevier. (17 original papers)

Margulis, L. 1970. *Origin of eukaryotic cells*. New Haven and London: Yale Univ. Press. (Symbiotic origin of eukaryotes; biology, biochemistry, palaeontology)

Rutten, M. G. 1971. *The origin of life by natural causes*. Amsterdam: Elsevier. (All aspects of chemical evolution)

Walter, M. R. (ed.) 1976. *Stromatolites*. Amsterdam: Elsevier. (Many original papers)

Individual references

Barghoorn, E. S. 1971. The oldest fossils. *Scient. Am.* **224** (5), 30–42. (Fully illustrated work on Precambian evolution)

Cloud, P. 1976. Beginnings of biospheric evolution and their biogeochemical consequences. *Paleobiology* **2**, 351–87. (Recent work on late chemical – early biological evolution)

Crimes, T. P. 1976. The stratigraphical significance of trace fossils. In *The study of trace fossils*, R. W. Frey (ed.), 109–30. Berlin: Springer-Verlag. (Trace fossil evidence on beginnings of metazoan life)

Glaessner, M. F. 1961. Precambrian animals. *Scient. Am.* **204**, 72–8. (Ediacara fauna)

Glaessner, M. F. 1962. Precambrian fossils. *Biol. Rev.* **37**, 467–94. (Ediacara fauna)

Glaessner, M. F. and M. Wade 1966. The late Precambrian fossils from Ediacara, South Australia. *Palaeontology* **9**, 599–628. (Ediacara fauna; technical descriptions)

Glaessner, M. F. and M. Wade 1971. *Precambridium:* a primitive arthropod. *Lethaia* **4**, 71–8. (Ediacara fauna)

Knoll, A. H. and E. S. Barghoorn 1974. Precambrian eukaryotic organisms: a reassessment of the evidence. *Science* **190**, 52–4. (Casts doubt upon fossil eukaryotic algae)

Matthews, S. C. and V. Missarzhevsky 1975. Small shelly fossils of late Precambrian and early Cambrian age: a review of recent work. *Quart. J. Geol. Soc. Lond.* **131**, 289–304. (Microfossil evidence on explosion of life in early Cambrian)

Newell, N. D. 1963. Crises in the history of life. *Scient. Am.* **867**, 1–16. ('Catastrophism' as a determinant for major extinction periods)

Schopf, J. W. 1968. Microflora of the Bitter Springs Formation, late Precambrian, Central Australia. *J. Paleont.* **42**, 651–8. (Earliest known eukaryotes)

Schopf, J. W. and D. Z. Oehler 1976. How old are the eukaryotes? *Science* **193**, 47–9. (Defends Precambrian algae as eukaryotic)

Sylvester-Bradley, P. C. 1975. The search for protolife. *Proc. Roy. Soc. Lond. B* **189**, 213–33. (Chemical and early biological evolution)

Valentine, J. M. and C. A. Campbell 1975. Genetic regulation and the fossil record. *Amer. Scientist* **63**, 673–80. (Origins of metazoans)

Wade, M. 1968. Preservation of soft-bodied animals in Precambrian sandstone at Ediacara, South Australia. *Lethaia* **1**, 238–67. (Why the Ediacara fauna was preserved)

Wade, M. 1972. Hydrozoa and Scyphozoa and other medusoids from the Precambrian Ediacara fauna, South Australia. *Palaeontology* **15**, 197–225. (Technical descriptions)

Whittington, H. B. 1966. Phylogeny and distribution of Ordovician trilobites. *J. Paleont.* **40**, 696–737. (Extinction and subsequent revival of trilobites at Cambrian/Ordovician boundary)

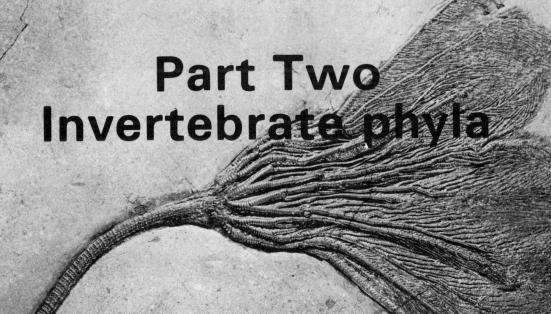

Part Two
Invertebrate phyla

4 Sponges, archaeocyathids and stromatoporoids

PHYLUM PORIFERA: SPONGES

The sponges are multicellular organisms but are not regarded as metazoans. They have only a few types of cell, and these are not really organised in tissues. There is no nervous system. In many ways they are of a grade of organisation in between protozoans and metazoans; hence they are sometimes known as **parazoans.** Apparently they derived from a protozoan ancestor but were an evolutionary blind alley, not ancestral to any metazoan.

The fine structure of sponges can often be used for identifying the species, but the external morphology may vary within fairly broad limits depending upon the environment. Sponges have remarkable powers of regeneration, and a living sponge squeezed through a silk net will re-form again on the other side. But individual cells removed from sponges, even though they may look just like some protozoans, will not live long; they are only viable as part of the sponge.

Sponges are generally sessile benthonic animals and are all filter feeders. A typical sponge (Fig. 4.1c) has an upright bag-shaped body with a central cavity (**paragaster**) opening at the top via an **osculum.** The outer surface of the sponge is perforated by numerous tiny holes (**ostia**), which lead to **incurrent canals** and thence to **chambers** within the sponge body. These chambers are lined by **collar-cells** or **choanocytes** (Fig. 4.2), each of which faces into the chamber. Exhalant passages lead from these to the central cavity. The collar-cells are the most important elements in the organisation of the sponge. Furthermore, these cells demonstrate the relationship between sponges and protozoans, for the single-celled or sometimes colonial protozoans known as the Choanoflagellida are in form virtually identical with sponge collar-cells. Some planktonic marine choanoflagellids, like sponges, construct basket-like capsules of geometrically organised rods of silica – a further indication of biological affinity.

Each collar-cell is a small globular cell with a cylindrical collar projecting from it, composed of fine pseudopodia. This collar encircles a solitary central

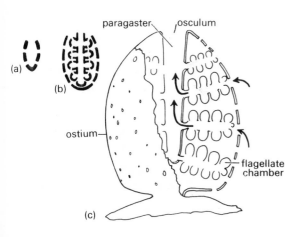

Figure 4.1 Elements of sponge morphology: (a) ascon type; (b) sycon type; (c) leucon (rhagon) type, showing passage of watercurrents. (Based on de Laubenfels, in '*Treatise*' Part (E).)

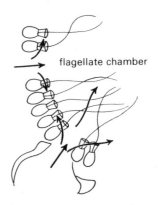

Figure 4.2 Wall of chamber showing incurrent pores with arrows showing current directions and collar cells (choanocytes).

flagellum. In life all the flagella beat continually, the tip of each whirling in a spiral motion. The current generated by all the flagellae draws water in through the ostia to the chambers and out to the central cavity and the osculum. Particulate organic matter then adheres to the sticky outside of the collar, which ingests it. Though the whirling of each flagellum is comparatively slow, the combined effect of all the flagellae operating together produces an efficient though low-pressure pump, which can pass through the sponge every minute a volume of water equal to the sponge's own volume. Sponges generally have other kinds of cells than the collar-cells. In one major group, Subphylum Gelatinosa (Fig. 4.4), the outer surface is covered with flattened epithelial cells (**pinacocytes**) of which those in the pore regions (**porocytes**) are perforated and can close off the pores by contraction if necessary. Each porocyte is stimulated to close individually by the presence of noxious substances in the water; the conduction of such stimulation to neighbouring cells may take place but in the absence of a nerve net is very limited. There are also amoeboid cells (**amoebocytes**) which wander through the sponge body and transfer nourishment from the choanocytes to other parts of the sponge.

Most sponges have a skeleton; this may simply be a colloidal jelly, but normally it consists of a horny material (**spongin**) or of calcareous or siliceous **spicules** (Figs 4.6, 4.7, 4.9, 4.12, 4.13) or of both. These spicules can be fossilised, especially when they are united so as to hold the body of the sponge together after death. Thus complete or partial sponge bodies, mats of spicules or isolated spicules are found in rocks extending back to the Cambrian and possibly further. Sometimes sponges use shell debris, sand or even the spicules of dead sponges to strengthen their skeletons.

There are three grades of organisation in sponges. In the simplest (**ascon**) grade (Fig. 4.1a) the sac-like body is merely a single chamber lined with choanocytes, with attendant epithelial cells and amoebocytes. **Sycon** sponges (Fig. 4.1b) have a number of grouped ascon-like chambers with a central opening, but the vast majority of sponges are of **leucon (rhagon)** type (Fig. 4.1c), in which a number of sycon-like elements open into the large central cavity (paragaster). In the larger modern sponges shrimps, ophiuroids or other commensal animals may live in

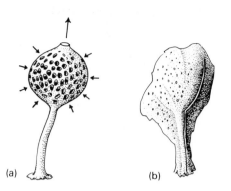

Figure 4.3 (a) Life form of a stalked still-water sponge; (b) life form of a 'rheophilic' sponge with wide lateral osculum directed away from current. (Redrawn from Carter, G. S. (1952) in *General zoology of the invertebrates* (London: Sidgwick and Jackson))

the paragaster, and such associations may be found fossil. Ascons are relatively small and not much larger than 10 cm high. Sycon and leucon sponges with their folded chambers exhibit obviously greater filtering efficiency and grow to a larger size.

There are about 1500 genera of modern sponges, of which some 80 per cent are marine (the rest are fresh water). The shallow marine genera show a remarkable tolerance of intertidal conditions. Sponges abound in the deep sea, especially the Hexactinellida or siliceous glass sponges, which are confined to quiet still waters. Sponge morphology is well adapted to the external environment, especially in such functional necessities as the separation of incurrent and excurrent water. Thus many deep water sponges are **stalked,** the incurrent water coming in near the stalk attachment and the excurrent waste water passing through the osculum (Fig. 4.3a). Sponges living in shallow water in which current directions are constant may have the osculum down one side, so that the broader fan-like surface faces into the current and the lateral osculum away from it (Fig. 4.3b). This not only makes use of the prevailing current but also prevents reuse of excurrent water. A parallel with this is found in bryozoans and crinoids.

CLASSIFICATION

Sponge taxonomy at the highest level is based upon the soft tissues, though only the skeleton is preserved

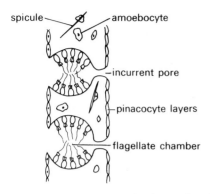

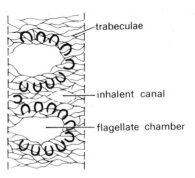

Figure 4.4 Wall structure of advanced gelatinosan ponge. (Redrawn from Reid, 1958)

Figure 4.5 Wall structure of hexactinellid (Nuda) of leucon grade. (Redrawn from Reid, 1958)

Figure 4.6 Various kinds of demosponge spicules (x 50 approx.). (Based on de Laubenfels, in '*Treatise*' Part (E))

Figure 4.7 Lithistid spicules (desmas) (x 25 approx.).

in the fossils. Two subphyla have been distinguished (Reid 1958–64) on the structure of the sponge wall. In Subphylum Gelatinosa (Fig. 4.4) the outer epthelial layer of flattened pinacocytes overlies a gelatinous middle layer (**mesenchyme**) in which the spicules are secreted by **scleroblasts** and wherein the amoebocytes wander. The innermost layer, in sponges of ascon grade, is that bearing the choanocytes; in more advanced grades these are invaginated within chambers.

The other subphylum, Nuda (Fig. 4.5), has neither pinacocyte layer nor mesenchyme; the choanocytes are borne in a network of syncitial filaments (**trabeculae**) but are likewise organised in chambers. Hence an accepted classification is:

PHYLUM PORIFERA (Cam.–Rec.): Multicellular animals with choanocytes and a skeleton of spongin and/or calcareous or siliceous spicules.

SUBPHYLUM 1. GELATINOSA: Porifera with pinacocytes and mesenchyme; choanocytes on the inner mesenchyme surface.

CLASS 1. DEMOSPONGEA (Cam.–Rec.): Gelatinosa of leucon grade with siliceous spicules and/or spongin and sometimes with foreign inclusions. Spicule rays usually diverge at 60° or 120°.

CLASS 2. CALCAREA (CALCISPONGEA) (Cam.–Rec.): Ascons, sycons or leucons with a skeleton of calcareous spicules.

SUBPHYLUM 2. NUDA: Porifera without pinacocyte cells or mesenchyme; choanocytes set in a network of protoplasmic threads.

CLASS HEXACTINELLIDA (HYALOSPONGEA) (L. Cam.–Rec.): Nuda, usually leucons, with spicular rays diverging at 90°.

Class Demospongea

Most fossil demosponges are represented by spicules only, the skeleton having collapsed. Such spicules may have either one single ray (**monaxon**) or four rays (**tetraxon**) diverging at 60° or 120° (Fig. 4.6). Ancient spicules, even of Cambrian age, can

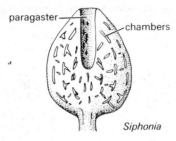

Figure 4.8 Cross section of *Siphonia*, a Cretaceous demosponge (x 0·75).

Figure 4.9 Sponge spicules: left-hand pair from calcisponges (x 100), right-hand pair from the Ordovician hexactinellid *Chancelloria* (x 20). (Redrawn from de Laubenfels, in '*Treatise*' Part (E))

Figure 4.10 *Raphidonema*, a Cretaceous calcisponge (x 0·5). (See also Fig. 4.15)

Figure 4.11 *Eusiphonella*, a 'compound' Cretaceous calcisponge (x 1 approx.).

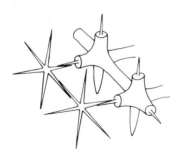

Figure 4.12 Part of skeleton of a dictyonine hexactinellid, showing overgrowth of contiguous hexact spicules by a siliceous envelope. (Redrawn from Reid, 1958)

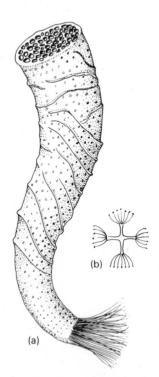

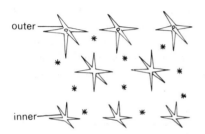

Figure 4.13 Construction of the hexactinellid skeleton with outer, middle and inner megascleres with scattered microscleres. (Redrawn from Reid, 1958)

Figure 4.14 (a) *Euplectella* (Hexactinellida), a Recent deep-sea glass sponge (x 0·5); (b) isolated microsclere of same (x 200).

normally be related to modern families. Only in Order Lithistida (Cam.–Rec.), are the spicules of different form. These knobbly or tubercular **desmas** (Fig. 4.7) are often so interlocked that the skeleton holds together after death. Lithistid genera are common from the Jurassic onwards, and such types as the Cretaceous *Siphonia* (Fig. 4.8) show details of the canals, chambers, paragaster and osculum when sectioned. Clionid sponges can penetrate into hard shell or limestone material, spreading out below the surface and secreting a spicular skeleton within the shell. Such sponges contribute greatly to the destruction of hard shell tissues and thus are of some geological importance.

Class Calcarea

The skeleton of calcareous sponges consists entirely of calcite spicules; there is neither spongin nor silica. The spicules are often of tuning-fork shape (Fig. 4.9). In the large Order Pharetronida the spicules form a closely packed mesh of different-sized 'tuning forks', giving a rigid and easily fossilised skeleton. Rich Jurassic calcareous sponge faunas are known, sometimes associated with reefs. Well-known Cretaceous genera include the vase-like *Raphidonema* (Figs 4.10, 4.15) and the digitate *Eusiphonella* (Fig. 4.11) which together form extensive sponge beds in the Lower Cretaceous of southern England.

Class Hexactinellida

Hexactinellids are of the normal sponge shape and may be anchored by an expanded flange or by a tuft of glassy fibres, especially in the deep sea forms (Fig. 4.14a). Bag-shaped, vase-like and dendritic forms are known, some with lateral oscula. The skeleton is of opaline silica, consisting of large (**megasclere**) and small (**microsclere**) spicules (Fig. 4.12, 4.13). These are well ordered, expressing cubic symmetry. The megascleres line the outer and inner walls and are here five-rayed, with the odd ray (**axon**) pointing inwards. Six-rayed megascleres are found in the central trabecular and choanocytic layers, together with the much smaller stellate or dumb-bell-shaped microscleres, not often found fossil. Often the megascleres unite as they grow together by complete fusion or by the formation of siliceous links between adjacent spicules. In such deep-sea glass sponges as

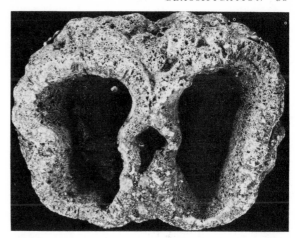

Figure 4.15 External appearance of an unusual twinned sponge, *Raphidonema;* from the Lower Cretaceous of Faringdon, England (x 0·75). (Photograph by courtesy of C. Chaplin)

Euplectella the whole skeleton becomes rigidly united, usually by the growth of a common sheath of silica enveloping adjacent parallel rays (Figs 4.12, 4.14). Here the skeleton forms a rigid lattice protected against torsion by spiral siliceous girders running round the outside at 45° to the axis.

Hexactinellids are classified mainly on skeletal characters – the structure of the mega- and microscleres, whether they unite and how they link – but since microscleres are seldom fossilised there are some difficulties with the taxonomy of fossil forms. Hexactinellids can be traced back to the Cambrian, and many now extinct groups flourished in the Upper Palaeozoic. These sponges were affected severely by the Permian extinctions but had recovered by the Jurassic, when many new taxa arose. Order Hexactinosida was very important in the Cretaceous but is now nearly extinct. Hexactinellids were abundant in the Tertiary on the continental shelves, but since most hexactinellids today live in the bathyal and abyssal zones either there has been a shift of habitat or the shelf forms have bcome extinct and only descendants of former deep-sea faunas survive.

GEOLOGICAL IMPORTANCE

Sponges are of little stratigraphical value, though they have been used as markers at specific horizons.

Their main importance geologically lies in that they have been a source of biogenic silica and that they are an element, though not an important one, in reef fabrics. Hexactinellids through time seem to have been mainly confined to the 'off-reef' facies in deeper waters, whilst demosponges and calcareous sponges preferred shallower waters. Ordovician lithistid sponges, especially in the carbonate facies of North America, were quite important as frame builders in stromatoporoid, bryozoan (Rigby 1971) and algal reefs, sometimes being as much as half the total reef volume. From time to time in the Palaeozoic there were quite extensive localised developments of sponges, such as rich Silurian sponge faunas of Tennessee and the Devonian and Mississippian glass sponges of New York and Pennsylvania, but these were not associated with reefs.

In the Permian the sponge faunas associated with the Texan reef complex are very well known. Here the calcisponges derived from Carboniferous shelf-living faunas became increasingly important on the patch reefs and finally on the barrier reefs, whilst hexactinellids as usual preferred the off-reef marginal facies. Some Triassic sponge reefs occur well developed in the Alps, whilst the Jurassic sponge reefs of southern Germany, very well known, contain sponges as primary frame-builders. Here the sponges are mainly siliceous thin-walled forms, often preserved in life position, forming small sponge-mounds, which occasionally grew to great size and united to produce large masses. But these sponges did not raise themselves far above the substratum, forming thin flat biostromes rather than true reefs. Other organisms are rarely associated with these Jurassic biostromes, being unable to provide suitable anchorage.

Most Cretaceous and Tertiary sponges, though locally abundant, do not seem to have been reef-related.

PHYLUM ARCHAEOCYATHIDA

The archaeocyathids are a very early group of calcareous fossils found mainly in Lower Cambrian carbonate facies and persisting only until the lower Middle Cambrian (Amgian). Though they are usually regarded as a distinct phylum (the only phylum ever to have become extinct) they are generally thought of as being of a grade of organi-

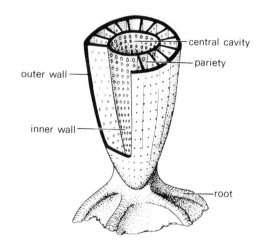

Figure 4.16 Archaeocyathid morphology modelled on *Ajacicyathus* (Cam.) (x 1·5). (Modified from Hill in '*Treatise*' Part (E))

sation comparable to that of the Porifera, to which they may bear a real biological affinity. They lived in shallow waters, often associated with stromatolites and forming thickets, localised bioherms or biostromes, or as isolated individuals. They have been considered to be the first 'reef-forming' animals, but they are usually subordinate in these reefs to stromatolitic algae, growing in clumps or scattered on the surface.

Archaeocyathids are found in all continents, but the best-known faunas are in the USSR, South Australia and western North America. They are absent from the British Isles and northern Europe. Where they occur in continuous sequence, as in Russia, they have proved to be of considerable stratigraphical value.

Archaeocyathids (Fig. 4.16) are normally solitary organisms bearing a superficial resemblance to corals or sponges. The skeleton or **cup** in a typical example such as *Ajacicyathus* (Class Regulares) is no more than a pair of inverted cones, one inside the other, forming **outer** and **inner walls** which are connected by vertical radial partitions (**parieties**) and separated by an annular cavity (the **intervallum**). A large central cavity seems analogous to the paragaster of sponges. The lower part of the cup is usually expanded into a basal flange with root-like holdfasts.

The outer and inner walls are perforated by numerous holes arranged in longitudinal rows; the

Figure 4.17 Septa connected by cross-partitions (synapiculae) in *Metafungia* (Cam.) (x 1·5). (Redrawn from Hill in '*Treatise*' Part (E))

parieties are sparsely perforate. In most archaeocyathids the outer wall has small pores whilst those of the inner wall are much larger. The microstructure of the cup is of polymicrocrystalline calcite, usually altered by diagenesis.

Whilst *Ajacicyathus* is of relatively simple construction, other archaeocyathids may have perforated walls and parieties, and also accessory structures crossing the intervallum (Fig. 4.17). These include radial rods, perforate transverse plates (**tabulae**) and imperforate arched plates (**dissepiments**). These elements tend to be concentrated towards the base of the cup. A few genera do not have an inner wall; some of these are very flattened cones, nearly discoidal in shape. Outgrowths from the outer wall are quite common and include tubular rods, tubercles or simply dense irregular masses; the latter apparently grow as a defensive mechanism where neighbouring specimens contact the individual in overcrowded conditions. Though most archaeocyathids are conical and solitary some rare colonial forms are known, with individuals being either arranged in chains or branched and shrub-like.

The usual size of archaeocyathids is 10–25 mm in diameter and up to 50 mm high, though specimens can reach a height of 150 mm and a few giants exceed this range.

Most archaeocyathids, like *Ajacicyathus*, belong to Class Regulares. The other class, Irregulares, has fewer members, distinguished by irregular pore structures on both walls and often an irregular outline of the cup as well.

SOFT PARTS, ORGANISATION AND ECOLOGY

The soft tissues of archaeocyathids are quite unknown, and though there is a general resemblance to sponge organisation no direct evidence exists. One useful pointer to their biological 'grade', however, is their capacity to regenerate the cup where damaged, which points to a quality of organisation and 'individual integrity' at least as good as that of the Porifera.

Most authorities agree that archaeocyathids were filter feeders, actively pumping water through the cup and straining off the food particles, perhaps with choanocyte-like cells, but the filtering function is not obvious in Irregulares. Flume tank experiments using model archaeocyathids (Balsam & Vogel 1973) suggest that 'passive flow' without flagellar pumping may have been important. The models were fixed in a flume tank in a laminar current whose strength could be varied. In this current a velocity gradient was established, generating a secondary flow from the outer wall through the intervallum and inner wall to the central cavity and out through the osculum. Short stout archaeocyathids were found to be better adapted to rapid current velocity than tall slender ones. Apparently the conical shape of archaeocyathids, and in particular the large excurrent opening, was well adapted for the generation of such passive flow. Though the passive generation of currents may well have been important, it is unlikely that the archaeocyathids abandoned the useful flagellar pumping system devised by their presumed protozoan ancestors, and feeding may have been a dominantly active process economically boosted by the passive flow component.

Archaeocyathids have been regarded by many as 'nature's first attempt to make a multicellular skeletal organism'. But the success of sponges, their ecological competitors, which arose independently from protozoans, may well have been an important factor in their early extinction.

Nearly all archaeocyathids come from carbonate shelf sediments deposited in warm seas. They adhered to the substrate by holdfasts or similar devices. They are most commonly associated with algal stromatolites; however, they also occur between and around the algal reefs and in bedded limestones with trilobites, brachiopods and hyolithids, but very rarely with sponges.

There is evidence from their occurrence in wave-smashed tumbled blocks and their common association with algae that they were depth-limited, the optimum being 20–30 m, individuals becoming

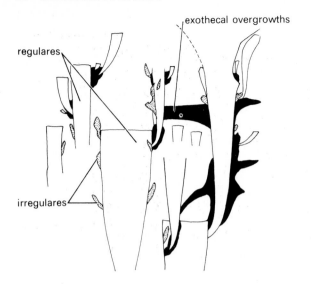

Figure 4.18 Archaeocyathid relationships in Cambrian life assemblages from South Australia (reconstructed from serial sections). (After Brasier, 1976)

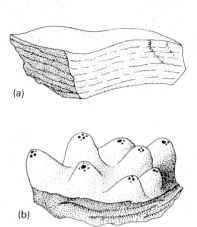

Figure 4.19 (a) *Stromatopora* (Dev.) – fractured part of colony showing laminae (x 0·75); b) upper surface of *Stromatopora* showing mamelons with openings of astro-rhizae (x 1). (Based on Lecompte in '*Treatise*' Part (F))

smaller and fewer down to 100 m. Their mode of growth is shown from a life assemblage of mid Lower Cambrian age from Australia (Fig. 4.18.) (Brasier 1976). In this fauna many species of Regulares were found clustered together, growing in the same general direction and reaching heights of up to 90 mm. Irregulares are present but fewer. The larvae always settled on dead archaeocyathids and usually on the outer wall near the top. Where the juveniles attached they fixed themselves by **exothecal outgrowths.** Irregular masses of exothecal material formed where individuals of the same and different species come into contact suggest competitive interactions – perhaps the oldest known case in the fossil record. In this fauna a micrite layer surrounds each cup, the work of the problematic encrusting organism *Renalcis*, which bound the archaeocyathid community to-gether and so enabled it to be preserved as a life assemblage.

DISTRIBUTION AND STRATIGRAPHICAL USE

Archaeocyathids are found in all continents, though their distribution is patchy; none are reported from the British Isles. Their abundance in the Lower Cambrian of the USSR has enabled very precise time ranges to be worked out for over 100 genera and used stratigraphically most successfully. The oldest known species come from the Tommotian stage in Russia. They diversified in the subsequent Adtabanian, reached an acme in the Botomian, and declined greatly in the Lenian stage which terminated the Lower Cambrian. Species are few in the Middle Cambrian, and only one possible Upper Cambrian survivor has been recorded from Antarctica. There is some evidence of provinciality, with three or more provinces in the Botomian, limiting intercontinental correlation using archaeocyathids. Even so, the Russian work has shown the potential stratigraphical value of this remarkable extinct phylum, whose zonal use will undoubtedly increase as the faunas become better known.

PHYLUM STROMATOPOROIDEA

Stromatoporoids are calcareous masses of layered and structured material found in carbonate sequences of Cambrian to Cretaceous age and dominant in the Silurian and Devonian. Palaeozoic stromatoporoids, especially those of the Ordovician to Devonian, were important reef formers. Though common enough and

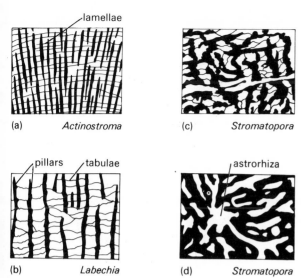

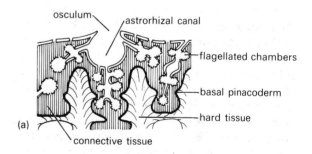

Figure 4.20 Structure of different stromatoporoids in section (x 10 approx.): (a)–(c) vertical sections; (d) horizontal. (Based on Lecompte in '*Treatise*' Part (F))

Figure 4.21 Wedge diagrams showing hard tissue in (a) *Stromatopora* and (b) *Actinostroma*. (Redrawn from Stearn, 1966)

greatly studied their zoological nature has only recently been clarified, and some of the Mesozoic 'stromatoporoids' may not in fact be stromatoporoids at all.

They bear some resemblance to compound tabulate corals, being of irregular layered form and forming rounded masses or thin flat sheets, occasionally cylindrical or discoidal (Fig. 4.19a, b). Where the upper surface is preserved it normally shows a pattern of polygonal markings. It also may have small swellings (**mamelons**) at intervals and frequently – a most characteristic feature of stromatoporoid organisation – stellate grooves known as **astrorhizae.**

The structure of stromatoporoids may be studied in vertical and tangential sections. It appears at first sight to be less defined than in corals, but the dominant structures are clear enough. A few selected types serve to illustrate the range in morphology. In *Stromatopora* (Ord.–Perm.) (Figs 4.20c, d; 4.21a) the **coenosteum** (colonial skeleton) in vertical section shows stout upright **pillars,** joined at intervals and traversed by thin horizontal **laminae,** though the differentiated structure is often lost and only a reticulate pattern of anastomosing elements remains. There may also be **tabulae** making partitions between chambers. With upward growth (Fig. 4.20d)

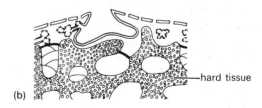

Figure 4.22 Restored morphology of stromatoporoids modelled upon sclerosponges: (a) *Actinostroma* (Cam.–L. Carb.); (b) *Stromatopora* (Ord.–Perm.) in which the hard tissue is composed of carbonate spherulites. (Based on Stearn, 1972)

new materials were secreted by the cells in contact with the upper coenosteal surface, thus roofing over the successive astrorhizal canals. In *Stromatopora* the calcareous material is constructed of carbonate spherulites in contact with one another, but this is unusual.

Actinostroma (Cam.–L. Carb.) (Figs 4.20a, 4.21b 4.22a) is somewhat similar, though the pillars are straighter and more slender and joined by horizontal radial processes in a kind of laminar network. Astrorhizae are not common and are absent altogether in some species.

Labechiella (Sil.–Dev.) has strong pillars and thin horizontal tabulae and possesses rare astrorhizae, whilst *Labechia* (Fig. 4.20b) (Ord.–Carb.) has very prominent tabulae between the pillars, convex upwards. In other genera horizontal laminae, often with cellular partitions, are well defined. These structural elements – pillars, laminae, tabulae, astrorhizae, etc. – are found variously developed in different stromatoporoids; sometimes one kind of structure develops much at the expense of others.

The original composition of stromatoporoid coenostea is unknown; it may have been calcite or aragonite. Some authorities (e.g. Stearn 1966) have suggested an originally aragonitic skeleton which is usually much altered differentially by diagenesis. Stearn attempted to distinguish primary from secondary microstructures and by so doing helped to remove the source of much taxonomic confusion. Kazmierczak (1976) however, who believes stromatoporoids to be algal, has suggested an originally calcitic coenosteum in which most of the structures preserved are primary.

BIOLOGICAL AFFINITIES

Stromatoporoids have been considered by different authorities to be hydrozoans, sponges, encrusting Foraminifera, bryozoans, algae or an extinct phylum with no modern counterparts. The suggestion that they are hydrozoans, which has been very popular, was based upon the presence of astrorhizae, which are somewhat similar to the hydrorhizae of hydrozoans, and of vertical zooidal tubes as in the modern reef-building hydrozoan *Millepora,* which in many other ways resembles stromatoporoids.

The recent discovery by Hartman and Goreau (1970) of some peculiar encrusting sponges in the Pacific and off Jamaica sheds further light on the problem. These have been placed in a new class of the Porifera, Class Sclerospongia, which contains only six genera. They have an aragonitic layered skeleton covered with organic tissue in which siliceous spicules are embedded. The outer surface of the tissue is pierced with two sizes of pores: small incurrent ostia, and larger widely-spaced excurrent oscula which join with a stellate system of canals, often running in grooves within the outer layer of the hard material and bearing a striking resemblance to astrorhizae. Water is sucked in through the ostia by choanocytes lying in linked chambers and is expelled through the oscula. The resemblance between the skeletons of sclerosponges and stromatoporoids is so close that some authorities have placed them all together in the same class, and a tentative soft-part anatomy has been reconstructed for stromatoporoids (Fig. 4.22). The Palaeozoic genus *Chaetetes*, usually regarded as a tabulate coral, may also be placed here. Evidently all stromatoporoids grew, as do the sclerosponges, on the surface and in the upper layers of the coenosteum, sealing off abandoned chambers with dissepiments, tabulae and other units. If the stromatoporoids really are Porifera, then they differ from all others in having no spicules of any kind. Hence their referral to sponges is not beyond question, and indeed Kazmierczak's suggestions of an algal origin bear serious consideration.

GEOLOGICAL IMPORTANCE

Many Palaeozoic stromatoporoids were reef builders, and since they had geological requirements similar to those of reef-building tabulate corals the two are often found together as frame builders. Massive 'ballstones' and solid unbedded stromatoporoid masses are found in clear water sediments, whilst the flattened and laminar stromatoporoids tend to be more common in impure limestones.

Very large reef-building stromatoporoid masses are often found to have marginal invaginations of sediment and inclusions of sediment within the coenosteum, indicative of cessations of growth for a time. Tongues of marginal sediment may be found on one side of the stromatoporoid mass, suggesting banking up of sediment on the lee side. The growth of such masses is often asymmetrical, and they often seem to lean over into the current, from which, of course, their food came. The most spectacular Palaeozoic reefs built largely by stromatoporoids are those of the Silurian of Götland, where they form huge reef masses up to 20 m high and 200 m in diameter. In these barrier reefs stromatoporoids were dominant and corals and other elements subsidiary.

In the Wenlock of the Welsh Borders a giant coral–stromatoporoid reef, with subsidiary bryozoans, formed a marginal barrier between an outer deeper-water zone in which reefs were absent and an inner shallow-water region where smaller reefs abounded. Mesozoic stromatoporoids, which were in many ways dissimilar to those of the Palaeozoic, were not generally reef formers.

BIBLIOGRAPHY

Books, treatises, symposia

Moore, R. D. (Ed.) 1955, '*Treatise*' Part (E). Archaeocyatha and Porifera. (The most recent comprehensive account|of sponges)

Teichert, C. (ed.) 1972. '*Treatise*' Part (E). (revised volume) Archaeocyatha. (Updated section)

Individual references

Balsam, W. L. and S. Vogel 1973. Water movement in archaeocyathids: evidence and implications of pressure flow in models. *Paleontology* **47,** 979–84. (Functional morphology)

Brasier, M. 1976. Early Cambrian intergrowths of archaeocyathids, *Renalcis* and pseudostromatolites from South Australia. *Palaeontology* **19,** 223–45. (Successive generations of archaeocyathids in life position; bibliography)

Hartman, W. D. and T. E. Goreau 1970. Jamaican coralline sponges; their morphology, ecology, and fossil representatives. *Zool. Soc. Lond. Symposium* **25,** 205–43. (Calcareous, archaeocyathid-like sponges)

Kazmierczak, J. 1976. Cyanophycean nature of stromatoporoids. *Nature* **264,** 49–51. (Evidence that stromatoporoids are algal)

Reid, R. E. H. 1958–64. *The Upper Cretaceous Hexactinellida*, 1–154. Palaeontogr. Soc. Monogr. (Classification, growth, morphology)

Rigby, J. K. 1971. Sponges and reef and related facies through time. *Proc. North Amer. Paleont. Convention, Chicago, 1969* (J), 1374–88. (The most recent treatment of sponge reefs)

Stearn, C. W. 1966. The microstructure of stromatoporoids. *Palaeontology* **9,** 74–124. (Distinction of primary from secondary microstructure)

Stearn, C. W. 1972. The relationship of the stromatoporoids to the sclerosponges. *Lethaia* **5,** 369–88. (See text)

Stearn, C. W. 1975. The stromatoporoid animal. *Lethaia* **8,** 89–100. (Sponge affinities of stromatoporoids)

5 Cnidarians

Corals, sea anemones, jellyfish and the small colonial hydroids are all representatives of Phylum Cnidaria. They were formerly grouped with the ctenophores ('sea-gooseberries' or 'comb-jellies') (Fig. 5.3a) in Phylum Coelenterata, but cnidarians and ctenophores are now regarded as separate though related phyla. Cnidarians are the simplest of all true metazoans, but they are of an evolutionary grade higher than sponges since their cells are properly organised in tissues which are normally constructed on a radial plan. The body is **diploblastic,** having cells organised in two layers only (Fig. 5.1a–c). These are the outer **ectoderm** and inner **endoderm,** which have no body cavity between them but only a jelly-like structureless layer. In this **mesogloea** runs a simple nerve net. The mesogloea is sometimes invaded by cells from the two primary layers. The single body cavity (**enteron**) has only one opening, the mouth, which also serves as an anus and is normally surrounded by a ring of tentacles. It is lined by endoderm which is sometimes infolded to form radial partitions or **mesenteries,** increasing the area over which digestion may take place, for the primary function of the endoderm is digestive. The cells of the endoderm ingest food by means of amoeboid pseudopodia; alternatively they can extend flagella into the enteron to stir its contents.

The cells of the ectoderm are more highly differentiated. Of these the principal types are the **musculo-epithelial** cells – columnar cells with contractile fibres – and there are also sense cells leading to the nerve net below. In addition there are ectodermal stinging cells, the **nematocysts** (Fig. 5.1c, e). These are large cells with a sealed central cavity containing poisonous fluid, within which lies a tightly coiled elongated tubular thread carrying barbs and stylets inside it. A small sensory hair on the outside of the nematocyst, the **cnidocil,** is sensitive to vibrations in the water, and when a small organism passes close to the cnidarian it triggers the nematocyst to discharge. When this happens the central cavity is intensely compressed by strong muscle fibres and the seal breaks; the thread is shot out with great force, turning inside out as it extends so as to expose the spiral barbs, which penetrate the prey so that it is injected and killed by the poison. Batteries of nematocysts are found on the tentacles, which all discharge together. The captured prey, held fast by the threads, is conveyed to the mouth as the tentacle then bends into it.

Cnidaria are characterised by a life cycle in which successive generations are of different kinds. This **alternation of generations** or **temporal polymorphism** is typical of the less specialised Cnidaria but may be suppressed entirely in the more 'advanced' kinds. Two types of individual, the fixed **polyp** (Fig. 5.1a) and the free **medusa** (Fig. 5.1b), alternate successively so that the polyp gives rise asexually to medusae, which reproduce sexually so that their zygotes produce polyps, and so on (Fig. 5.1d).

A polyp is a sedentary animal with a cylindrical body and an upward-facing mouth surrounded by a ring of tentacles. The mesogloea is comparatively thin. Medusae are free-swimming inhabitants of the plankton, often very small; their orientation is inverted relative to that of polyps. The medusoid body is discoidal, with the mouth prolonged centrally into a downward-facing funnel or **manubrium.** The mesogloea is very thick, and below it the enteron extends through four or more radial canals which join a peripheral circular canal, forming a kind of circulatory system for dissolved food material. The tentacles hang down freely beneath the outer rim, each usually carrying at its base an organ of balance or statocyst.

Since medusae are the sexual as well as the dispersal phase of the cycle they carry the four gonads, one of which is located along each of the primary radial canals. When the gametes are ripe

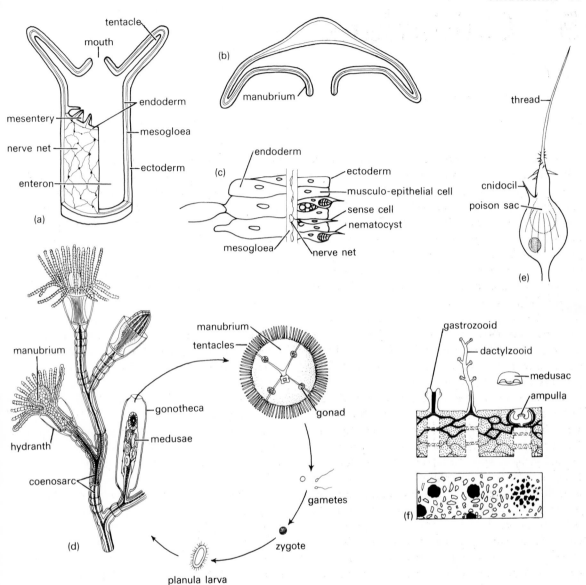

Figure 5.1 (a) Schematic diagram showing cnidarian hydroid cut longitudinally; (b) cnidarian medusoid with thickened mesogloea above the manubrium; (c) body wall of cnidarian showing details of diploblastic structure; (d) *Obelia* – morphology and life cycle (x 25): hydroid phase (left), medusoid (right) from below; (e) discharged nematocyst, showing spiral barbs near the base of the thread; (f) *Millepora* (Cret.–Rec.) – longitudinal section (above) and surface view (below) (x 10); the small medusa has been recently budded off from the ampulla. ((d) modified from Borradaile L. *et al* (1956) *The Invertebrate*. (Cambridge: C.U.P.); (f) based on Boschma in '*Treatise*' Part (F)).

they are shed by rupture of the gonad wall, and fertilisation takes place in the sea water. The zygote then develops into a small ciliated **planula larva.** This has at first a solid core of endoderm, which eventually develops a central cavity, and settles down to become the next polyp generation.

Medusae move by rhythmic pulsation of the bell-like body through highly modified musculo-epithelial cells. They normally die after reproducing.

Though alternation of generations seems to have been basic to the more generalised Cnidaria, such as most of Class Hydrozoa, it has been modified or suppressed in others. In Class Scyphozoa the medusoid phase is dominant and the polypoid phase very reduced, whilst in Class Anthozoa the medusoid has been eliminated entirely and the polypoid phase has become the sexual generation.

A description of the major divisions of the Cnidaria (simplified) follows.

PHYLUM CNIDARIA

The Cnidaria (Precam.–Rec.) are metazoans with radial or biradial symmetry. They have body walls of ectoderm and endoderm separated only by mesogloea; a single body cavity (enteron) with mouth but no separate anus; a simple nerve net; and no separate excretory or circulatory systems. Nematocysts are present. They may be solitary or colonial, and they are often polymorphic with alternate polyps and medusae. Many have calcareous or organic skeletons. There are three classes:

CLASS 1. HYDROZOA (L. Cam.–Rec.)
CLASS 2. SCYPHOZOA (Precam.–Rec.): Most large jellyfish.
CLASS 3. ANTHOZOA (Cam.–Rec.): Corals, sea-anemones, gorgonians, sea-pens.

Class Hydrozoa

The Hydrozoa are usually polymorphic, with radial and often tetrameral symmetry. Several hydrozoan orders embrace a wide range of forms:

ORDER 1. HYDROIDA (Cam.–Rec.): Have colonial sessile polyps and planktonic medusae e.g. *Obelia.*
ORDER 2. TRACHYLINA (?Jur.–Rec.): Free medusae only, the polyp stage having been suppressed entirely.
ORDER 3. SIPHONOPHORA (Ord.–Rec.): Large and complex floating colonies, lacking hard parts, with some division of labour amongst the members of the colony.
ORDERS 4 and 5. MILLEPORINA and STYLASTERINA (Tert.–Rec.): Have calcareous skeletons and are reef formers.
ORDER 6. SPONGIOMORPHIDA (Trias.–Jur.): Form massive colonies with radial pillars united by horizontal bars, and structures resembling the astrorhizae of stromatoporoids (q.v.).

Order Hydroida

Hydroids are rather generalised hydrozoans in which there is a chitinous external skeleton (**perisarc).** The order includes three suborders: Eleutheroblastina (free solitary naked forms, the pond *Hydra* being one), Calyptoblastina and Gymnoblastina. Representatives of the two latter orders are occasionally found fossil. *Obelia* (Suborder Calyptoblastina) (Fig. 5.1d), a standard example of a colonial hydroid, has a hollow root-like structure for attachment from which branching tubes arise, giving rise to polyps (**hydranths**) alternately on each side. The perisarc is annular in places, especially below the cups (**hydrothecae**) which surround each polyp. Though each polyp is an individual, the polyps are all connected together by a tubular system (**coenosarc**) through which the enteron continues. Each polyp has many tentacles and a prominent bulbous funnel or manubrium upon which the mouth is set. Towards the base of the colony are cylindrical **gonothecae:** flask-shaped structures with a constricted aperture. Within each of these is a central stem arising from the coenosarc, and from this the medusae form, being budded off continually from the upper end of the stem as they mature. These escape through the aperture and thereafter swim freely and grow until they are old enough to reproduce sexually. Gonothecae are typical of the Calyptoblastina; in the Gymnoblastina the medusae form directly from the coenosarc without being enclosed, and there is no hydrotheca.

Fossil chitinous hydroids are well known in the Ordovician and have been recorded from several localities.

The calcareous tubes of several Mesozoic and

Figure 5.2 *Millepora;* a recent specimen from Barbados (x 0·35). (Photograph by courtesy of C. Chaplin)

Tertiary serpulid tube-worms (e.g. *Parsimonia*) are often found infested by a colonial organism *(Protulophila)* which has been interpreted as a hydroid buried in the outer part of the calcareous tube (Scrutton 1975) (Fig. 5.3b). This has two components: polyp chambers, and stolonal networks which connect them in a regular scale-like pattern. The polyps grew around the rim and were probably incorporated into the tube as the serpulid grew, forming polyp chambers opening through semicircular apertures. This hydroid–serpulid association was probably symbiotic or commensal, in the same way that several species of the modern hydroid *Proboscidactyla* are symbiotic with tubular sabellid worms.

Protulophila would have been able to feed on some of the supply of food particles drawn in by the feeding arms of the worm, whilst the serpulid in turn probably derived some protection from the nematocyst batteries of the hydroid. It must have been a successful association, for the degree of infestation is sometimes as high as 95 per cent and infested serpulids are found in beds ranging from the Middle Jurassic to the Pliocene, 170 million years in all.

Orders Milleporina and Stylasterina

The milleporines and stylasterines are hydrozoans which superficially resemble corals and may be quite important in some modern reefs. *Millepora* (Figs 5.1f, 5.2) has a thick laminar calcareous skeleton of many vertical tubes with cross-partitions connected by thin horizontal ramifying canals. The soft tissues invest only the upper part of the skeleton, degenerating in the older lower layers as the skeleton builds up. There are three types of vertical tubes, each with its own **zooid** arising from the surface and capable of retracting into the tube: the **gastrozooids,** which are purely manubrium-like mouths; the **dactylozooids,** which are elongated tentacles with batteries of nematocysts; and the **ampullae,** which produce medusae. All these are connected by soft tissues running through the horizontal canals. This kind of colony illustrates an effective 'division of labour' and consequent modification of the zooids.

Millepora in its natural reef habitat is very variable in form. The millepores of Barbados (Stearn & Riding 1973) can be grouped in four forms:

(a) encrusting dead corals on gorgonians in thin sheets less than 1 cm thick;
(b) 'boxwork', i.e. forming thick near-vertical blades joined along their edges to form box-like subangular upward-opening cavities up to 30 cm high;
(c) bladed, i.e. with erect blades buttressed by other blades, making colonies up to 50 cm high;
(d) branching, up to 20 cm high, with delicate or stubby branches.

It is possible that the boxwork, bladed and branching forms are distinct non-intergrading biospecies, but the encrusting form grades into the others and is probably an environmental growth form of the other species, being most common in shallow water or encrusting gorgonians. The habitat preferences of the other species seem to be governed by water turbulence, boxwork forms being the strongest and bladed the most delicate.

Class Scyphozoa

The Scyphozoa are free-swimming medusae, entirely marine, and usually with radial symmetry based on a tetramerous plan. The sessile polyp stage (Fig. 5.3c) is called a **scyphistoma,** which buds by **strobilation.** This means that towards its upper end it is continually growing and being divided by transverse grooves, so that a pile of small medusae form one on top of the other and are released in turn to break free

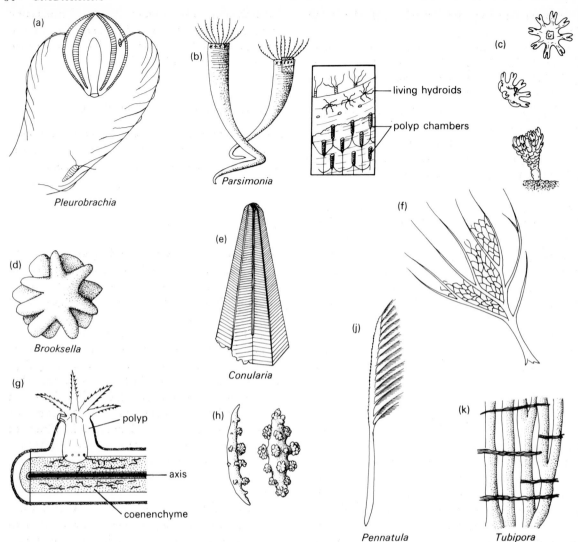

Figure 5.3 (a) A ctenophore, showing four of the comb-like fused plates of cilia, enteron and tentacles, which have just caught a copepod (x 1); (b) a Cretaceous serpulid worm infested by the symbiotic hydroid *Protulophila* (x 1). Right: detail of rim of the serpulid with living hydroids; the worm tube below is exfoliated to show the arrangement of the polyp chambers linked by connecting stolons (Based on Scrutton, 1975); (c) strobilating scyphozoan scyphistoma releasing ephyra larvae; (d) internal mould of a Cambrian protomedusan (x 0·75); (e) representative of a Cambrian–Permian fossil group of possible cnidarian affinities (x 0·75); (f) a Recent gorgonian (Octocorallia), with some of the anastomosing interstitial branches removed; (g) section through gorgonian stem with 8-armed polyp set into organic coenenchyme with central axis; (h) gorgonian spicules (x 1); (j) a Recent sea-pen (Pennatulacea) (x 0·25); (k) *Tubipora* (Recent), lateral view (x 2). ((d) Based on Walcott; (c) on Slater; (f) on Bayer: all in '*Treatise*' Part (F)).

as **ephyra larvae:** the juveniles of the adult jellyfish.

Few jellyfish are found in the fossil record. Of these, some are found as compressions in very fine sediment (e.g. *Peytoia* of the Burgess Shale) whilst others, including the many Ediacara species, are preserved as sand infillings of the gut cavity. *Brooksella* (M. Cam.) (Fig. 5.3d) shows the original lobes of the gut, likewise preserved by sand infilling, but as these differ in form from those of modern jellyfish *Brooksella* is sometimes placed in a separate class, the Proto-medusae.

Some steeply pyramidal chitinophosphatic fossils, the Conulata (Cam.–Trias.) (Fig. 5.3e) (conulariids), which have a quadrate cross-section and often marked herringbone ridges down the sides, have also been considered by some authorities as belonging to Class Scyphozoa. This is based upon the reported presence of tentacles preserved below the pyramidal structure in rare instances and upon the resemblance in shape of one rather flattened genus *(Conchopeltis)* to a medusa. On the other hand, most conulariids do not look at all like jellyfish and may belong to an extinct phylum.

Class Anthozoa

The Anthozoa are polyps with no trace of a medusoid stage. Anthozoans (corals, sea-anemones, gorgonians and sea-pens) are solitary or colonial, entirely marine cnidarians. The polyps produce gametes which develop directly into planulae larve after fertilisation and thence to a new generation of polyps; the medusoid generation has been eliminated entirely. They resemble hydrozoan polyps in many respects though are often very large. They always have a tubular **gullet** or **stomodaeum** leading down into the enteron, which hydrozoan polyps do not have. The exterior itself is divided by radial partitions (mesenteries) whose number and morphology is important in the subdivision of the class.

Those that secrete hard parts, and especially the corals, are of great geological importance, often forming thick beds in the Palaeozoic and sometimes, in association with stromatoporoids, reef-like masses. From Tertiary time onwards true coral reefs of vast thicknesses have formed, and corals living as reef formers are probably more important now than at any other time.

Anthozoans are grouped in three subclasses: Ceriantipatharia, Octocorallia and Zoantharia. Only the last of these, which includes the corals, will be considered in detail.

SUBCLASS CERIANTIPATHARIA

Ceriantipatharia are solitary or colonial polyps which for morphological reasons are placed apart from other groups. They are virtually unknown as fossils.

SUBCLASS OCTOCORALLIA

Octocorallia (Cret.–Rec.), poorly known as fossils, are especially represented by the gorgonians (Order Gorgonacea) common in many modern coral reefs (Fig. 5.3f–h). The colony forms a flat fan of anastomosing branches. The skeleton consists of horny branching tubes each of which may have a central calcified core. From the outer surface of the branching tubes project the many small polyps **(autozooids),** each of which has eight tentacles, characteristic of the subclass. These autozooids are set in a thick gelatinous outer coating over the branching tubes, within which are set innumerable calcareous **spicules** which help to support the autozooids. These are the only part to preserve as fossils, and only from them can the former presence of gorgonians be inferred.

Heliopora (Order Coenothecalia) has a massive aragonitic skeleton, bright blue in colour. It is an important reef former in the Indo-Pacific province. Its similarity to the tabulate *Heliolites* has suggested that the latter is a Palaeozoic octocoral, but this resemblance is far more likely to result from convergent evolution.

The red octocoral *Tubipora*, the 'organ pipe' coral (Order Stolonifera), has polyps inhabiting long horny tubes (Fig. 5.3k) and often forming large colonial masses. The sea-pens (Order Pennatulacea) (Fig. 5.3j) are another kind of octocoral. Here the colony has the form of a feather, the base of which is set in mud whilst the feathery upper branches are lined with autozooids. The presence of *Rangea* in the late Precambrian of the Ediacara fauna, interpreted as a sea-pen, testifies to the success of the pennatulaceans in ancient times.

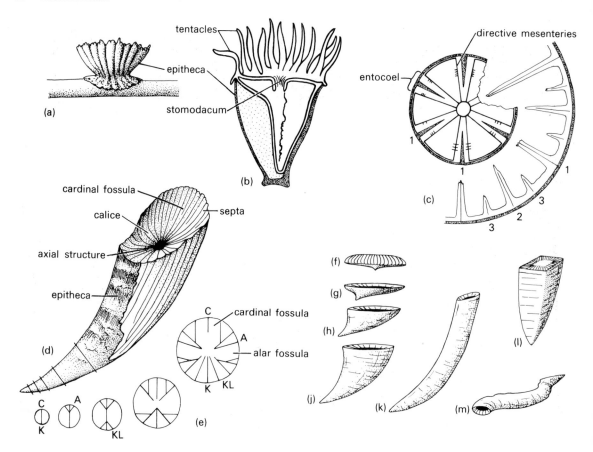

Figure 5.4 (a) Young *Caryophyllia*, a Recent scleractinian coral attached to the tube of the worm *Ditrupa* (drawn from Wilson, 1976); (b) *Caryophyllia*, vertical section; (c) transverse section through young (centre) and older (right) scleractinian coral; 1 – proseptum; 2, 3 – metasepta; (d) *Amplexizaphrentis* (Carb.) with part of epitheca removed (x 2); (e) septal emplacement shown by serial sections taken through marked level; C – cardinal septum; K – counter; KL – counter lateral; A – alar. (f)–(m) Shapes of coralla: (f) discoidal; (g) patellate; (h) turbinate; (j) trochoid; (k)

SUBCLASS ZOANTHARIA: CORALS

The structure of the Zoantharia is exemplified by the Upper Jurassic to Recent solitary coral *Caryophyllia* (Order Scleractinia). Recent species are of cosmopolitan distribution, living on various substrates and at depths of 0–2750 m. *C. smithii* lives in cool temperate waters off western Scotland (Wilson 1975, 1976). In regions of strong tidal currents it is found attached to pebbles or boulders, but in weaker currents on the outer shelf it can live in a variety of substrates; the most common form of attachment in sand patches is to calcareous tubes of the worm *Ditrupa* (Fig. 5.4a).

The polyp of *Caryophyllia*, which is some 2–3 cm across (Fig. 5.4b), is many-tentacled and has a stomodaeum leading down to the enteron, which is divided biradially by numerous mesenteries. Each of these has frilled free edges. The soft basal tissues secrete an aragonitic cup or **corallum** which is short and horn-shaped. It has an outer wall (**epitheca**) within which are numerous radially arranged **septa** between which lie the mesenteries. The first-formed **prosepta** (Fig. 5.4c) are larger and more pronounced than the **metasepta** intercalated between them.

When preserved as a fossil the aragonite is often altered to calcite or dissolved completely leaving a negative mould in the matrix. Aragonitic skeletons are known as far back as the Triassic, but not all

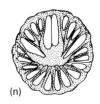
(n)

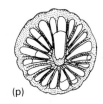
(p)

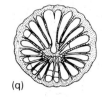
(q)

(r)

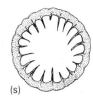
(s)

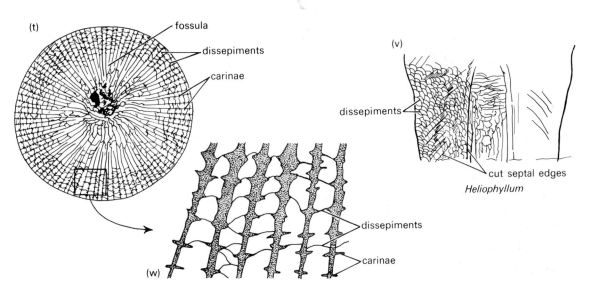

(t) fossula
dissepiments
carinae

(v)
dissepiments
cut septal edges
Heliophyllum

dissepiments
carinae
(w)

cylindrical; (l) pyramidal; (m) scolecoid; (n)–(s) *Amplexizaphrentis delanouei* (Carb.) group – transverse sections: (n) *A. delanouei;* (p) *A. parallela;* (q) *A. constricta;* (r) *A. disjuncta* (early); (s) *A. disjuncta* (late) (not to scale); (t) *Heliophyllum* (Dev.) transverse section showing near-radial symmetry (x 1·5); (v) vertical section (x 1.5), (w) enlargement of peripheral zone showing relationships of septa and dissepiments and two types of carinae. ((f)–(m) Redrawn from Hill in '*Treatise*' Part (F); (n)–(s) based on Carruthers, 1910)

Scleractinia are aragonitic. In *Caryophyllia*, as in all Scleractinia, the septa are laid down in multiples of six. When the coral is very young the corallum is simply a **basal plate** with six small prosepta inserted between the mesenteries. Later the metasepta appear, and there may be two or three generations of these so that the whole becomes multiseptate. A slight shift of their spacing throughout growth as the corallum becomes elliptical imparts the characteristic biradial symmetry. Throughout growth new material is added to the exposed edges of the septa and to the epitheca so that the coral expands as it grows.

In broad and general terms the simple structure of *Caryophyllia* applies to most living and fossil corals, but there are major differences in skeletal structure, septal arrangement and mode of colony formation. Four orders have been defined:

ORDER 1. RUGOSA (?Cam. – L. Trias.)
ORDER 2. TABULATA (?Cam. – Perm.)
ORDER 3. SCLERACTINIA (Trias.–Rec.)
ORDER 4. HETEROCORALLIA (U. Dev. – L. Cam.)

Order Rugosa

Morphology

The Rugosa are an almost exclusively Palaeozoic group of solitary and colonial corals. Primitively they show bilateral symmetry, arising because the numerous metasepta are inserted in four loci alone. Such bilaterality is clear in many Rugosa though in others it is obscured by the proliferation of septa. Two examples are chosen to show the basic structure of rugose corals. The widespread Carboniferous *Amplexizaphrentis* (Superfamily Cyathaxoniicae) (Fig. 5.4d, e) is a small solitary 'horn-coral', so called because of its shape.

The outer horn-shaped part of the corallum is covered by a thin calcareous skin, the epitheca, which extends most of the way from the tip to the upper (distal) surface or **calice,** where the skeletal elements filling the inside of the corallum are exposed. These internal elements form the basis of classification of the Rugosa and are of two main types: the vertical elements (septa, **axial structure**), and the horizontal elements (**tabulae),** which will be considered separately. (**Dissepiments,** which are important horizontal elements in some Rugosa, are lacking in *Amplexizaphrentis.)*

The septa are thin vertical plates arranged in a characteristic biradial pattern, which develops to maturity throughout the ontogeny of the coral. Their manner of insertion can be studied by investigating the ontogeny of the coral from the early stages; this is usually done by making serial sections normal to the axis, from the tip to the calice, and arranging them in a successional series. This provides a full record of how the coral grows, since structures once formed by accretion are not resorbed but retain their original form through the life of the coral.

When *Amplexizaphrentis* is very young a single proseptum divides it (Fig. 5.4e), soon becoming separated into **cardinal (C)** and **counter (K)** prosepta. Two other pairs of prosepta follow: the **alar (A)** adjacent to the cardinal septum, and the **counter-lateral (KL).** Even at this stage some bilaterality is evident, which becomes more pronounced when the metasepta are inserted in four quadrants only, on the 'cardinal' side of the alar and counter-lateral septa. Through differential growth the counter-lateral and alar septa move towards the counter septum to make room for new metasepta. Eventually, when more metasepta have been inserted, short minor septa (**second-order metasepta)** are laid down between the first-order metasepta. Around the cardinal septum there is a **cardinal fossula** where metasepta are not inserted, and especially during the intermediate growth stages

there are lateral (**alar) fossulae** as well. These are visible in the adult *Amplexizaphrentis*, but in large Rugosa where the major and minor septa have proliferated greatly the fossulae are so compressed as to be hard to distinguish.

In *Amplexizaphrentis*, as in most other Rugosa, two other skeletal elements are (a) the **epitheca,** and (b) **tabulae,** which are flat horizontal plates forming a floor to the cavity in which the polyp resided. The major septa join together in a central boss formed of updomed tabulae.

Heliophyllum (Subfamily Zaphrenticae) (Fig. 5.4t–w), a large North American rugosan, is sometimes found as poorly developed colonies. In this coral there are so many septa that it is not easy to see the four main lines of emplacement. The spacing of the septa is more or less equidistant, and the symmetry is virtually radial. There are some structural elements not found in *Amplexizaphrentis*, especially the dissepiments, which are concentrated in a broad marginal zone or **dissepimentarium.** These are small curving plates located between the septa and set normal to them, inclined downwards at about 45° to the epitheca. The major septa may reach the centre, in which updomed tabulae are the most prominent components. Another characteristic feature is presence of **carinae:** short 'yard-arm' bars projecting laterally from the septa.

Though bilateral symmetry has been masked by radial structure the cardinal fossula can still be seen where the dissepimentarium narrows and the axial ends of neighbouring septa are shortened and curve round it.

Range of form and structure in Rugosa

Form, type and habit of corallum. Rugosa are either solitary or colonial (**compound),** and either form may exist even within closely related genera of the same family. In solitary Rugosa (Fig. 5.4f–m) the usual shape is a curved horn: the **ceratoid** shape of *Amplexizaphrentis*. If the 'cone' has expanded very fast a flat **discoidal** shape may result; with progressively less abrupt expansion **patellate, turbinate, trochoid** and ceratoid forms will be the product. **Cylindrical** corals are virtually straight-sided, except for the first-formed part; **scolecoid** forms are irregularly twisted cylinders; **pyramidal** types have sharply angled sides; whilst **calceoloid** genera,

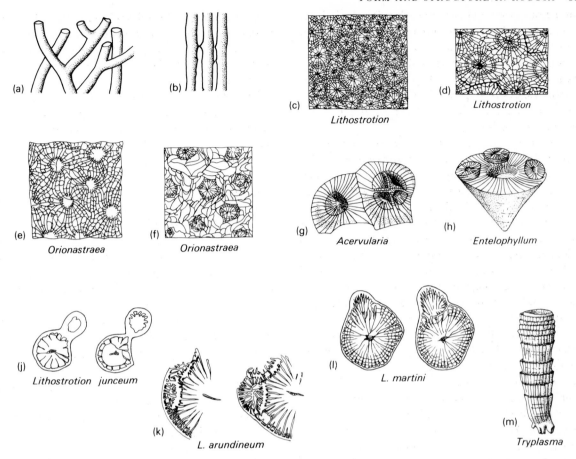

Figure 5.5 Form of corallum in Rugosa: (a) fasciculate dendroid; (b) fasciculate cerioid; (c) massive cerioid (x 0·6); (d) massive astraeoid (x 1); (e) thamnasteroid (x 1·5); (f) aphroid *(Orionastraea)* (x 1). Modes of increase: (g) quadripartite axial increase in *Acervularia* (Sil.) (x 1 approx); (h) peripheral increase by offsets in *Entelophyllum* (Sil.) (x 1 approx); (j) lateral increase in *Lithostrotion junceum* (Carb.), two successive stages (x 5); (k) lateral increase in *L. arundineum* (Carb.) (x 5); (l) lateral increase in *L. martini* (Carb.) (x 5); (m) rejuvenescence in *Tryplasma* (Sil.–L. Dev.) each 'shelf' represents a successive episode at rejuvenescence (x 0·75). ((a), (f), (g), (h), (m) Redrawn from Hill in '*Treatise*' Part (F), (j)–(l) redrawn from Jull, 1965).

oddest of all, possess a curved corallum with one flattened side and a hinged lid or **operculum.**

Colonial Rugosa (Figs 5.5a–f, 5.6) are those in which a single skeleton (corallum) is produced by the life activities of numerous adjacent polyps each contributing a **corallite** to the whole. The habit of the colony is **phenetic,** i.e. may be greatly influenced by the environment, but colony type, which includes the morphology and relations of the corallites, is more rigidly defined genetically. In the **fasciculate** type the corallites are cylindrical but not in contact. There are two kinds of fasciculate morphologies: **dendroid,** with irregular branches; and **phaceloid,** in which the corallites are more or less parallel and sometimes joined by connecting processes. **Massive** corals are those in which the corallites are so closely packed as to be polygonal in section. Four kinds of massive corals are distinguished:

(a) **cerioid** (e.g. *Lithostrotion*), in which each corallite retains its wall;
(b) **astraeoid** (e.g. other species of *Lithostrotion*), in which the corallite walls are wholly or partially lost but the septa stay unreduced;

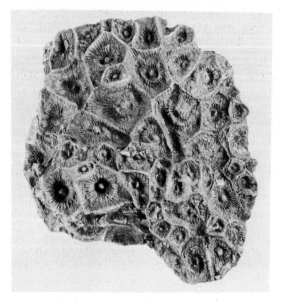

Figure 5.6 *Lonsdaleia floriformis* (L. Carb.) (Bathgate, Scotland): small colony showing both axial and lateral increase with a few small encrusting *Syringopora* (x 0·75). (Photograph by courtesy of Jeremy Jameson)

Figure 5.7 Fine structure of septa in Rugosa: (a)–(c) trabecular, fibro-normal and lamellar tissue; (d) trabecular and lamellar tissue in *Syringaxon* (Sil.–Dev.); (e)–(g) fibre fascicle arrangement in (e) *Keriophyllum* (M. Dev.); (f) *Palaeosmilia* (Carb.); (g) *Dinophyllum* (Sil.); (h) dominance of fibronormal tissue in *Amplexizaphrentis* (Carb.); (j) dominance of trabecular tissue in *Keriophyllum* (M. Dev.). (Mainly redrawn from Wang, 1950, and Kato, 1963)

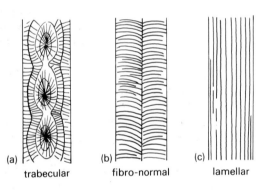

(a) trabecular (b) fibro-normal (c) lamellar

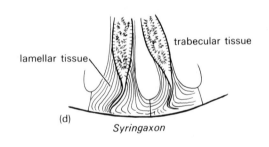

(d) *Syringaxon*

(e) *Keriophyllum* (f) *Palaeosmilia* (g) *Dinophyllum*

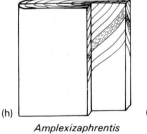 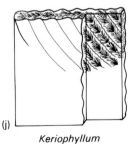

(h) *Amplexizaphrentis* (j) *Keriophyllum*

c) **thamnasteroid** (e.g. *Orionastraea*), in which the septa of adjacent corallites are confluent and often sinuous or twisted;

d) **aphroid** (e.g. *Orionastraea*), where the septa are reduced at their outer ends so that neighbouring corallites are united by a zone of dissepiments alone.

Fine skeletal structure of septa (Fig. 5.7a–j). The basal ectoderm has numerous radial invaginations, each lying above a septum which it secretes by crystallisation. The skeletal elements are formed of calcium carbonate, some or all of which may have been initially calcite. According to Wang (1950) and other authors, the process of crystallisation results in the production of three main components: **trabecular, fibro-normal** and **lamellar tissue,** the latter usually investing structures made of trabecular or fibro-normal tissue. When septa consisting of fibro-normal tissue are examined in section a median dark line is seen bisecting each septum, and projecting from it there are narrow parallel calcite fibres arranged normal to this median line, though radially around it at the tip of the septum. Trabecular tissue consists of parallel or fan-like rods arranged in a kind of palisade in the plane of the septum. Each **trabecula** is composed of serially arranged radiating whorls of tiny fibres (**fibre fascicles**). These fascicles are secreted in linear series from a continually producing centre of calcification; the trabecula grows as long as calcification goes on.

The relative use of fibro-normal and trabecular tissue in rugosan septa is very variable, as is the arrangement of trabeculae within the septum. In some genera septa are entirely fibro-normal and in others there is only trabecular tissue; **semitrabecular** septa have a trabecular median part and fibro-normal peripheral regions. Many other septal types have been described, depending upon the arrangement of trabeculae and the fibre fascicles within them. Wang has noted, amongst types of fibre fascicle construction, simple widely spaced fascicles (e.g. *Keriophyllum*), grouped fascicles (e.g. *Palaeosmilia*) or fascicles in continuous series (e.g. *Dinophyllum*). In *Tryplasma* the trabeculae are discontinuous as a result of intermittent calcification from the same centre. There are also many kinds of trabecular grouping and orientation in the septa. They are, however, generally directed upwards and inwards from the wall, subparallel and fan-like, but sometimes diverging from the wall in a complex and irregular manner. They commonly turn upwards axially and may be involved in the formation of axial structures.

Dissepiments, tabulae and epithecae are formed of fibro-normal tissue. Carinae are trabecular, formed either of single swollen trabeculae or of compound trabeculae, branched or in bundles.

Lamellar tissue, which usually surrounds the septa and other skeletal elements in some corals, was first described as a primary calcitic material. However, many recent workers (Kato 1963, Jell 1969, Sorauf 1971) have argued that lamellar tissue is probably secondary after aragonite. Some authorities have even suggested that the whole rugose skeleton was originally aragonite and that trabeculae are diagenetic artifacts. Sandberg (1975), however, who studied ultrastructural and chemical changes in aragonitic Pleistocene to Recent bryozoans converting to calcite, has shown that the results of such diagenesis do not in any way resemble the forms taken by trabecular and fibro-normal tissue. Hence he has contended that all the preserved parts of rugose coral skeletons must originally have been calcite and that most of the fine skeletal structure actually seen is primary. For the moment, therefore, the issue is unresolved.

Since the gross morphology of corals is so highly variable and has been subject, in so many independent stocks, to repeated convergent evolution, there is a clear need in taxonomy to find more 'stable' characters to use as the basis of classification. Whilst microstructure has proved useful here along with other characters, its interpretation is still the subject of much discussion and it is by no means a universal key. Microstructure has proved of some value in evolutionary studies, Kato (1963), for instance, suggesting a general stratigraphical change from lamellar to fibro-normal tissue.

Types of septa. In the Rugosa septal microstructure and arrangement may vary greatly, and those differences are of high taxonomic value. Septa are normally laminar, but perforated septa do occur in some genera where the trabeculae are not closely joined. Usually they are straight, sometimes zigzag. Commonly septa are not of uniform thickness throughout their length. In *Kodonophyllum* (Fig. 5.8q), for instance, the outer parts of the septa are so greatly

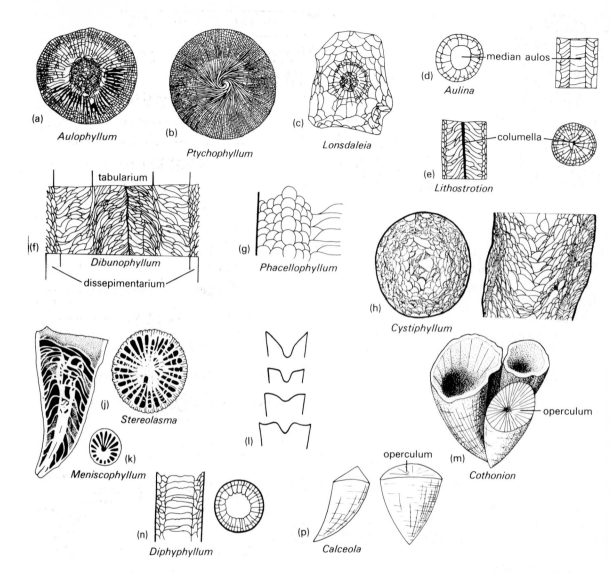

Figure 5.8 Morphology of Rugosa: (a) *Aulophyllum* (Carb.) with axial strucure and partial stereozone of thickened septa (x 1); (b) *Ptychophyllum* (Sil.) with axial vortex (x 0·75); (c) *Lonsdaleia* (Carb.) with lonsdaleoid dissepiments or presepiments (x 1·25); (d) *Aulina* (Carb.) with median aulos (x 1·6); (e) *Lithostrotion* (Carb.)–vertical section showing dissepimentarium and tabularium (x 1·5); (f) *Dibunophyllum* (Carb.), showing location of central tabularium and dissepimentarium; (g) *Phacellophyllum* (Dev.) vertical section of dissepimentarium with horseshoe-shaped dissepiments (x 4); (h) *Cystiphyllum* (Sil.) showing broad dissepimentarium (x 0·75); (j) *Stereolasma* (Dev.)—vertical and transverse section (x 1·5); (k) *Meniscophyllum* — transverse section (x 1·5); (l) diverse calical shapes in various Devonian species of Spongophyllidae (diagrammatic); (m) *Cothonion* (M. Cam.), the earliest genus referred

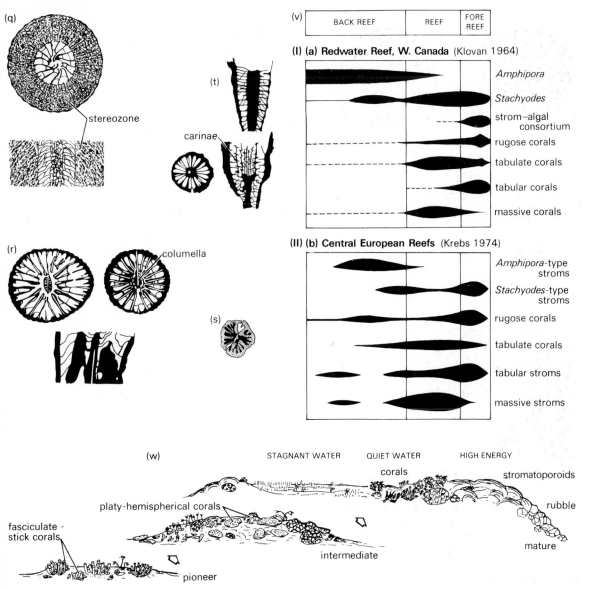

(q) *Kodonophyllum* stereozone

(t) carinae

(r) columella

(s)

(v)

| BACK REEF | REEF | FORE REEF |

(I) (a) Redwater Reef, W. Canada (Klovan 1964)

Amphipora
Stachyodes
strom–algal consortium
rugose corals
tabulate corals
tabular corals
massive corals

(II) (b) Central European Reefs (Krebs 1974)

Amphipora-type stroms
Stachyodes-type stroms
rugose corals
tabulate corals
tabular stroms
massive stroms

(w) STAGNANT WATER   QUIET WATER   HIGH ENERGY
corals
stromatoporoids
rubble
platy-hemispherical corals
fasciculate - stick corals
mature
intermediate
pioneer

to the Rugosa, with one operculum in place (x 1); (redrawn from Jell & Jell, 1976); (n) *Diphyphyllum* (Carb.) showing downturned outer edges of tabulae forming a tube (x 1·5); (p) *Calceola* a Devonian solitary rugose coral with operculum (x 0·5); (q) *Kodonophyllum* (Sil.) with septal stereozone (x 1·5); (r) *Lophophyllidium* (Carb.–Perm.) two transverse and one vertical section (x 2); (s) *Plerophyllum* (Dev.–?Trias.), one of the last of all Rugose corals (x 1·5);

(t) *Metriophyllum* (Dev.) with carinae (x 0·75); (v) distribution of Rugose corals, stromatoporoids and other reef-associated organisms in Devonian reef facies (redrawn from Klovan, 1964, and Krebs, 1974); (w) palaeoecological succession of a Devonian reef, from pioneering to climax stages (based on Copper, 1974). (Mainly redrawn from Hill in '*Treatise*' Part (F), except where indicated)

thickened that they are contiguous, forming a well-marked **stereozone**. In *Aulophyllum* (Fig. 5.8a) half of the coral only usually has medially thickened septa, so the stereozone is incomplete. The sides of rugosan septa are usually smooth but are often provided with **flanges** or small **denticulations** where the trabecular tips show through. In *Orionastraea* and some other aphroid genera the outer parts of the septa are replaced by dissepiments, whilst adjacent corallites of thamnasteroid types have confluent septa. *Amplexus* illustrates a type of coral in which septa are imperfectly developed, only being properly formed immediately above each 'calicular' surface. Carinae may be of various kinds as in *Heliophyllum* and are especially well displayed in *Metriophyllum* (Fig. 5.8t).

Axial structure. Very many Rugosa have a central structure (Fig. 5.8a–f), which may be one of three basic kinds. An **axial vortex** is formed by the joined axial ends of the major septa, slightly twisted as in *Ptychophyllum*. A **columella** is simply a dilated end of the counter septum and is representative of *Lithostrotion, Lophophyllidium* and other related genera. An **axial column,** as in *Aulophyllum*, is a broad axial zone formed of **tabellae,** i.e. small incomplete tabulae within the axial complex. In *Lonsdaleia* and *Dibunophyllum* the axial column in section appears web-like and is likewise composed largely of tabulae with septa crossing it; in the latter genus it is divided by a central **septal plate** formed by the fused cardinal and counter septa. *Aulina* has a very simple axial column, the **aulos,** which is a vertical tube truncating the inner ends of the septa and traversed by horizontal tabulae.

Tabulae and dissepiments (Fig. 5.8f–h). Tabulae are transverse plates which may be flat, convex or concave. They usually occupy a central space or **tabularium,** and if there is an axial complex they join with it. Tabulae may be replaced by a number of smaller plates called tabellae. Tabulae in the Cyathaxoniicae, the most primitive superfamily, generally extend right across the corallum, but in most Rugosa they terminate externally in a **marginarium,** which is either a septal stereozone or a dissepimentarium, and hence do not reach the edge.

Dissepiments, the small plates usually found towards the edge of the corallum, lie peripheral to the tabularium and like the tabulae are constructed of fibro-normal tissue. They are of various kinds. They may be simply inclined or slightly swollen plates dipping towards the axial region, but in *Phillipsastrea, Thamnophyllum* and relatives, for instance, there evolved an extraordinary range of dissepimental types, including horseshoe-shaped ones, which have proved useful in classification.

Dissepiments are not found in the earliest rugose genera. On their first appearance in late Ordovician streptelasmatinids and columnariids they are confined to a thin peripheral dissepimentarium. In later genera, however, they become more important structural components, most noticeably in types such as *Lonsdaleia*. Rarely, as in *Cystiphyllum*, the dissepimentarium may become very wide and, together with tabellae with which dissepiments may intergrade, make up virtually the whole skeleton, since the septa are reduced to separate trabeculae developed only on the upper surface of the tabellae and dissepiments.

Calice. The calice is that part of the coral which was in life in contact with the basal ectoderm of the polyp. It is in effect a mould of the secretory surface. It is not commonly seen in specimens preserved in limestones unless these have been weathered, but in individuals collected from shales calicular surfaces can often be clearly seen. Usually there is an annular calicular platform surrounding a deep central depression (calicular pit). Some genera with an axial complex have a marked calicular boss where it emerges to the surface, but this is not always present. The form of the calice is quite variable even within families, as shown in Figure 5.8l which illustrates various forms found in the Spongophyllidae in which there are conical, bell-shaped, saucer-shaped and everted types.

Asexual budding and the production of new corallites (Fig. 5.5g–l). When a compound coral grows the first-formed **protocorallite** gives rise asexually to offsets, which may form in a number of ways. The term **budding** is used for the soft parts, **increase** for the skeleton. Detailed studies have been made in the Lithostrotionidae and many other groups of the three main methods of increase and budding:

(a) **Axial increase.** In this mode the corallite splits by fission into two or more daughter polyps within the calice. Such budding is inevitably parricidal (i.e. destroys the parent). The upper surfaces of many compound corals can often be seen with several corallites undergoing axial increase at approximately the same time. As the colony continues to grow the parent corallite is buried and the daughter corallites grow to full size. This is the least common of modes of increase in the Rugosa, being very rare in *Lithostrotion*.

(b) **Peripheral increase.** Here small offsets arise round the parent corallite. This may or may not be a parricidal system; usually the daughter arises from the parent's side without killing it. It is quite common in fasciculate genera and has been reported in a few *Lithostrotion* species.

(c) **Lateral increase.** This is the most common method of increase in the Rugosa and is not parricidal. In *Lithostrotion* it has been well investigated (Jull 1965) but has proved to be so variable as to be of limited taxonomic value at the generic level. Three distinct types of lateral increase have been distinguished in *Lithostrotion* using serial sections. In one type (e.g. *L. junceum*) the lateral bud arises fully external to the parent calice and is initially aseptate; septa only form later. In a second type, typical of species with a narrow dissepimentarium (e.g. *L. arundineum*), the new corallite arises from near the periphery and, though septate from the start, does not inherit septa from the parent. In a third type, in species such as *L. martini* with a wide dissepimentarium, the daughter arises from well within the parent calice and inherits its initial septa, especially those of the outer wall, from the ends of the parent septa. Cerioid species of *Lithostrotion* increase in a manner analogous to the third type, but the daughter remains attached to the parent and is polygonal from the start.

Rejuvenescence (Fig. 5.5m). In solitary cylindrical forms the corallite is often found to be constricted at irregular intervals, leaving a broad or narrow shelf where the septate older calice is exposed. Above this it may expand again, before once more being constricted. These constricted bands probably represent periods of famine during which the polyp resorbed its own tissues in order to stay alive and shrank away from the edges, becoming smaller whilst retaining its

form. The next period of increased food supply permitted growth to begin once more; this is rejuvenescence. Starvation and regrowth, however, are only one explanation. In some species rejuvenescence may be especially strong, so that sharp rims are formed with deep and almost slit-like contractions between.

Evolution in the Rugosa

The Rugosa have relatively few phylogenetically useful characters. Furthermore, they were very plastic in their evolutionary potential and notoriously subject to homeomorphic trends in separate stocks. Yet however common such iterative trends may have been, the guiding principle controlling their evolution was always towards a strong and firm skeleton. Thus many of the observed structures can be interpreted functionally; for instance, the complication of the trabeculae increased skeletal strength, the carinae held the polyp more firmly and the axial column gave greater support in the central region.

There is thus some functional meaning in all the trends that have been observed, though it is not always clear what the function was.

Overall pattern of evolution. The first known 'corals' from the Middle Cambrian of New South Wales, Australia, include the genus *Cothonion* (Fig. 5.8 m) (Jell & Jell 1976) which has a conical septate corallum and a biradial operculum. This has been tentatively referred to the Rugosa. Undoubted Rugosa do not appear until the Blackriveran (M. Ord.) in North America. They spread rapidly and evolved in a series of five major episodes. During each of these one particular faunal group was dominant, and subsidiary elements at the time were: (a) small persistent bradytelic stocks; (b) early members of a later dominant stock; and (c) remnants of a formerly dominant stock. The main phases are as follows:

(a) In the Middle Ordovician to Lower Llandovery period small solitary corals dominated, mainly without dissepiments and with only feeble development of trabecular tissue. The Streptelasmatina and Columnariina were dominant.

(b) The Upper Llandovery to Lower Devonian period saw the first really prolific burst of Rugosa and the exploitation of the reef habitat, by both solitary and colonial corals, though the Rugosa were subsidiary components of most reefs relative to stromatoporoids and tabulate corals. Even so, at generic level the Rugosa outnumbered the Tabulata in the Middle Silurian and thereafter kept ahead.

A dissepimentarium was independently evolved in all three suborders, and seems to have been associated in some way with the great success of the Rugosa at this time. The Columnariina were dominant, but other groups also flourished including a curious short-lived family Calostylidae, which had perforate septa. (Perforate septa are also known in the Scleractinia, but the Calostylidae were not ancestral to them.)

(c) There were relatively few Lower Devonian corals, and not many of the Silurian genera lasted into the Devonian. A new burst of evolutionary diversity in both simple and compound corals took place in Middle to Upper Devonian times, with digonophyllids (Suborder Cystiphyllina) and phillipsastraeids being numerically important. These are all large colonial forms with a wide dissepimentarium and long septa with trabecular fans. But by the late Devonian coral faunas had become impoverished and the dominant Phillipsastraeidae and the last Cystiphyllina had died out.

(d) The Lower and Middle Carboniferous saw the climax of rugosan development with the Superfamily Zaphrenticae dominant (Family Aulophyllidae, Family Lithostrotionidae) and a whole variety of solitary and colonial forms evolving rapidly. The independent development of an axial structure was characteristic of virtually all groups, and the microstructure became very complex. Some primitive Cyathaxoniicae still continued. By this time the Tabulata were declining whilst the Rugosa were relatively more successful. Late Middle Carboniferous and Upper Carboniferous forms were few in number, and most families had died out by the Artinskian.

(e) The final radiation of the Rugosa was in the Permian, based mainly on the surviving Cyathaxoniicae, in which the protosepta were strongly developed. Such genera as *Plerophyllum*

(Fig. 5.8s) and the columnariid *Waagenophyllum* were dominant. The last known Rugosa are known from the lowermost Triassic. The early Scleractinia may have been derived from one of these later Cyathaxoniicae; on the other hand, many authorities consider that the Scleractinia derived from a soft anemone stock and are not descended from the Rugosa.

Evolutionary trends. Rugose corals had only a limited evolutionary potential; hence it was inherently likely that the same kind of structure would be evolved several times over. According to Lang (1923), who first tried to establish such repetitive 'trends', in Carboniferous corals, 'each character follows in varying degrees and at varying rates one of a comparatively few possible developments pointing to the existence . . . of limited tendencies . . . repeated in each lineage'. In taxonomy, which becomes confused by such trends, the only way to separate and identify the genera is by very detailed study of morphology, fine structure and ontogeny.

In general terms the development of a marginarium (stereozone or dissepimentarium) is perhaps the most persistent and universal of all trends in the Rugosa, and since this happened in all three suborders independently in the later Llandovery, just prior to the first success of the Rugosa as reef builders, it seems to have been important and fundamental, though it is not entirely clear how. A number of other trends fundamental to all the Rugosa have also been distinguished.

The phylogeny of the Rugosa, however, is still too poorly known to allow trends to be properly documented except in a few specific cases.

Evolutionary studies of trends within Family Lithostrotionidae have shown that the parallels are so close in several descendent stocks that species originally lumped in the same genera were actually derived polyphyletically. Thus '*Diphyphyllum*' (Fig. 5.8n) is a composite genus of which several species were originally derived from an ancestral *Lithostrotion* stock. In all of these the central tabulae have become so shaped that the outer down-turned edges have formed a single central tube, the advantage being that a stronger axial structure results. Similarly '*Orionastraea*', which has a weak or absent columella and a tendency for septa to be withdrawn from both the axis and the periphery, is evidently polyphyletic and has had to be split since it included morphs that are of similar form but derived from different ancestral species.

Apart from these trends, there is a cerioid trend in some Lithostrotionidae and in addition a particularly interesting trend which involves the loss of dissepiments through time. In the early forms (e.g. *Lithostrotion praenuntius*) there are six circlets of dissepiments, reducing in the descendent *L. martini* to four and then to two. In addition the septa generally become fewer and the tabulae more widely spaced. The evolutionary principle governing the origin of the new types is probably paedomorphic since the descendent adults tend to resemble ancestral juveniles. Such morphology would seem to be especially appropriate for an environment in which rapid growth with minimal expenditure on skeletal elements was desirable. Likewise, the advantage of food sharing between the linked polyps in the 'orion-astraeoid' aphroid morphologies, and the fact that in these all the available surface space is covered by polyps, would lead to greater efficiency of the corallum as an integrated unit.

Each of these trends therefore, expressed polyphyletically, has its own advantage suitable for particular ecological conditions.

Microevolution. There are relatively few well-documented accounts of small-scale evolutionary change in the Rugosa. One of the best known is that of Carruthers (1910) who worked on the *Amplexizaphrentis delanouei* group in Scotland and England. The species group actually ranges as far as Belgium. His collections were made from thin limestones | in | an alternating | limestone–shale–sandstone–coal sequence, all well located stratigraphically. In studying the populations in ascending sequence he noted that most characters (shape, size, tabular spacing, etc.) tended to remain constant but that there seemed to be successive changes in the shape of the cardinal fossula and the length of the major septa (Fig. 5.4n–s). The oldest beds contain *A. delanoeui s.s.*, which has septa meeting in the centre and a large cardinal fossula expanded axially. A variant of this, *A. parallela*, with a parallel-sided fossula, occurs in the same beds. It was probably derived paedomorphically.

A short-lived side branch led to the small *A. lawstonensis*. In younger beds *A. delaneoui* and *A.*

constricta are rare and replaced largely by *A. constricta* in which the axial end of the cardinal fossula is constricted to a keyhole shape. In the youngest limestones there occurs *A. disjuncta;* this form passes through a *constricta*-type morphology in its juvenile development, but thereafter septa retreat away from the centre.

Though this has been regarded as a classic study it is severely limited in two respects. One is that since the corals occur only at certain horizons within a highly variable sedimentary sequence, it was not possible to establish continuity of variation. Secondly, the study was done within a limited area alone, and it is not clear how far the sequence can be substantiated in other regions. And since different 'species' coexisted in time *(A. delanouei and A. parallela)* and there is some evidence of the same species being found at different horizons in different areas, the pattern of evolution was probably very complex and is perhaps best regarded as the expression of a trend within a species group rather than as a matter of linear descent. In functional terms the morphology of *A. constricta* makes for a stronger axial region than that of *A. delanouei*, but that of *A. disjuncta* is clearly weaker. Perhaps the need for rapid growth, as would be allowed by septal reduction, may in this case have offset the axial weakness.

So many factors seem to have been involved in the evolution of the *A. delanouei* group – including trending, response to local environmental conditions, paedomorphosis, allopatry and polyphyly – that the evolutionary pattern now seems to be so complex as to elude analysis altogether.

Classification of the Rugosa

The following classification, erected upon the various structural features already described, is based upon that in the *Treatise*. It is largely founded on megascopic features; knowledge of microstructure is not yet advanced enough to be of much use. The inevitable effect of parallel trends and convergent evolution, however, creates taxonomic problems which if resolved may lead to a radical rearrangement of the group.

ORDER RUGOSA (?Cam./Ord. – L. Trias.): Solitary or colonial corals with the major septa inserted in four quadrants only relative to the protosepta. Epitheca normally present, and up to three orders of septa. Dissepiments occur in more advanced groups; tabulae normally present. First- and second-order septa well developed; third-order septa very rare.

SUBORDER 1. STREPTELASMATINA (Ord.– L. Trias.): Solitary or colonial. Marginarium either a septal stereozone or a dissepimentarium constructed of small globose interseptal dissepiments. Tabulae domed.

SUPERFAMILY 1. CYATHAXONIICAE (Ord. – L. Trias.): Small and solitary. Usually without dissepiments, with only a narrow septal stereozone and with outwardly declined tabulae. e.g. *Cyathaxonia, Syringaxon, Amplexizaphrentis.*

SUPERFAMILY 2. ZAPHRENTICAE (Ord.–Perm.): Solitary or colonial. Usually with a dissepimentarium or stereozone and with conical or domed tabulae. e.g. *Heliophyllum, Dibunophyllum, Lithostrotion, Phillipsastrea, Streptelasma.*

SUBORDER 2. COLUMNARIINA (Ord.–Perm.): Usually colonial, rarely solitary. Marginarium variable (absent, septal stereozone, or with lonsdaleoid dissepimentis). Septa thin. Tabulae flat, depressed or sagging axially, sometimes domed. e.g. *Lonsdaleia, Spongophyllum, Waagenophyllum.*

SUBORDER 3. CYSTIPHYLLINA (Ord.–Rec.): Solitary or colonial. Septa with large and often complex trabecular structure, or sometimes absent. Marginarium a stereozone or wide dissepimentarium with small globose dissepiments. Tabulae present. e.g. *Cystiphyllum, Calceola, Goniophyllum.*

Ecology of the Rugosa

Solitary Rugosa had no really effective means of anchorage on the sea floor. They did not normally cement themselves to the substratum though some have cicatrices of attachment or root-like 'talons' which may have helped to fix them. They seem to have preferred soft substrates, but since relatively few 'hardgrounds' are found preserved it is not known how far they could colonise such hard substrates. The horn-like shape would give a reasonable degree of support if the coral were to be half sunk in mud, convex side down. This would allow the polyp to be exposed above the sea floor and would enable it to stay in the same general attitude whilst the coral grew. There is some evidence that Rugosa such as *Aulophyllum* could grow like this, orientated with respect to currents. Commonly, however, in any population some or all of the coralla are twisted; they must have toppled over and then grown up again.

The solitary *Calceola* lay on the sea floor, convex side down with its operculum open to 90°. Presumably it could shut it in case of emergency.

Compound Rugosa did not normally fix themselves to the sea floor either, though rare examples of hardgrounds encrusted by Rugosa are known. In softer substrates the weight of the colony in the mud would no doubt give some stability. Since they were uncemented they were environmentally restricted and could not build proper reefs. This seems to have been their most severe evolutionary limitation, from which, in spite of their 230-million-year history, they were never able to escape. Thus they are very rarely found within the limits of strong wave action and so were never really involved as frame builders in the algal and coral–stromatoporoid reefs of the Palaeozoic (Wells 1957).

Rugose coral faunas were very sensitive to environmental conditions, and the form of both solitary and colonial species was strongly influenced by ecological factors (Wells 1937).

Separate facies faunas can be distinguished, adapted for particular circumstances and normally found with characteristic associations of brachiopods and tabulates. Three such facies faunas were noted by Hill (1937) working mainly with Scottish Carboniferous successions. These are:

(a) The *Cyathaxonia* fauna. Usually found in black or greenish calcareous shales, though sometimes in bryozoan reefs also, these corals are all small and solitary with dissepiments poorly developed. This facies fauna is very long-ranged and certainly antedates the Carboniferous; indeed the earliest of all rugose coral faunas seem to have been of this type, though occurring in mudstones. It is usually found with tabulates, small brachiopods and trilobites. As represented typically in the Carboniferous the sea floor was generally muddy, with decaying organic matter, but must have been fairly well oxygenated.

(b) The *Caninia*–clisiophyllid fauna. This fauna, found first in the Tournaisian but probably older, consists of large solitary genera with dissepiments. Clisiophyllids gradually took over from caniniids with time. These faunas were adapted to shallow limestone seas, well lighted and oxygenated but with little terrigenous material.

(c) Colonial coral faunas. Compound rugose corals with dissepiments may occur in small and scattered 'reefs'; they are represented in bedded limestones as well. These were first considered to be reef faunas analogous to modern reef corals, but since the Rugosa were not really reef formers they are best considered as just a group of colonial corals. These colonial corals evolved from the *Caninia*–clisiophyllid fauna.

Hill's three facies associations have generally been well substantiated, though the Cyathaxonia fauna is very much separate whilst the other two are sometimes intergrading and ecologically less well defined. Thus it is not uncommon to see rapid alterations of faunas (b) and (c) in vertical sequence, as in the Carboniferous of the west of Ireland where, in addition, thickets of fasciculate lithostrotionids are found with the large solitary *Aulophyllum* growing within the thickets.

More recent work has concentrated upon the distribution of rugose and other corals in Palaeozoic reefs. Klovan (1964) and Krebs (1974) have charted the facies associations of corals and other reef organisms in Devonian reefs of western Canada and central Europe respectively (Fig. 5.8v). In these reefs colonial Rugosa tend to be confined to the quieter fore-reef facies around the flanks of rigid frame-reefs; they do not normally occur within the reef itself nor, except in localised patches, within the back-reef facies. Solitary Rugosa are likewise confined but are numerically insignificant.

Tabulate corals, especially those with a creeping or foliaceous habit, were somewhat less restricted but could not compete with stromatoporoids and algae in the turbulent zone of most active reef growth. The hard skeletons of both rugose and tabulate corals, however, seem to have contributed to establishing a firm foundation for later reef growth of algae and stromatoporoids. Whilst the broad pattern of distribution of rugose and tabulate corals with respect to ancient reefs has been generally confirmed (e.g. Jamieson 1971), a more detailed picture of the various coral–stromatoporoid facies associations is emerging. Scrutton (1977), for example, who worked upon an evolving reef complex in south-western England, has shown how the genera in stromatoporoid reef limestones differ from those in biohermal, bioclastic and bedded dark limestones.

Copper (1974) (Fig. 5.8w) has described a generalised palaeoecological succession of a Devonian reef from pioneer through intermediate to a mature reef system. Devonian reef systems were evidently more extensive and elaborate than those of the Silurian but by the end of the Devonian had collapsed. A fuller treatment of the theme has been given by Walker and Alberstadt (1975).

On a global scale, evidence is also accumulating on the biogeographical distribution of rugose corals (Oliver 1976).

Order Tabulata (Fig. 5.9)

The tabulate corals are entirely Palaeozoic, and though they appear a little earlier than the Rugosa they have an otherwise similar time range. They are always colonial, never solitary, and usually their corallites are small. Invariably they have prominent tabulae, but other skeletal elements, in particular the septa, are reduced or absent.

Having relatively few structural elements, the Tabulata are of comparatively simple construction. *Favosites* (Sil.–Dev.) (Fig. 5.9a), for example, has elongated thin walled prismatic polygonal corallites with horizontal tabulae extending right across the corallum. The septa are reduced to short and somewhat irregular spines. All the walls are perforated by numerous **mural pores,** connecting the corallites. In *Favosites*, as in tabulates generally, elements of 'skeletal microstructure' (tabulae and walls) are fibro-normal. Trabeculae are scattered in tabulates. In *Favosites* they are normal to the surface of the tabulae but inclined upwards and outwards along the axis of the septa. In other genera septal spines are not necessarily trabeculate. Some genera (e.g. *Trabeculites*) have trabeculate walls, and *Halysites* has trabeculae in the walls.

Range in form and structure of the Tabulata

Form of corallum. The corallum (colonial skeleton) is built up by individual polyps which may or may not be directly connected to each other.

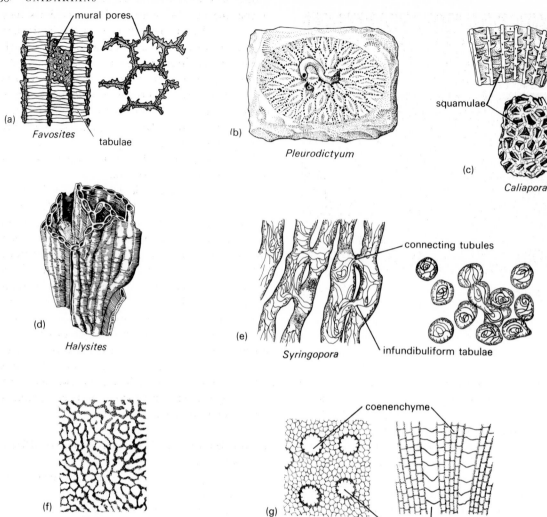

Figure 5.9 Morphology of Tabulata: (a) *Favosites* (Sil.–Dev.) – vertical and transverse sections showing tabulae, reduced spines and mural pores (x 4); (b) *Pleurodictyum* (Dev.) (internal mould) with commensal worm, (x 1·5); (c) *Caliapora* Dev.) – vertical and transverse section showing shelf-like squamulae (x 4); (d) *Halysites* (Ord.–Sil.) – external view of colony showing cateniform growth (x 0·75); (e) *Syringopora* (Sil.–Carb.) – vertical and transverse section showing infundibuliform tabulae and connecting tubules (x 3.5); (f) meandroid growth in *Chaetetes* (Ord.–Perm.) (this may not be a tabulate coral) (x 0·75); (g) *Heliolites* (Sil.–Dev.) – transverse and vertical section with corallites embedded in coenenchyme. (All redrawn from Hill and Stumm in '*Treatise*' Part (F), other than (g) redrawn from Woods, H. (1896) *Palaeontology* (Cambridge: C.U.P.).)

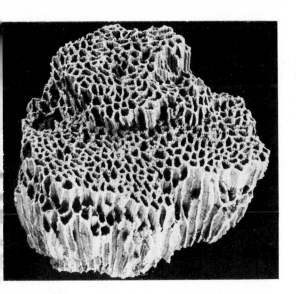

Figure 5.10 *Halysites* – a Silurian colony cleared of matrix showing cateniform growth (x 1). (Photograph by courtesy of C. Chaplin)

Cerioid forms (e.g. the Favositina) (Fig. 5.9a) have polygonal corallites all in contact. **Cateniform** colonies (e.g. the Halysitina) (Figs 5.9d, 5.10) have elongated corallites joined end to end in wandering palisades. Fasciculate tabulates (e.g. *Syringopora*) (Fig. 5.9e) have cylindrical corallites which may be dendroid or phaceloid and may be provided with connecting tubules. **Auloporoid** genera have a branching (**ramose**) tubular structure and often an encrusting creeping (**reptant**) habit; others are erect. Finally, **coenenchymal** (Fig. 5.9g) types, character-istic of the more advanced tabulates (e.g. the Heliolitina), have no dividing walls between the corallites but instead a common mass of complex tissue, the coenenchyme, originating as a margi-narium and forming a dense calcareous mass in which the corallites are embedded. As in the Rugosa, colony form, which refers to the relations of the corallites, must be distinguished from colony habit; thus *Thamnopora* is a ramose but cerioid favositid. Other tabulates may be **massive, foliaceous** (in which habit the coral forms thin overlapping laminar sheets) or **creeping.**

Skeletal elements. Tabulae, the most important skeletal elements, are always present and commonly traverse the corallite horizontally. Sometimes they are replaced by smaller tabellae. In the fasciculate genus *Syringopora* (Fig. 5.9e) the tabulae are funnel-shaped (**infundibuliform**) and run through the tubules that connect the branches. Septa in tabulates are rarely more than short spines, commonly twelve in number, but in some cases they do reach the centre of the corallite. In *Caliapora* (Fig. 5.9c) and other genera of the Favositina they may be replaced by small shelves (**squamulae**) which jut inwards at intervals from the corallite wall.

Where there is a marginal zone, as in the more advanced tabulates, it can be of two kinds. It may be only a thickened zone of annular lamellae as in *Striatopora*, constricting the aperture. In the other kind the growth of the marginarium is accompanied by the loss of the corallite walls, resulting in a coenenchyme. This can be constructed in various ways and evidently evolved independently in several stocks; tabulae, trabeculae borne on dissepiments and trabeculae closely packed or sometimes organised into tubes may all have a part in it. The production of a coenenchyme is one of the more important evolutionary trends in the Tabulata, another being the development of mural pores or connecting tubules for interconnection of the corallites.

Asexual budding and growth of the corallites. The offsets may grow in various ways:
(a) axially (parricidally) through the growth of dividing walls;
(b) peripherally, i.e. erupting in the coenenchyme;
(c) laterally, in which each bud originates by outgrowth from the parent and thereafter grows intermurally.

Evolution and ecology of the Tabulata

Possible tabulates are recorded from the Cambrian, though the earliest undoubted Tabulata are known in the Chazyan of North America, somewhat ante-dating the arrival of the first Rugosa. By Trenton times they had spread to many parts of the world and the first Heliolitina and Halysitina had appeared, whilst the Sarcinulina and Lichenariina were already on the wane. The Favositina from their first appear-ance in the Middle Ordovician became a very dominant group throughout the Silurian and Devonian, though the Heliolitina were also important.

With the Carboniferous the Favositina, which had undergone many changes, lost their dominance and became subordinate to the Auloporina, Syringoporina and Chaetetina (if the latter are indeed tabulates). These remained the important groups until the end of the Permian.

Larger tabulates are found in coral–stromatoporoid reefs and were relatively important, though they were not really frame builders since they had no proper means of attachment. Smaller tabulates tended to occur in deeper waters, and fasciculate genera usually lived in quieter environments. Ordovician and Lower Silurian tabulates of small size are often found with early solitary Rugosa, forming a characteristic association in calcareous mudstones. In later rocks also the two are frequently associated, and evidently the Rugosa and Tabulata had similar habitat preferences. Generally they lived in similar conditions to those of modern corals, but they are not normally found in very high energy environments.

Tabulates are not of great stratigraphical value, but they do sometimes occur in useful marker bands. The Lower Devonian *Pleurodictyum* (Fig. 5.9b), a small domed tabulate always found with a commensal tubiculous worm in the centre of the colony, is characteristic of sandy facies, by contrast with other corals. Along with other fossils of the same facies it has been used successfully to tell directions of faunal migration across Europe.

Classification of the Tabulata

ORDER TABULATA (Ord.–Perm.): Invariably colonial. Generally composed of small corallites with tabulae; septa (usually twelve in number) reduced to vertical rows of spines or absent.

There are seven or eight suborders which are quite clearly circumscribed and easily distinguished. The position of SUBORDER CHAETETINA is not clear since some evidence suggests that these may be hydrozoans or stromatoporoids rather than corals. The following classification is much abbreviated.

SUBORDER 1. CHAETETINA (Ord.–Perm.): Slender aseptate corallites with no mural pores. e.g. *Chaetetes*.

SUBORDER 2. LICHENARIINA (L. Ord. – L. Sil.): Massive colonies, Corallites prismatic with sixteen or more septa and with mural pores in horizontal rows. e.g. *Lichenaria*.

SUBORDER 3. SARCINULINA (U. Ord. – M. Sil.): Massive coenenchymal colonies, with coenenchyme enclosing horizontal spaces and formed largely of tabular and septal extensions. Up to twenty-four septa. e.g. *Sarcinula*.

SUBORDER 4. FAVOSITINA (M. Ord.–Perm.): Variform colonies with slender corallites having mural pores and short spinose septa. e.g. *Favosites, Alveolites, Michelinia, Pleurodictyum*.

SUBORDER 5. AULOPORINA (Ord.–Perm.): Corals with creeping or erect habit, fasciculate, with tabulae widely spaced or absent. e.g. *Aulopora*.

SUBORDER 6. SYRINGOPORINA (U. Ord.–Perm.): Large erect coralla, dendroid or phaceloid, with cylindrical corallites connected by horizontal tubules. Septa often absent; tabulae horizontal or funnel-shaped. e.g. *Syringopora*.

SUBORDER 7. HALYSITINA (M. Ord. – U. Sil.): Cateniform colonies of imperforate coralla having in some cases small vertical tabulate tubules between the large tubes. e.g. *Halysites*.

SUBORDER 8. HELIOLITINA (M. Ord. – M. Dev.): Massive colonies with twelve-septate corallites embedded in coenenchyme. (These are sometimes classified apart and have some similarity to certain octocorals.) e.g. *Heliolites*.

Order Scleractinia (Fig. 5.11)

All post-Lower Triassic corals are included in Order Scleractinia. All of them, like *Caryophyllia* (p. 66), secrete an aragonitic exoskeleton in which the septa are inserted between the mesenteries in multiples of six (Fig. 5.4a–c). After the first six prosepta grow, successive cycles of six, twelve and twenty-four metasepta are inserted in all six quadrants. Through their pattern of septal insertion the Scleractinia are immediately distinguished from the Rugosa. Each of the earlier cycles is complete before the next cycle grows, though in some cases this system breaks down in the higher cycles.

The skeleton originates as a thin basal plate from which the septa arise vertically. The polyp then secretes an epitheca growing up from the basal plate as a cone, over which the lower rim of polyp hangs as an **edge zone.** Within the cup there may be dissepiments as well as septa. In general, the structure of both simple and compound scleractinians is light and porous, rather than solid as in the Rugosa. In several other respects other than the primary septal plan the Scleractinia differ from Palaeozoic groups; these can be mentioned briefly.

Range in form and structure of scleractinians

Type and habit. Scleractinians may be solitary or compound. In solitary scleractinians the form of the corallum depends on the relative rates of vertical and peripheral growth once the basal plate has been secreted. Where peripheral growth has been dominant a discoidal form results (Fig. 5.12a), often with everted septa in cases where these have grown rapidly. Perhaps the commonest kinds are those with turbinate or conical coralla in which the axis is straight (Fig. 5.11a), though horn-shaped and cylindrical types are common.

In colonial Scleractinia, as with other colonies, the corallites are interconnected as a result of repeated asexual division by the polyps. Colony type is genetically defined and refers mainly to the relationships of the corallites. As with the Rugosa there are dendroid, phaceloid, thamnasteroid and, rarely, aphroid types. In cerioid scleractinians the walls of adjacent polygonal corallites are closely united and of dissepimental or septal origin, by contrast with the walls of cerioid rugosans, which are epithecal. **Plocoid** types of corallites have separated walls but united by dissepiments. Ramose types of creeping habit, paralleling the tabulate *Aulopora,* are not uncommon.

In addition there are two other types which do not correspond to any rugosan type:

(a) the **meandroid** type (Figs 5.11b, 5.12b) in which corallites are arranged in linear series with the cross-walls absent, and confined within the lateral walls which run irregularly over the surface like the convolutions of a human brain;
(b) the **hydnophoroid** type (Fig. 5.11c) in which the centres of the corallites are arranged around little hillocks or **monticules.**

In habit scleractinian colonies may be branching (ramose), massive, encrusting and creeping (reptant) or foliaceous. Sometimes adjacent colonies of the same species fuse to form a single colony.

Septa and associated structures. Septa are formed of aragonitic trabeculae (simple or compound) normally arranged in a fan-like system and often with a denticulate upper surface, but laminar tissue is unknown (Fig. 5.11e). Usually the trabeculae are united, though sometimes, paralleling the Palaeozoic Family Calostylidae, the trabecular framework is loosely united and may be perforated. Such **fenestrate** septa are more important in scleractinians than rugosans. The grouping of trabeculae and their structure are important stable characters for taxonomy. They form initially from aragonitic spherulites in an organic matrix (Sorauf 1972).

The septa originate between the mesenteries, which are also the site of digestion, excretion and gonad development. These mesenteries have a double layer of endoderm separated by a thin sheet of mesoglea. Muscles are arranged on one side of the mesentery only (Fig. 5.4c), and the mesenteries are grouped in pairs with the pleated muscle blocks facing each other. The space within such a pair is the **entocoel,** whilst between (the muscles facing away from each other) is the **exocoel.** But two pairs of mesenteries on opposite sides, the **directive mesenteries,** have the muscle pleating reversed, giving a bilateral symmetry to an otherwise radial coral. If the coral is elliptical the long axis extends in the directive plane.

Septa are thus entocoelic (**entosepta**) or exocoelic (**exosepta**) in origin; usually the first two cycles are of entosepta, the rest of exosepta. In some scleractinians there may be a peculiar growth system producing vertical pillars (**pali**) along the inner edges of some of the entosepta (Fig. 5.11 g–m).

If a columella forms it is always of septal origin and may begin as pali; it is usually a central rod or a dividing plate formed from a proseptum.

The septa are often connected by cross-bars called **synapticulae** (Fig. 5.11d) which grow towards each other from the walls of adjacent septa and eventually fuse, perforating the mesenteries in the process.

Other primary structures. The thin basal plate is semitransparent and firmly adherent to the substratum; it may later be thickened by secondary deposits. The epitheca is not developed in many scleractinians but if present may consist of chevron-like crystallites of aragonite. Dissepiments, like those

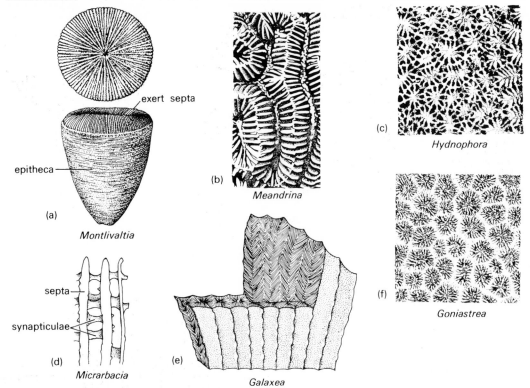

Figure 5.11 Scleractinian morphology: (a) *Montlivaltia* (Jur.) well-preserved specimen (x 1). (b) *Meandrina*, upper surface of a Recent meandroid coral; growth pattern resulting from intratentacular budding (x 1). (c) *Hydnophora* (Rec.), upper surface of hydnophoroid coral; growth pattern resulting from extratentacular (circummural) budding (x 2). (d) Synapticulae in *Micrarbacia* (Cret.–Rec.). (e) Trabecular structure in *Galaxea* (Mio.–Rec.). (f) *Goniastrea* (Eoc.–Rec.), upper surface showing distomodaeal budding, i.e. division of a single stomodaeum into two (x 2). (g)–(m) Origin of pali: (g) Sextant of solitary scleractinian. (h) Same divided into layers: (j)–(m) Transverse of polyp and corallum at each of these levels: (n) *Lophelia* (Oligo.–Rec.), a dendroid deep-water caryophylline (x 1). (p) Scleractinian epithecal wall with internal layers of stereome. (q) *Galaxea* polyp/coral

of the Rugosa, are secreted by the base of the polyp. As the latter grows it moves upwards, detaching a small part of the base from the calice at a time and forming a dissepiment to enclose each space. Where the fine skeletal structure of dissepiments have been studied (Sorauf 1972) dissepiments have been shown to have two layers: a first-formed primary layer, overlain by vertically aligned clusters or aragonitic spherulites. Endothecal dissepiments are confined within the corallite, but in some colonial scleratinians the corallites are united by a common spongy tissue, the coenosteum, which may be formed partially of exothecal dissepiments and partially by rods and pillars (**costae**)**,** as in *Galaxea* (Fig. 5.11q).

Tabular dissepiments are flattish plates extending across the whole width of the corallite or confined only to its axial part (thus resembling tabulae in rugosans). **Vesicular dissepiments** are small arched plates, convex upwards and overlapping, and usually inclined downwards and inwards from the edge of the corallite.

Secondary structures.   Stereome (Fig. 5.11p) is an adherent layer of secondary tissue which may cover the septal surface. It is composed of transverse bundles of aragonitic needles. Stereome normally thickens the epitheca internally as well, but in some cases its function is taken over by primary thickening of the septa, by dissepiments or by synapticulae.

In compound scleractinians individual corallites are usually separated by a complex perforated tissue: the **coenosteum.** This may consist entirely of

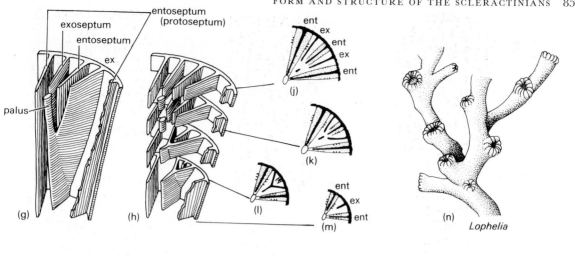

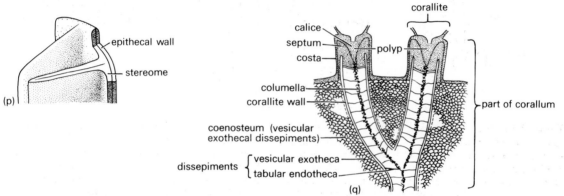

relationships in a coenostial scleractinian. (Redrawn from Wells, 1956 except (n) in '*Treatise*' Part (A) and (a) redrawn from *British Mesozoic Fossils* (1962). (London: British Museum).)

exothecal dissepiments (tabular) or other material, but it provides a support for numerous canals linking the individual corallites and binding the living tissues together in a functionally cohesive mass.

Asexual reproduction and colony formation. The form of the colony in the Scleractinia is largely determined by the mode of budding combined with relative growth rates. The following types have been distinguished:

(a) **Intratentacular budding.** Here the polyp divides by simple fission across the stomodaeum, each daughter (or which there may be two, three, four or more) retaining part of the original stomodaeum and regenerating the rest. The products of such **polystomodaeal** budding remain linked and except in very rare instances never really separate as individuals. Meandroid corals (Fig. 5.11b) are clearly the results of this kind of budding.

(b) **Extratentacular budding** (Fig. 5.11f). In this type new stomodaea are produced outside the tentacular ring of the parent. These extratentacular buds soon separate and do not remain linked. The colony that results from such budding (whether branching or massive) thus consists simply of numerous separate individuals and is not integrated functionally in the same way as are the products of intratentacular

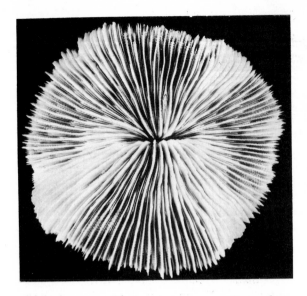

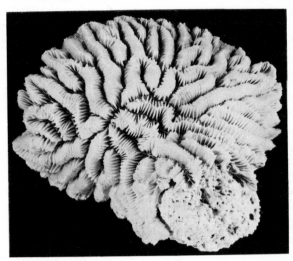

Figure 5.12   (a) *Fungia*, a large recent solitary scleractinian showing several orders of septa (x 0·5); (b) *Meandrina*, a recent colonial coral from Barbados with typically meandroid structure resulting from intratentacular budding (x 0·5) (c.f. Fig. 5.11b). (Photograph by courtesy of C. Chaplin)

budding. There may, however, be some nervous or chemical linkage if the soft tissues are in contact, even if the enterons are not united.

Evolution in scleractinians

The oldest known scleractinians are Middle Triassic in age. They may have been derived from a group of soft anemones or from a group of late Palaeozoic cyathaxonians. On their first appearance they are already differentiated into a number of important families. **Hermatypic** in type, they formed small patch reefs, may have possessed zooxanthellae and were confined to warm shallow well-lighted waters (p. 87). By the late Triassic they were expanding fast in many parts of the world, and many new families arose. This pattern persisted through the Lower Jurassic but with the first true deeper-water **ahermatypic** corals (suborder Caryophylliina) emerged at about that time, though these were of limited importance. By the Middle Jurassic scleractinians were on the increase everywhere, and patch reefs erupted wherever conditions were suitable. In the Tethyan Ocean the first large reefs developed, persisting until a general setback in the early Cretaceous. Later the large reefs and coral banks developed, and whilst some of the older families died out others arose. The success of reef corals fluctuated in the Cretaceous, and the ahermatypes became really important for the first time. By the later Cretaceous both hermatypic and ahermatypic corals were becoming of distinctly modern type, and by the end of the Eocene the dominant groups were much like those of the present since the formerly important Mesozoic families (e.g. the Montlivaltiidae) had disappeared. During the late Tertiary the two main coral faunal provinces of the present – the Indo-Pacific and western North Atlantic provinces – had become distinct, and reef-building corals flourished on an unprecedented scale. Many reefs were killed by the lowering of sea level in the Pleistocene, but their remains have provided stable surfaces for the regrowth of coral since their submergence. Comparatively few genera were affected by the conditions of the Pleistocene, so that Pliocene coral faunas are effectively the same as those of the present.

Most families of scleractinians show evolutionary trends, especially from solitary to compound coralla and, in colonial types, from phaceloid through plocoid and cerioid to meandroid form. This probably increased the degree of colonial integration (p. 92).

Ecology of Scleractinians

Scleractinians fall into two main ecological categories: hermatypic and ahermatypic.

Hermatypic corals are those in which the endodermal cells are replete with symbiotic algae (dinoflagellates or zooxanthellae). These algae exist in enormous numbers, and they are so essential to the metabolism of the coral supplying it with nutrients and oxygen that hermatypes can only flourish in the photic zone, at depths normally less than 50 m (rarely 90 m). Since the corals have firm anchorage they grow successfully and most abundantly within the zone of wave action, at depths less than 20 m.

Hermatypes generally need a minimum water temperature of 18 °C and flourish best between 25 and 29 °C and at normal salinity. Hence most reef corals, which are hermatypic, are restricted to shallow well-lighted warm water. In addition they need a good supply of oxygen and generally prefer a firm non-muddy substrate for otherwise the planulae will not settle. There are, however, hermatypes living and building reefs in muddy regions. Light is perhaps the most important of all the limiting factors, though a reef can be killed by tidal emergence during heavy rain, during a hurricane for instance, and may take twenty years to re-establish itself.

Ahermatypic corals have no zooxanthellae and do not have the same environmental restrictions. They are found at depths up to 6000 m but are most abundant down to 500 m, and though they can survive temperatures below 0 °C they flourish best between 5 and 10 °C. Furthermore, they can live in total darkness. Over two-thirds of these 'deep sea corals' are solitary and do not form reefs, though the colonial dendroid caryophyllid *Lophelia* (Fig. 5.11n) forms deep water banks or thickets at the edge of the western European continental shelf.

The earliest (Triassic) scleratinians were evidently hermatypes, though perhaps not so adapted for the reef habitat as those of the present since they formed only small reefs. Ahermatypic corals developed from many hermatypic stocks by the slow spread of certain types into colder deeper waters. The first known ahermatypes were the Jurassic caryophyllids whose descendants retained this habit, but representatives of many other groups also become ahermatypic, so that the hermatype/ahermatype division today cuts across many families and even genera.

Classification of the Scleractinians

In classifying the Scleractinia, as with the Rugosa and Tabulata, it has been necessary to find stable taxonomic characters. Septal structure can be used for defining subordinal categories whilst mode of colony formation is of value only at the familial level. Habitat and dimensions at any level of classification are of little value.

ORDER SCLERACTINIA (M. Trias.–Rec.): Solitary or colonial corals with mesenteries paired and the septa arranged in multiples of six. Basal plate gives rise to septa and epitheca.

  SUBORDER 1. ASTROCOENIINA (M. Trias.–Rec.): Normally colonial hermatypes with small corallites. Septa of up to eight trabeculae, spinose to laminar. e.g. *Acropora, Thamnasteria*.

  SUBORDER 2. FUNGIINA (M. Trias.–Rec.): Solitary or colonial with perforate septa linked by synapticulae. Corallites usually large. Mainly hermatypic. e.g. *Fungia, Cyclolites, Isastraea*.

  SUBORDER 3. FAVIINA (M. Trias.–Rec.): Laminar septa with fan-like trabeculae. Dissepiments present but synapticulae rare. Mainly hermatypic. e.g. *Favites, Montlivaltia, Thecosmilia*.

  SUBORDER 4. CARYOPHYLLINA (Jur.–Rec.): Normally solitary and ahermatypic. Septa laminar with simple trabeculae in fan system; rare synapticulae. e.g. *Parasmilia, Caryophyllia*.

  SUBORDER 5. DENDROPHYLLINA (U. Cret.–Rec.): Solitary or colonial. Septa laminar but irregularly perforate. Wall of swollen synapticulae. Mainly ahermatypic. e.g. *Dendrophyllia*.

Coral reefs

Recent coral reefs (Stoddard 1969, Ladd 1971) are confined to tropical seas and are best developed in the Indo-Pacific area and in the western North Atlantic. The Indo-Pacific reefs, where maximally developed (Melanesia to South-east Asia), have ninety-two recorded genera and subgenera and over 700 species, whilst the coral fauna of the North Atlantic is smaller with only thirty-four genera and sixty-two species. The North Atlantic fauna seems to have been derived initially from a Pacific source since Oligocene reefs of the West Indies have many of the same corals, but it has since developed in isolation, particularly since the

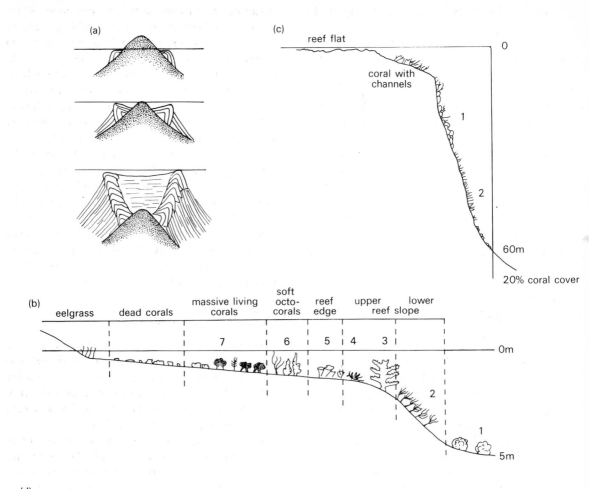

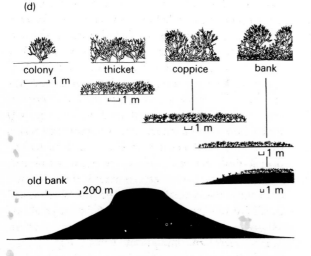

Figure 5.13 Coral reefs: (a) growth stages in the development of an atoll by volcanic subsidence according to Darwin: fringing reef (top), barrier reef (middle), atoll (bottom). (b) Section through edge of growing reef showing zonation. Manauli reef off N.W. Ceylon. The coral genera are represented diagrammatically: 1 – *Symphyllia recta*; 2 – *Acropora formosa*; 3 – *A. hyacinthus*; 4 – *A. humilis*; 5 – *Montipora foliosa*; 6 – Soft corals: *Alcyonaria, Gorgonaria, Antipatharia*; 7 – *Glavia, Pocillopora, Goniastrea, Porites* (Simplified from Mergner and Scheer, 1974, *Atoll Res. Bull.*). (c) Coral zonation on growing edge of Arno Island (Marshall Islands): 1 – *Porites* dominant; 2 – *Pachyseria* dominant (simplified from Sachet and Dahl, 1974, *Atoll Res. Bull.*). (d) Stages in development of deep-water coral bank (redrawn from Squires, 1964).

Americas were linked by the Panama isthmus in the Pleistocene.

In **structural reefs** the scleractinians have built up massive wave-resistant structures, often to great thicknesses, where the bulk of the material is contributed by corals, even though calcareous algae may play an important part in reef structure. Such a large rigid structure built up by generations of corals is the habitat of a host of diverse organisms, the whole forming one of the most complex ecosystems known. In coral reefs productivity, calcium metabolism and carbonate fixation (3500 g/m² per year) are very high. The range of habitats is such that there is a high degree of specialisation in the associated fauna, especially fishes, molluscs and arthropods; these are found in zoned and localised niches in various parts of the reef.

In **reef communities,** however, though equally confined to tropical seas the corals merely live together without building up a rigid structure.

Types and genesis of coral reefs

Structural coral reefs today belong to three major types: **fringing reefs, barrier reefs** and **atolls.** Fringing reefs develop along shorelines, especially those of volcanic islands. Barrier reefs are formed some distance from the shore. Atolls are circular or horseshoe-shaped and usually orientated with respect to the prevailing wind, with a central shallow lagoon in which small patch reefs may develop. The Great Barrier Reef (Maxwell 1968) off eastern Australia is a special case, being a linear feature parallel with the fault-bounded margin of the continent, where coral growth was initiated on the edges of an upraised fault.

It is now well known that the three major types represent successive stages in the growth of coral around a sinking volcanic island (Fig. 5.13a). The growth of corals has kept pace with subsidence, whilst broken reef talus has formed cascading fans going down to depth on the outside. This theory, originally suggested by Darwin, has been confirmed by deep borings into atolls. At Eniwetok atoll, bored in 1951, basalt was encountered at about 1·25 km below the surface. The limestone overlying the basalt was Eocene in age. Thus Eniwetok is a limestone pillar 1·25 km thick standing on a deeply drowned volcanic island some 3·2 km high on the ocean floor. The

maximum rate of subsidence had been about 50 cm per year. Confirmatory seismic and drilling evidence from other atolls suggests that this is the common pattern. The now sunken volcanic islands arose at 'hot spots' on a submarine ridge crest and have been carried away into deeper waters (as on a conveyor belt) as the ocean floor spread away from the ridge.

Darwin's theory of atoll formation, though now vindicated, did not take into account Pleistocene changes of sea level, which were far reaching since the reefs were exposed and the corals killed off. The exposed surfaces were eroded to some extent and often fretted into a karst topography. Hence the veneer of corals which has grown since the end of the Pleistocene has re-established itself on an anomalous topographic surface, and it is because of this that the form of a modern reef is not simply a reflection of constructional activity.

Growth and zonation of corals

Corals on reefs have few natural enemies other than coral-eating fishes and the starfish *Acanthaster*. The latter has recently wreaked havoc on some Pacific reefs, for after the depredations of the starfish the dead coral becomes covered with an algal film which prevents coral planulae from settling.

Because of the steep environmental gradients within the reef complex the corals themselves are ecologically zoned with specific reference to the shoreline (Fig. 5.13b, c). Each species occupies a particular habitat and is not found elsewhere. Within particular habitats there is a kind of 'pecking order' in corals; dominant species may actually eat other species of the same genus by mesentery extrusion and digestion. Different species inhabit the windward and leeward sides of the island. There is, furthermore, a great change both in species and in their growth forms, from the upper part of the fore-reef slope across the reef flats to the central lagoon with its shell sand floor and patch reefs. In some cases massive rounded corals such as *Diploria* tend to inhabit the surf zone, whilst the stout branching *Acropora* (the 'stag's horn' coral) lives at greater depths out of the range of surf action. In other areas massive corals are commoner in quiet waters, and *Acropora* occurs in high energy zones. The apparent structural strength of corals is therefore only one factor amongst many that control ecological zonation; ability to compete for space and

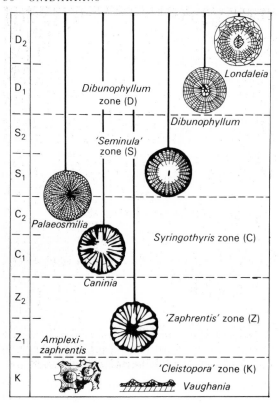

Figure 5.14 British Lower Carboniferous coral–brachiopod zones; based on the original work of Vaughan and showing time ranges of various coral genera. The *Seminula* zone is based on the ' eponymous brachiopod (redrawn from Tasch, 1973).

to cope with sediment influx may be likewise important.

Zonation varies from reef to reef depending on various factors. The presence of resistant algal ridges, for instance, can greatly affect the pattern of zonation. Not only the scleractinians but also the gorgonians (in the Caribbean reefs) and octocorals (in the Indo-Pacific reefs) are ecologically zoned, and so are the various bivalves and other molluscs that form specialised communities in the very varied reef habitats (Yonge 1974).

The growth rate of corals varies greatly with the environment and the season but may reach 10 cm per year in branching corals and 1 cm per year in the more massive colonies. Reef growth as a whole may reach a maximum of 2·5 cm per year, but in the past it has generally been less.

Within reefs there is not only continual construction by the scleractinians and encrusting algae, but also a constant attrition through wave action and, equally important, destruction by boring organisms and coral-eating fishes. These produce a great quantity of fine comminuted coral sand, which fills up the lagoon and is transported to deeper waters.

Deep-water coral banks

Deeper water corals (ahermatypes such as *Lophelia* and *Dendrophyllia*) may form structures differing from those of their shallow water counterparts. Some of these deep water corals are solitary; others are colonial and dendriform.

The development of these structures (Squires 1964) takes place in four stages (Fig. 5.13d). At first there is a single **colony.** When this colony is joined by others it becomes a **thicket** a few metres across, in which the members may be all of the same species or of different species. This newly developed environment attracts other organisms, e.g. fishes, molluscs and crustaceans, giving it a new ecological character. As the thicket spreads and matures, skeletal debris accumulates from broken and bored coral, providing a new substratum which again attracts other animals. This stage is now a **coppice** and may be several metres across. Eventually, with the accumulation of more debris and further coral growth, a **bank** develops: a topographic entity with a core of solid skeletal debris, a covering mat of more open debris and a capping of live coral. The proportion of living to dead coral decreases with time.

The exposed sequence studied by Squires were Tertiary in age, but similar banks are forming today. The study of fossil thickets, coppices and banks can yield valuable palaeoecological data.

Geological uses of corals

Corals as stratigraphical indicators

Corals are generally too long-ranged for use as zone fossils. Nevertheless they have been used where no shorter-ranged fossils are available, especially in the Carboniferous. In 1905 Vaughan, working on the Lower Carboniferous of the Bristol region, England, erected a zonal scheme based upon the first appearance of corals and brachiopods (Fig. 5.14). This scheme, though crude in terms of time range, has proved to be effective and to be applicable, with minor local modifications, to other parts of Britain and Europe.

Vaughan believed that the faunal sequence he erected reflected evolutionary lineages, but in fact the corals really occur as assemblage biozones with each of the main cycles bringing in a new migrant fauna.

The correlations are somewhat inexact and are being supplemented or even replaced where possible by a stratigraphy based on goniatites, conodonts and foraminiferids. New stages based upon transgressive cycles in the Lower Carboniferous are now standard, but much refinement is still needed.

Coral zonation in the European Devonian is increasingly proving of value.

Corals as geochronometers

In well-preserved specimens of many rugose and scleractinian corals the epithecal surface shows fine growth ridges, some 200 per centimetre. Each of these represents a growth increment: the former position of the rim of the calice. These fine growth ridges are often grouped in prominent bands of annulations between which the epitheca is constricted. It is normally recognised that the fine growth ridges represent daily growth increments, whilst the banding is monthly and the broader and more widely spaced annulations are yearly. Wells (1963) first established in the scleractinian *Manicina* that the growth ridges were diurnal, and since the Recent *Lophelia pertusa*, from Norwegian fjords, has twenty-eight growth ridges (presumably diurnal) per band, each band corresponding to a lunar cycle, the monthly banding hypothesis seems well established.

Wells (1963), working on Devonian corals with annual rather than monthly annulations, counted the number of growth ridges per annulation in a number of species. Though the results were limited by the preservation of the corals he concluded, mainly from observations on *Heliophyllum*, *Eridophyllum* and *Favosites*, that there were an average of 400 days in the Devonian year. This figure corresponded well to astronomical estimates that the Earth's rotation has been slowing down through tidal friction by about 2 seconds per 100 000 years. Assuming that the Earth's annual circuit round the sun was the same length, which seems to be fairly well substantiated, there

Figure 5.15 Banding on the epitheca of the Middle Devonian rugose coral *?Heliophyllum* sp. (x 3). (Copyright photograph by courtesy of Dr C. T. Scrutton)

must have been more days in the Devonian year, but they were shorter (Scrutton & Hipkin 1973).

Annual banding is not often found in corals, having only been recorded in specimens that originally lived in waters where there were seasonal fluctuations. It is normally easier to work with the Rugosa that show the growth ridges grouped in monthly bands. These are more abundant and do not have to have a complete epitheca. Many Recent corals have monthly breeding cycles during which carbonate deposition is inhibited, and this may have been the case with the Rugosa.

The monthly banded Middle Devonian corals studied by Scrutton (1965) (Fig. 5.15) had an average of 30·6 growth ridges per monthly band, and if the figure of 399 (the astronomical estimate for length of the Devonian year) is divided by the average growth-ridge count per band, the result is consistently 13. Hence it seems that thirteen bands each of about 30·6 growth ridges were laid down by these corals in the Middle Devonian year.

The use of corals as geochronometers is important as it has provided a consistent check upon estimates of the slowing down of the Earth's rotation, based on a biological rather than a physical system. If more becomes known about the growth rates of rugose corals, then the use of corals as geochronometers may become even greater than it is now.

Corals as colonies: the limits of zoantharian evolution

Colonies have been defined as 'groups of individuals structurally bound together in varying degrees of skeletal and physiological integration; all genetically linked by descent from a single founding individual' (Coates & Oliver 1973).

Such colonies (e.g. cnidarians, bryozoans, graptolites) can exhibit a wide range of organisation and integration. At one extreme the zooids budded off from the parent may be completely independent, whilst at the other end of the scale the individuals of the colony may become linked and co-ordinated so that the whole colony can function as a single unit. In some cases zooids may become highly specialised, promoting division of labour amongst the members of the colony.

Advantages of the colonial state presumably are mainly connected with greater efficiency in protection and stability as well as in reproduction, feeding and respiration. Thus a colony of asexually reproducing individuals can soon develop a large and effective biomass, with a strong and stable skeleton in which there is the potential for integration and co-operation between zooids.

In the octocorals (Bayer 1973) there are two distinct functional types: (a) those which secrete a massive scleractinian-like skeleton with very little integration; and (b) those (the vast majority), including pennatulaceans, in which integration varies from the very simple, through numerous intermediate stages, to very complex colonies with the various zooids specialised for different functions. In the most extreme cases not only are interzooidal canals linking the enterons retained, so that food captured by a few zooids can be shared by the rest of the colony, but also there are specialised 'siphonozooids' organised for pumping water through the coenenchymal canals, to ensure that adequate supplies of oxygenated water for respiration reach all members of the colony. Such **polymorphic** zooids function together but cannot function apart, and the colony acts as a 'superindividual'. Whilst this level of integration in the octocorals evidently originated very early in their history (if the interpretation of *Rangea* as a Precambrian pennatulacean is correct), the zoantharians have never fully reached the same levels of colonial organisation since their zooids are not polymorphic.

Throughout the Phanerozoic colonial corals have outnumbered solitary types, though in the case of the Rugosa solitary corals were more abundant than colonial ones. Colonial Rugosa vary in level of integration between (a) those phaceloid colonies in which individuals are functionally unconnected, and (b) aphroid colonies where the corallite walls have gone and zooids were probably connected together by linked enterons. This is the highest level attained in the Rugosa.

Tabulate corals likewise did not achieve a high degree of colonial integration, though mural pores in *Favosites* suggest enteron connections. The heliolitids, however, had a coenosteal system paralleling that of coenosteal scleractinians, but for some reason which is not clear they were not especially successful. They too perhaps were limited in growth rate because the skeleton was massive rather than light and porous like that of scleractinians.

Whilst cerioid and phaceloid forms are known amongst scleractinians, a generally higher level of integration is apparent in the preponderance of meandroid and coenosteoid types. Meandroid genera have a **polystome** system in which hundreds of mouths with their adjacent tentacles connect in each 'valley' with a single elongated enteron. In coenosteal Scleractinia the enterons of adjacent polyps are also all connected, and skeleton building has become a co-operative venture. But even the most advanced scleractinians, with the possible exception of a very few rare genera, do not exhibit polymorphism.

Whilst rugose and tabulate corals were able to make some headway towards an integrated colonial system, they were generally less successful in this than the Scleractinia. But their comparative failure as reef builders was probably due not so much to this as to certain other features of their organisation. The Rugosa and to some extent the Tabulata had slow-growing and rather solid and heavy skeletons in which a great deal of calcium carbonate was deposited in internal structures. They lacked an edge zone and could not attach themselves firmly to the substratum. It is not known whether Palaeozoic corals had zooxanthellae in their tissue; if they did not, one of the great disadvantages possessed by reef-building scleractinians was denied them, and this would have affected their growth rate, ability to get rid of waste material and other factors.

But the strong, yet light and porous, stable, fast-growing and at the same time increasingly integrated skeletons of the scleractinians permitted the development of the coral reef environment, so laying down the framework for some of the most complex and productive of all ecosystems that have ever evolved.

BIBLIOGRAPHY

Books, treatises, symposia

Maxwell, W. G. H. 1968. *Atlas of the Great Barrier Reef.* Amsterdam: Elsevier. (Illustrated atlas)

Moore, R. C. (ed.) 1956. *'Treatise'* Part (F). Coelenterata. (Old but valuable)

Individual references

Bayer, F. M. 1973. Colonial organisation in octocorals. In *Animal colonies*, R. Boardman, A. Cheetham and W. J. Oliver (eds), 69–93. Pennsylvania: Dowden, Hutchinson & Ross. (Integration of colonies)

Carruthers, R. G. 1910. On the evolution of *Zaphrentis delanouei* in Lower Carboniferous times. *Quart. J. Geol. Soc. Lond.* **66,** 523–36. (Classic, though now dubious, evolutionary study)

Coates, A. G. and W. A. Oliver 1973. Coloniality in zoantharian corals. In *Animal colonies*, R. Boardman, A. Cheetham and W. J. Oliver (eds), 3–27. Pennsylvania: Dowden, Hutchinson & Ross. (Integration of colonies)

Copper, P. 1974. Structure and development of early Palaeozoic reefs. In *Proc. Second Int. Symposium Coral Reefs*, A. M. Cameron (ed.), 365–86. Brisbane: Great Barrier Reef Committee. (Changes in reef structure from incipient to maturity)

Garwood, E. J. 1913. The Lower Carboniferous succession in the north-west of England. *Quart. J. Geol. Soc. Lond.* **68,** 449–596. (Zonation using coral-brachiopod fauna)

Hill, D. 1937. *A monograph of the Carboniferous rugose corals of Scotland*, 1–78. Palaeontogr. Soc. monogr. (Taxonomy and coral community structure)

Jamieson, E. R. 1971. Paleoecology of Devonian reefs in western Canada. *Proc. North Amer. Paleont. Convention; Chicago, 1969* (J), 1300–40. (Reef environments)

Jell, J. S. 1969. Septal microstructure and classification of the Phillipsastraeidae. In *Stratigraphy and palaeontology* (essay in honour of Dorothy Hill), K. S. W. Campbell (ed.), 50–73. Canberra: Australian National Univ. Press. (Composition of skeletal elements and chemistry)

Jell, P. A. and J. S. Jell 1976. Early Middle Cambrian corals from western New South Wales. *Alcheringa* **1,** 181–95. (The oldest known corals)

Jull, R. K. 1965. Corallum increase in *Lithostrotion*. *Palaeontology* **8,** 204–25. (Various methods of growth)

Kato, M. 1963. Fine skeletal structures in Rugosa. *J. Fac. Sci. Hokkaido Univ. Ser. 4* **11,** 571–630. (Microstructure)

Klovan, J. E. 1964. Facies analysis of the Redwater Reef Complex, Alberta, Canada. *Bull. Can. Pet. Geol.* **12,** 1–100. (Distribution of corals within a reef)

Krebs, W. 1974. Devonian carbonate complexes of central Europe. In *Reefs in time and space*, L. F. Laporte (ed.), 155–208. Tulsa: Soc. Econ. Min. Paleont. (Distribution of corals within reefs)

Ladd, H. S. 1971. Existing reefs: geological aspects. *Proc. North American Paleont. Convention, Chicago, 1969.* (J). 1273–1300. (Good summary and bibliography)

Lang, W. D. 1923. Trends in British Carboniferous corals. *Proc. Geol. Ass.* **34,** 120–36. (An early analysis of evolutionary trends; actually of limited value)

Oliver, W. J. 1976. Presidential address: biogeography of Devonian rugose corals. *J. Palaeont.* **52,** 365–73. (Coral faunal provinces)

Sandberg, P. A. 1975. Bryozoan diagenesis: bearing on the nature of the original skeleton of rugose corals. *J. Paleont.* **49,** 587–606. (Detailed microstructural study with good bibliography)

Scrutton, C. T. 1965. Periodicity in Devonian coral growth. *Palaeontology* **7,** 552–8. (Monthly banding and the length of the Devonian year)

Scrutton C. T. 1975. Hydroid–serpulid symbiosis in the Mesozoic and Tertiary. *Palaeontology* **18,** 225–74. (*Parsimonia* and *Protulophila*)

Scrutton, C. T. 1977. Reef facies in the Devonian of eastern South Devon, England. *Mem. Bur. Rech. Geol. minièr.* **89,** 125–35. (Faunal variation with facies)

Scrutton C. T. and R. G. Hipkin 1973. Long-term changes in the rotation rate of the Earth. *Earth Sci. Rev.* **9,** 259–74. (Partially based on coral banding)

Sorauf, J. E. 1971. Microstructure in the exoskeleton of some Rugosa (Coelenterata). *J. Paleont.* **45,** 23–32. (Scanning electron micrographs)

Sorauf, J. E. 1972. Skeletal microstructure and microarchitecture in Scleractinia (Coelenterata). *Palaeontology* **15,** 88–107. (Scanning electron micrographs)

Squires, D. F. 1964. Fossil coral thickets in Warrarapa, New

Zealand. *J. Paleont.* **38,** 904–15. (Development of thickets, coppices and banks)

Stearn, C. W. and R. Riding 1973. Forms of the hydrozoan *Millepora* on a Recent coral reef. *Lethaia* **6,** 187–200. (Living milleporines)

Stoddard, D. C. 1969. Ecology and morphology of Recent coral reefs. *Biol. Rev.* **44,** 433–98. (Valuable reference work with bibliography)

Taylor J. D. 1968. Coral reef and associated invertebrate communities (mainly molluscan) around Mahe, Seychelles. *Phil. Trans Roy. Soc. Lond. B* **793,** 129–206. (Coral/molluscan ecology)

Vaughan, A. 1905. The palaeontological sequence in the Carboniferous limestone of the Bristol area. *Quart. J. Geol. Soc. Lond.* **61,** 181–307. (Classic study using coral/brachiopod zonation)

Walker, K. R. and L. P. Alberstadt 1975. Ecological succession as an aspect of structure in fossil communities. *Palaeobiology* **1,** 238–57. (Reef structural development and ecology with increasing maturity)

Wang, H. C. 1950. A revision of the Zoantharia Rugosa in the light of their minute skeletal structure. *Phil. Trans Roy. Soc. Lond. B* **234,** 175–246. (Classic study, now rather superseded)

Wells, J. W. 1937. Individual variation in the rugose coral species *Heliophyllum helli. Palaeont. Americana* **2,** 1–22. (Environmental factors in variation)

Wells, J. W. 1963. Coral growth and geochronometry. *Nature* **197,** 948–50. (Annual banding and the slowing of the Earth's rotation)

Wells, J. W. 1957. Annotated bibliography: corals. In *Treatise on marine Ecology and Palaeoecology*, 2. J. Hedgpeth (ed.), 773–82. Mem. Geol. Soc. Amer. No. **67.** (Invaluable source work)

Wilson, J. 1975. The distribution of the coral *Caryophyllia smithii* S & B on the Scottish continental shelf. *J. Mar. Biol. Ass. UK* **55,** 611–25. (Precise documentation)

Wilson, J. B. 1976. Attachment of the coral *Caryophyllia smithii* S & B to tubes of the polychaete *Ditrupa arietina* (Muller) and other substrates. *J. Mar. Biol. Ass. UK* **56,** 291–303. (Documentation)

Yonge, C. M. 1963. The biology of coral reefs. In *Adv. Mar. Biol. 1,* F. S. Russell (ed.), 209–60. London, George Allen & Unwin. (Valuable summary and bibliography)

Yonge, C. M. 1968. Living corals. *Proc. Roy. Soc. Lond. B* **169,** 329–44. (Invaluable reference)

Yonge, C. M. 1974. Coral reefs and molluscs. *Trans Roy. Soc. Edin.* **69,** 147–66. (Environmental niches in reefs)

6 Bryozoans

Bryozoa, because they are small and often delicate, tend to be less familiar than most invertebrate groups preserved fossil. Yet they are common in sedimentary rocks and abundant in the sea today; at least 3500 living and 15 000 fossil species are known. The fronds of flat seaweeds such as *Laminaria* are often covered with the lacy calcareous skeletons of the bryozoan *Membranipora*, each colony being composed of hundreds of individuals.

All bryozoans are colonial, most are marine, and in the majority of cases the units or zooids of the colony secrete tubes or boxes of lime partially encasing the soft parts (e.g. Fig. 6.2). Each zooid is basically cylindrical, has a ring of tentacles and at first sight seems to resemble a small cnidarian polyp. But the zooids are coelomate, having a freely suspended gut with both mouth and anus, and are unquestionably of a higher grade of organisation. They were recognised as being distinct from cnidarians in 1820, and the name Bryozoa was given formally to them in 1831 by Ehrenberg, one year after the name Polyzoa had already been given by the Irish naturalist J. Vaughan Thompson. However, since both Thompson and Ehrenberg included in the phylum, as understood by them, certain other invertebrate groups, the true bryozoans are sometimes (and especially in American usage) known as Phylum Ectoprocta. The term Bryozoa, being in more common use, is here retained.

Two genera, one simple in construction and one more complex, are described here to show the basic morphology. Both of these belong to the largest and most successful marine class: the Gymnolaemata.

TWO EXAMPLES OF LIVING BRYOZOANS

BOWERBANKIA (FIGS 6.1–6.5)

Bowerbankia (Order Ctenostomata) 6.1a–c has a creeping cylindrical **stolon** from which numerous bottle-shaped zooids arise in clusters.

Each living zooid is housed in a skeleton (membranous and unpreservable in this case) called the **cystid.** (The term **zooecium,** which has sometimes been used as an alternative to cystid, is now restricted to the preservable skeleton only.) The whole colony may be called a **zoarium.** When feeding, the zooid extends its **lophophore,** a ring of ten ciliated tentacles, into the surrounding water. These tentacles converge on a central mouth from which the gut descends into the body of the zooid. This gut is U-shaped, hanging down in the coelom, and has an oesophagus, stomach and intestine which opens by an anus close to the mouth but outside the ring of tentacles. The tentacles attach to the body by an eversible tentacle sheath. In *Bowerbankia* and its relatives the sheath is protected by a collar, but this is not represented in other bryozoa. Between mouth and anus is a major ganglion from which nerves run to all parts of the body. From the base of the stomach a thread, the **funiculus,** connects to a main **funicular tube** running along the stolon and connecting all the zooids.

The tentacles can be either extruded for feeding (Fig. 6.1c) or retracted entirely within the body for protection (Fig. 6.1d). In the resting zooid the tentacle sheath becomes invaginated, with its surface facing inwards and surrounding the tentacles rather than facing outwards below them. Eversion of the tentacles is accomplished quite simply by compression of the body wall by transverse muscles; this raises the hydrostatic pressure of the coelomic fluid so that the tentacles have to emerge. A large retractor muscle, attached to the tentacular base and accompanied by longitudinal muscles, pulls the tentacles within the body when danger threatens, and the zooid is finally closed off by a circular sphincter muscle just below the collar. The collar then folds inwards in a series of pleats and completes the closure of the zooid.

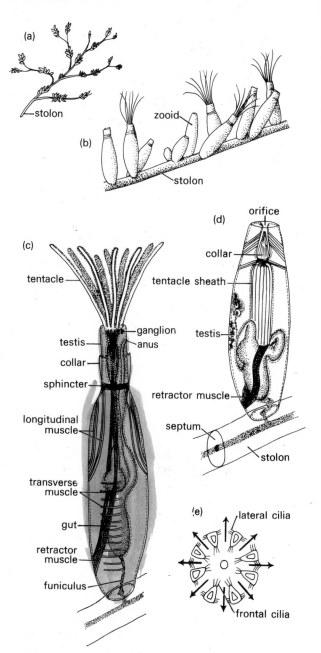

Figure 6.2 *Tricephalopora pustulosa*, Cretaceous cribrimorph (x 25 approx.). (Photograph by courtesy of Dr G. P. Larwood)

Figure 6.1 *Bowerbankia*, a Recent ctenostome: (a) appearance of colony (x 1), (redrawn from Bassler, 1953 in '*Treatise*' Part (A)); (b) section of zoarium with extended and retracted zooids (x 7); (c) zooid with tentacles extended (x 35 approx); (d) zooid with tentacles retracted (x 35 approx); (e) section through tentacle crown, arrows showing current directions. ((b)–(e) redrawn from Ryland, 1970)

When the zooid feeds, the tentacles are extended in an erect funnel by hydrostatic pressure of the coelom. They are quadrate in cross-section and have **ciliary tracts** on each side, those of adjacent tentacles nearly touching. Another tract of 'frontal cilia' on each tentacle faces the mouth, being more strongly developed towards the base. When these cilia beat downwards in a co-ordinated metachronal rhythm, currents are generated which pass straight down the funnel towards the mouth and out between the tentacles (Fig. 6.1e). The food particles (mainly phytoplankton) in the incoming stream are strained off in the process and may be conveyed to the mouth mainly by cilia. The operation of the lophophore and its structure are decidedly similar to those of brachiopods; hence the Bryozoan and Brachiopoda are assumed to be related and are grouped together with phoronids in 'Superphylum' Lophophorata.

There are no separate excretory, circulatory or respiratory organs. The colony grows from an initial zooid, the **ancestrula,** by growth of the stolon (itself a series of modified zooids) and by asexual budding (Fig. 6.3). New colonies, however, are produced sexually. The zooids are hermaphrodite, but the ovary and testis may develop at different times. The ovary is a cluster of several egg cells, which are released one at a time into the tentacle sheath where

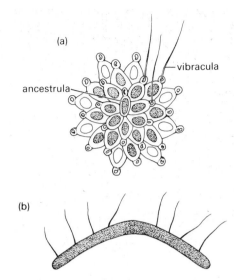

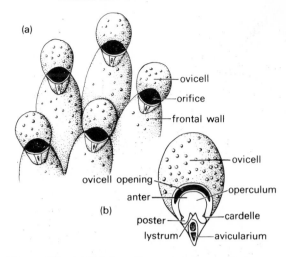

Figure 6.3 (a) *Cupuladria* (Mio.–Rec.) – Young zoarium showing growth from central ancestrula; blank zooecia are the last formed (x 10); (b) probable life position of zoarium. (Modified from Lagaiij, 1963)

Figure 6.4 (a) *Smittina*, a Recent ascophoran cheilostome, with opercular structures missing (x 35) (modified after Bassler in '*Treatise*' Part (A)); (b) enlargement of anterior end with operculum in place.

they are fertilised. Each then develops into a **trochophore larva** whilst still in the tentacle sheaf. When fully mature the larva swims away; meanwhile the zooid degenerates. A new egg is released only when the last fertilised egg has developed into a larva. Sperm developed within the testis is seen in many bryozoans to make its way out into the sea via tiny pores in the tentacles.

SMITTINA (Figs. 6.4a, b; 6.5a, b)

Smittina (Order Cheilostomata, Suborder Ascophora) is an example of one of the more complex bryozoans. The zoarium is encrusting, and the individual cystids are arranged like flat elliptical boxes radiating away from the ancestrula or first-formed zooid. They are all constructed of calcium carbonate. Mature cystids may develop distal **ovicells:** swollen spherical structures in which the fertilised eggs develop into larvae. In different cheilostomes ovicells may lie within the cystid, overlap onto the next one distally or be embedded in its posterior wall; alternatively they may be wholly or partially separated. Directly behind the ovicell is the **orifice** or opening to the zooid. This is keyhole-shaped and closed by an **operculum,** hinged at its narrowest points or **cardelles,** so that

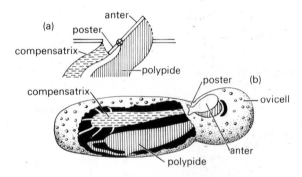

Figure 6.5 Structure of ascophoran cheilostome: (a) dissected lateral view with operculum opening up; (b) median section of opened operculum hinged by cardelles.

when the operculum opens about this fulcrum the distal part (**anter**) rises and the proximal part (**poster**) sinks. When the **polypide** is retracted, it lies entirely within the cystid and the orifice is closed. When it emerges, it comes through the distal part of the orifice as the anter lifts up. Connected to the proximal part of the orifice below the poster is a sac, the **compensatrix,** which is suspended from the body wall by many **radial muscles.** This compensatrix is concerned with polypide extrusion. When the radial muscles contract, the compensatrix

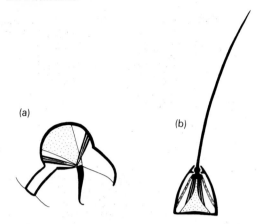

(a)

(b)

Figure 6.6 (a) Avicularium in lateral view with different sets of muscles to open and close; (b) vibraculum, musculature for lashing shown. (Redrawn from Ryland, 1970)

expands and the poster is depressed so that water enters the sac. As the compensatrix swells, the polypide is displaced and has to emerge from the anter because of the hydrostatic pressure. The polypide can be pulled back in again when the retractor muscle contracts. This causes the compensatrix to be evacuated hydrostatically (since the radial muscles are by this time relaxed) and the operculum shuts.

The **frontal wall** (i.e. upper surface) is complex in structure with a regular sculpture, secreted in several layers. Cheilostomes are strongly polymorphic, as is shown by the specialised organs **avicularia** and **vibracula** which are both modified zooids.

The avicularia (Fig. 6.6a) are attached to the upper surface at species-specific locations. Each resembles a bird's head in shape and contains a single modified polypide. It has a hinged chitinous mandible which opens and snaps shut in a constant motion, discouraging both predators and settling larvae.

The polypide is reduced to a rudiment, and the main internal soft-part structures are the paired antagonistic muscle sets with which the mandible snaps. The mandible is infrequently preserved fossil, but since its edge generally fits the avicularium, its form is quite well reflected by the preservable skeleton. Avicularia may be sessile or, in the most extreme form pedunculate, i.e. mounted on a short stalk. The least modified sessile avicularia are said to

be **vicarious;** these are slightly smaller than normal zooecia and replace them in the colony at regular intervals. **Interzooidal** avicularia occur between zooids and are reduced in size, whilst the much smaller **adventitious** and usually pedunculate avicularia are found mainly in cheilostomes in which they may occur anywhere on the frontal wall. Ryland (1970) has discussed their possible evolution in some detail.

Vibracula (Fig. 6.6b) possess a long whip-like bristle (**seta**) projecting from a sessile basal chamber which contains only the muscles. The seta swings on a pair of opposing pivots just above its lower end. The muscles are attached below the pivots. When triggered into action the contraction of the muscles causes the seta to lash violently. This stimulates neighbouring vibracula, and the whole ensemble will strike hard against any alien object, discourage settling larvae or winnow away sediment. In lunulitiform bryozoans, vibracula may be modified as 'legs' on the underside of the colony, enabling it to move.

CLASSIFICATION

PHYLUM BRYOZOA (POLYZOA) (Ord. – Rec.): Sessile colonial coelomates, normally marine, rarely freshwater, consisting of small linked zooids, usually with a calcareous or more rarely an organic, exoskeleton. Zooids have a tentacle-bearing retractile lophophore and a U-shaped gut with the anus outside the tentacular ring. Colonies arise from an ancestrula (or rarely a **statoblast**), encrusting, creeping, erect or in chains, polymorphic in some groups.

CLASS 1. PHYLACTOLAEMATA (Mesozoic–Rec.): Non-calcareous fresh-water bryozoans. Zooids with horseshoe-shaped lophophores; statoblasts produced as resting buds. Twelve genera only. e.g. *Plumatella*, *Cristatella*.

CLASS 2. STENOLAEMATA (Ord.–Rec.): Calcified marine bryozoans, usually non-operculate, with an extensive fossil record. Tentacle extrusion in living forms is brought about by muscular action forcing coelomic fluid into the proximal part of the zooid displacing the tentacles distally. 550 genera.

ORDER 1. CYCLOSTOMATA (Ord.–Rec.): Zoaria

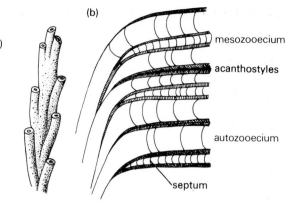

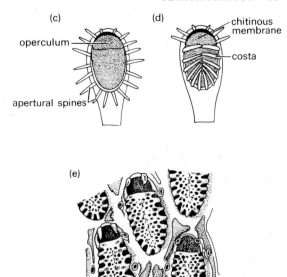

Figure 6.7 (a) *Crisia*, a Recent cyclostome (x 25); (b) *Dekayella*, an Ordovician trepostome in section (x 20). (c–d) Possible origin of Cretaceous cribrimorphs from membraniporoids: (c) ancestral membraniporoid type with radiating apertural spines; (d) closure and fusion of apertural spines to form costae covering frontal chitinous membrane; (e) *Pelmatopora gregoryi*, a Cretaceous cribrimorph, frontal view (x 25), ((c)–(e) Redrawn from Larwood, 1962).

of tubular zooecia possessing either circular apertures separated by pseudoporous frontal walls, or contiguous polygonal apertures. Interzooecial walls usually pierced by **mural pores.** e.g. *Crisia* (Fig. 6.7a), *Berenicea*, *Heteropora*.

ORDER 2. CYSTOPORATA (Ord.–Perm.): Similar to cyclostomes but with regions of curved **cystiphragms** separating zooecia which often have crescentic projections (**lunaria**) around their apertures. e.g. *Fistulipora*, *Ceramopora*.

ORDER 3. TREPOSTOMATA (Ord.–Perm.):'Stony bryozoans' forming massive zoaria with elongate **autozooecia** which are initially thin-walled but become thick-walled close to the zoarial surface where small **mesozooecia,** filled by closely spaced **diaphragms,** may intervene between autozooecia. Mural pores absent. e.g. *Monticulipora*, *Dekayella* (Fig. 6.7b), *Hallopora*.

ORDER 4. CRYPTOSTOMATA (Ord.–Perm.): Frequently net-like, occasionally cylindrical, zoaria with short autozooecia which are partly divided by a **hemiseptum** close to their distal aperture. Mural pores absent. e.g. *Fenestella* (Fig. 6.8a, b, d, e), *Archimedes* (Figs 6.8c, f), *Rhabdomeson*.

CLASS 3. GYMNOLAEMATA (Ord.–Rec.): Marine,

occasionally brackish, bryozoans which may be calified. Lophophores everted by muscular deformation of part of the body wall. Strongly polymorphic. 650 genera.

ORDER 1. CTENOSTOMATA (Ord.–Rec.): Zooids uncalcified, walls membranous or gelatinous, lacking ovicells, frequently **penetrant.** e.g. *Bowerbankia* (Figs 6.1a–e) *Alcyonidium*.

ORDER 2. CHEILOSTOMATA (Jur.–Rec.): Usually calcareous, with short box-like zooecia having a distal orifice closed by a hinged operculum. Avicularia and vibracula common. Embryos brooded in ovicells. e.g. *Smittina* (Figs 6.4a, b; 6.5a, b), *Membranipora*, *Cribrilina*, *Pelmatopora* (Fig. 6.7e), *Cupuladria* (Fig. 6.3a, b).

MORPHOLOGY AND EVOLUTION

Shortly after their first appearance in the early Ordovician the Bryozoa underwent a great burst of evolution, resulting in the establishment of the early Stenolaemata which by the Upper Ordovician were very abundant and diverse. The Stenolaemata are all extinct now apart from the Cyclostomata, but were the dominant class of Palaeozoic bryozoans. Of all the Stenolaemata the most important and abundant in

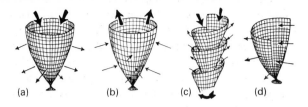

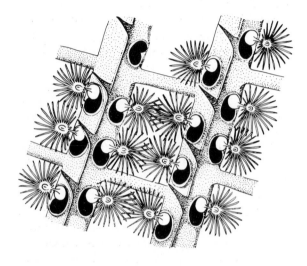

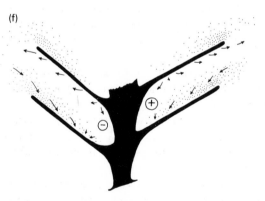

Figure 6.8 Morphology and direction of currents inferred in fenestellids: (a) Carboniferous fenestellid with inward-facing zooecia; (b) Silurian fenestellid with outward-facing zooecia; (c) *Archimedes* (Carb.–Perm.), a spiral colony; (d) fan-shaped fenestillid, using unidirectional current; (e) *Fenestella* (Carb.–Perm.): part of zoarium showing zooids reconstructed as extending into fenestrules (x 8 approx.); (f) *Archimedes:* interpretation of feeding by spiralling currents. (Redrawn from Cowen and Rider, 1972)

the Lower Palaeozoic were the 'stony bryozoans', the Trepostomata (Fig. 6.7b). They formed stick-like or globular calcareous zoaria up to 50 cm across. The zooecia are tubular and closely packed with cross-partitions and thickened distal parts. Sometimes the zooecial orifices are clustered in groups at the summit of small mounds (**monticules**) or in depressions (**maculae).** The zooecial walls are constructed of thin laminae, with the individual laminae at intervals forming meniscus-like diaphragms or cross-partitions. There is no communication through autozooecial walls. Between the large zooecial orifices (autozooecia) there are often smaller openings (mesozooecia); these suggest the former presence of some kind of smaller polymorphic zooid. Sometimes **acanthostyles** (rod-like spines of clear calcite) are visible at the intersections of zooecial walls.

The Trepostomata, which after their initial evolutionary burst became so abundant in the Ordovician and Silurian, declined thereafter as cryptostomes became dominant, and they finally died out in the Triassic.

The Cryptostomata, of which one of the most distinctive groups is the fenestellids, appeared in the Ordovician but reached their acme in the later Palaeozoic. Whilst they often had mesozooecia and acanthostyles, cryptostomes are short and box-like, and the colony may be delicate and lacy rather than large and heavy as in the trepostomes.

Fenestellid colonies (Figs 6.8a–d) grew from an ancestrula forming first a ring of zooecia and then a circlet of upright branches; these bifurcated at intervals giving the colony a cup-shaped form or in some species the form of a half-cup or fan. Whatever the form of the colony, the branches are subtriangular in cross-section and have rows of zooecia opening onto only two faces of the colony. The branches are connected by cross-bars with rectangular spaces or **fenestrules** between (Fig. 6.8e). The skeleton was probably secreted by a thin epithelium which extended over the whole surface, and as in some living cyclostomes the zooids probably shared a common coelom.

Since the growth of the bryozoan skeleton is a 'colonial effort', the colony may become highly integrated rather than remain as no more than an aggregate of individuals. The functional morphology of bryozoan colonies is potentially a fruitful field for research. Fenestellid colonies, to take an example, have been analysed functionally by Cowen and Rider

(1972). The arrangement of the zooecia and the form of the zoarium are clearly important in understanding how the colony operated as a whole. It has been postulated that the 'operational subunits' of the fenestellid colony are the fenestrules. In the main branches, the lophophores of equally spaced zooids on opposite sides of a median ridge are considered to have extended into the fenestrule forming a filtering net, the cross-bars giving support and dividing up the fenestrules. A combined feeding current set up by the zooids would draw water in through the fenestrules, enabling them to strain off all the food material with their lophophores, and would thus give high filtering efficiency.

The growth forms of fenestellid colonies may have been adapted for maximising filtering efficiency at the colonial level. Two kinds of cup-shaped colonies are known: those with zooecia facing outwards (common in the Silurian), and those with inward-facing zooecial apertures (more common in the Carboniferous). There are also fan-shaped (half-cup) and spiral colonies. The cup-shaped colonies would probably be equally efficient whether they drew in water from the top and exhaled it laterally through the fenestrules, or whether they sucked it in at the sides (like a sponge) and sent out an excurrent stream from the central cavity. Fan-shaped colonies (again as with certain sponges) would be the best shape for taking in water in a regime where a weak current was constantly flowing normal to the fan surface.

The peculiar *Archimedes* has a typically fenestellid network structure, but this describes a helical spiral around a thick central calcareous column. Such a bryozoan standing upright on the sea floor would be functionally efficient if the combined action of the lophophores generated a current stream coming in at the top and running spirally along the 'deck' to the base, as down a fairground helter-skelter (Fig. 6.8f). As the main stream travelled spirally downwards some of it would be sucked through the innumerable fenestrules and passed away in a centrifugal stream just below each deck. Whether the fenestrular system would have functioned in the manner postulated would depend largely upon whether the lophophores actually did extend into the fenestrule, and this remains unknown.

Cryptostomes outclassed the trepostomes in the Upper Palaeozoic, but by the end of the Permian they too were extinct.

The Cyclostomata (Fig. 6.7a) were the only group of fossilised bryozoans to cross the Palaeozoic/Mesozoic boundary (though probably the membranous ctenostomes did also). Since there was little competition they were able to dominate the Mesozoic scene. They had become very diverse and important by the Lower Cretaceous; indeed according to Ryland (1970) 'their zenith during this period constitutes one of the highlights in the history of the bryozoa'. They have declined since, though some survive to the present, and since they are the only living representatives of an ancient stock most of our conceptions of soft part morphology in Palaeozoic bryozoans are based upon them. The cyclostome zooid is quite like that of *Bowerbankia*, though with a few characteristic differences, and there is some degree of polymorphism.

Many cyclostome colonies seem to have been quite well integrated, and the colony rather than the individual, as with fenestellids, is the functional unit in terms of effective feeding.

The numerical decline of the Cyclostomata in the later Cretaceous and Tertiary seems to relate to the great contemporaneous expansion of the Cheilostomata, the last and perhaps the most successful of the bryozoan orders to arise. They probably originated not from cyclostomes but from ctenostomes, which were then in existence but whose fossil record is poor since they are generally uncalcified.

Order Cheilostomata is divided by Ryland into four suborders: Ascophora, Anasca, Gymnocystidea and Cribrimorpha. Ascophoran structure has been described in *Smittina*, though within this suborder there is much structural variation in the position of the ovicell and the orifice and in the sculpture of the upper surface.

The Anasca lack the compensatrix, and the polypide extrudes through the action of internal muscles on the flexible chitinous frontal wall. Both the Anasca and the Ascophora are very important today. Cribrimorpha (Figs. 6.2, 6.7c–e), however, after a brief though substantial expansion in the Upper Cretaceous (e.g. Larwood 1962) have now greatly declined. These are usually unilaminar encrusting forms with calcified side walls, though the primary frontal wall in which the aperture lies is of chitin. This chitinous wall is overarched by calcareous **costae** or ribs which form a secondary frontal wall and meet in the midline. These fuse, making a kind of porous cage over the primary wall – perhaps the most elaborate wall structures ever evolved in the bryozoans. The apertural region is usually protected by a semicircle of oral spines.

Cheilostomes, by their marked polymorphism of the avicularia and vibracula and by the connections between zooids, express a high degree of integration, which may have involved some modification of the structure of the colony to make maximum effective use of the feeding currents.

The geological history of bryozoans is incompletely known because of the poor record of the non-calcified forms. Nevertheless it is clear that particular groups were dominant at certain times. In the Lower Palaeozoic the trepostomes were especially important, whilst the acme of the cryptostomes was in the Upper Palaeozoic. After the Permian extinctions the cyclostomes, which had been present throughout the Palaeozoic, expanded vastly in the Jurassic and Lower Cretaceous, declining only when the cheilostomes arose to become the dominant bryozoans of the Upper Cretaceous and Tertiary. In Recent seas they are perhaps the most numerous lophophorates. In each of these groups the degree of colonial integration varies. It is claimed indeed that the colony rather than the zooid has been the unit of natural selection. Thus in Recent seas the specialisation of avicularia and other polymorphs can only be seen in terms of benefit to the colony as a whole. And polymorphism occurs in 75 per cent of all living cheilostomes, especially in those species living in predictable environments where sufficient food resources would allow the 'luxury' of non-feeding zooids.

ECOLOGY AND DISTRIBUTION

Bryozoans are abundant in all oceans with a maximum in the western Pacific. They are found at all depths from the shoreline down to the abyssal zone – the deepest record is over 8500 m – but they decrease in numbers and importance in the fauna with depth.

The controls of distribution are:

(a) Temperature, for though a few species are eurythermal most have restricted temperature ranges.

(b) Wave action, since the colonies are liable to damage.

(c) The availability of a hard substrate upon which the larvae can settle. This is especially important in limiting depth range, for deeper sea sediments are much finer than those of the continental shelves and deep sea oozes offer little prospect of a firm anchorage.

(d) Salinity. This is a fairly important control, and since waters off large river mouths have reduced salinity as well as much suspended sediment few bryozoans are found there. Even so there are a few euryhaline bryozoans which can withstand salinities of only 20 ‰, including the ubiquitous *Bowerbankia*.

Shallow-water bryozoans

Relatively few Recent bryozoans are intertidal since the high environmental energy and the problems of dessication between tides are too great for such delicate organisms. The sublittoral zone, however, has a wealth of bryozoans encrusting the rocks and seaweeds. They feed largely on the abundant phytoplankton of this zone and are especially common at depths between 20 and 80 m. The depth ranges of many bryozoan species from shallower waters are well known and they tend to form characteristic associations with other organisms, some of which have been remarkably persistent through time.

One species of particular ecological interest is the widespread anascan cheilostome *Cupuladria canariensis* (Fig. 6.3a, b), for its habitat and distribution in Recent seas and from the Miocene onwards has been very thoroughly researched (Lagaiij 1963). It lives at depths between about 5 and 500 m, but it is most common on continental shelves on a sand substrate in the Atlantic and East Pacific. It can tolerate temperatures between 12 and 31 °C, though it is normally confined within the 14 °C isocryme and to salinities between 27 and 37 ‰.

The larvae have strong sediment preferences, so colonies are always found where the particles (e.g. quartz grains, Foraminifera, glauconite pellets, shell fragments) are large enough to permit settlement, but not too large. *Cupuladria* can tolerate a certain amount of clay since the vibracula whip constantly and prevent the settlement of clay particles.

Colonies are always lunulitiform, i.e. have a concavo-convex lensoidal form, and probably rest on the bottom on the growing edge or are raised on vibracular setae. Many lunulitiform types can 'walk' on these setae across coarse unconsolidated sediments. Since the temperature limits of Recent colonies are well known, and assuming no change of habitat preference through time, *C. canariensis* has been able to be used as a good palaeotemperature gauge. Since it is common in Miocene and Pliocene sediments of the North Sea Basin, the temperature of the water during deposition of these sediments must have been at least 9 °C higher than it is today. The first appearance of this species is a good stratigraphical marker for the Lower Miocene, and on this criterion several suites of Tertiary sediments have been assigned to their correct system.

All this testifies to the great geological use that can be made of living species that occur in the fossil state and whose ecology and distribution have been investigated in minute detail.

In very general terms the proportion of erect and rigid species of bryozoans tends to increase with depth; encrusters prefer shallow water habitats. Even so, encrusters are found on hard substrates at all depths and erect forms are found in shallow waters, though many of these (e.g. *Flustra*) have flexible zoaria which are more likely to withstand wave damage. The very delicately branched forms prefer a habitat with weak or absent current action, whilst the discoidal or lunulitiform genera live successfully in loose moving rippled sands.

It may be possible to infer depth relationships in fossil bryozoan-bearing assemblages from colony form, but only in the broadest and most general sense and in conjunction with other criteria, for by itself the shape of a colony is not an unequivocal palaeoecological indicator.

Reef-dwelling bryozoans

Bryozoans have been quite significant as frame builders or as sediment binders in various kinds of reefs through geological time. In Ordovician to Devonian coral–stromatoporoid reefs they form a subordinate part of the reef fabric and assist in binding the sediment. Commonly they bridged gaps and allowed cavities to form below them, these often becoming filled with fine sediment. But so far no real evidence of ecological zonation has been found.

In the large Permo-Carboniferous algal reefs they contribute in a minor way, or on a localised scale sometimes more importantly to the reef framework. Fenestellids have been found in life position, projecting outwards from a steeply dipping reef face. They also occur in patch reefs as frame builders.

Modern coral reefs carry abundant faunas of encrusting bryozoans, which are in places significant frame builders. Strongly built thick-walled encrusters are found in regions of turbulent water, whilst the more delicate cribrimorphs live in sheltered cavities. The larvae of these types have strong habitat preferences and will only settle on particular substrates, usually coralline algae or dead skeletal material. Water turbulence seems to be the primary control of distribution within the multifarious habitats provided by the coral reef and bryozoans of various types, though almost all encrusters will flourish almost anywhere that is free of suspended sediment. Most species seem to have particular functional adaptations for such habitats, though many of these have not been investigated in detail.

Deep-water bryozoans

Most of the abyssal and bathyal Bryozoa of Recent seas are cheilostomes. The majority taken from depths of over 1000 m were attached to shells, pebbles and other hard surfaces, and they have been found at some 25 per cent of deep sea stations. A few species survive as mud surface dwellers as they have long root-like threads capable of holding them securely in soft sediment.

STRATIGRAPHICAL USE

Since Palaeozoic bryozoan genera tend to be long-ranged and facies-controlled their stratigraphical applications are usually poor. Species assemblages within given facies are regionally very useful for zonal purposes, however, especially in widespread carbonate shelf sediments. Some of the Cretaceous and Tertiary bryozoans seem to have limited vertical distribution, and as their time ranges become better known so their stratigraphical potential increases.

BIBLIOGRAPHY

Books, Treatises, Symposia

Larwood, G. P. (ed.) 1973. *Living and fossil bryozoans*. New York and London: Academic Press. (Many original papers; a standard reference work)

Moore, R. C. (ed.) 1953. '*Treatise*' Part (G). Bryozoa. (R. S. Bassler's original Treatise, somewhat dated)

Ryland, J. S. 1970. *Bryozoans*. London: Hutchinson. (Invaluable account of Recent and fossil bryozoans)

Individual references

Busk, G. 1859. *A monograph of the fossil Polyzoa of the Crag.* Palaeontogr. Soc. monogr. (Early description of a bryozoan-rich bed)

Cowen, R. and J. Rider 1972. Functional analysis of fenestellid bryozoan colonies. *Lethaia* **5,** 147–64. (See text)

Lagaaij, R. 1963. *Cupuladria canariensis* (Busk): portrait of a bryozoan. *Palaeontology* **6,** 172–217. (Morphology, distribution and ecology of a single species)

Larwood, G. P. 1962. The morphology and systematics of some Cretaceous cribrimorph Polyzoa (Pelmatoporinae). *Bull. Brit. Mus. Nat. Hist.* **6,** 1–281. (Excellent illustrations)

7 Brachiopods

Brachiopods are benthic marine animals whose soft parts are enclosed within a two-valved shell. They have some resemblance to bivalves in that they possess a hinged pair of **valves** and feed by drawing water into the shell and filtering off the food particles, but they are zoologically quite separate. The two valves are of different sizes but symmetrical about a **median plane,** by contrast with the equal-sized but inequilateral valves of the bivalves. Brachiopods are first found in the Cambrian and are very abundant in the fossil record, often being the commonest and most ubiquitous fossils in any shallow water deposit. In the Palaeozoic they were a very important phylum, and though they were decimated in the Permian some genera continued as the dominant benthos in localised areas during the Mesozoic where they may be vastly abundant, and by far the commonest fossils. Whilst their importance has since declined, they are, however, commoner today than formerly thought, especially in deep and cold waters. This may represent a real shift in environment through time. Well over 1600 fossil genera are known, though no more than seventy are living today albeit these are widely distributed and found at all depths in the sea. Many of the fossil genera have been found to be stratigraphically useful at various horizons.

MORPHOLOGY

Certain fundamental characters are common to all brachiopods. They all have a shell of two valves, usually fixed to the sea floor by means of a stalk or **pedicle** (though some are cemented, attached by spines or free-lying), and a complex food-gathering mechanism called the **lophophore.** The differences in composition and structure of these elements, however, immediately divide the brachiopods into two distinct classes:

CLASS 1. INARTICULATA (L. Cam.–Rec.): Brachiopods in which the valves are not hinged by **teeth** and **sockets.** They are usually chitinophosphatic in composition, though sometimes calcareous. They often have a fleshy muscular pedicle, though some groups have lost the pedicle and may fix themselves to the substrate by cementation. There are four orders, mainly of long-ranged groups (e.g. *Lingula*). Inarticulates are usually considered as less advanced than articulates.

CLASS 2. ARTICULATA (L. Cam.–Rec.): Calcareous-shelled brachiopods in which the valves are hinged by teeth in one valve and sockets in the other. The pedicle is made of a dead horny material and in some fossil genera appears to have atrophied. The articulates are much more diverse and abundant than the inarticulates. Though they appear at the same level in the Lower Cambrian, the main early radiation of the articulates was in the early Ordovician. Seven orders are found in the Palaeozoic of which three persist today.

Class Articulata

Morphology of three genera

MAGELLANIA

The recent brachiopod *Magellania* (Order Terebratulida) (Fig. 7.1) has been well chosen by many authors as a typical example of a modern articulate brachiopod. It shows clearly how the hard parts relate to the living anatomy; useful in interpreting the ubiquitous terebratulides of the Mesozoic.

Magellania lives epifaunally, attached by its pedicle to rocks and stones at depths of 12–600 m in Australian and Antarctic waters.

The two oval calcareous valves differ in size but are equilateral and divided by a single plane of symmetry. In standard orientation the 'upper' valve is smaller than the 'lower'. There has been some

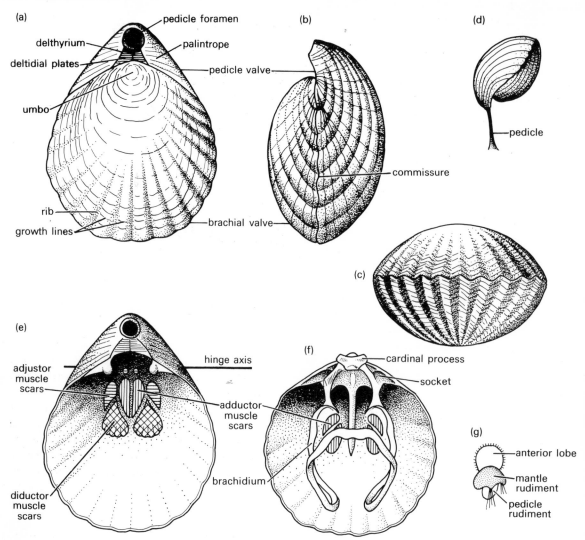

(a) pedicle foramen
delthyrium
deltidial plates
palintrope
pedicle valve
umbo
rib
growth lines
brachial valve

(b) commissure

(d) pedicle

(c)

(e) hinge axis
adjustor muscle scars
adductor muscle scars
diductor muscle scars
brachidium

(f) cardinal process
socket

(g) anterior lobe
mantle rudiment
pedicle rudiment

Figure 7.1 *Magellania flavescens:* (a) upper surface with brachial valve; (b) lateral view; (c) anterior view; (d) in life position showing pedicle attachment; (e) internal view of pedicle valve; (f) brachial valve ((a, b, e, f) x 2 approx), (g) Larva. ((a)–(f) Based on Davidson, 1851; (g) on Rudwick, 1970)

controversy as to what these two valves should be called; the 'upper' valve is variously known as the **brachial** or **dorsal valve,** the 'lower' as the **pedicle** or **ventral valve.** The terms 'brachial' and 'pedicle' refer to structures contained within the valves and thus are more technically correct. Furthermore, the conventional dorsal or ventral orientation is not necessarily a life orientation; in fact the brachiopod, when fixed by its pedicle, has both valves vertically held or has the pedicle valve uppermost. Hence the terms 'brachial' and 'pedicle' are adopted here as standard practice to designate the valves though the other system is admittedly simpler.

External morphology. The superficial resemblances of such brachiopods as *Magellania* to Roman oil lamps has long been drawn, hence the vernacular term 'lamp-shells'. The two valves are unequally biconvex. Each grows from the first-formed part, the **protegulum,** which later becomes part of the

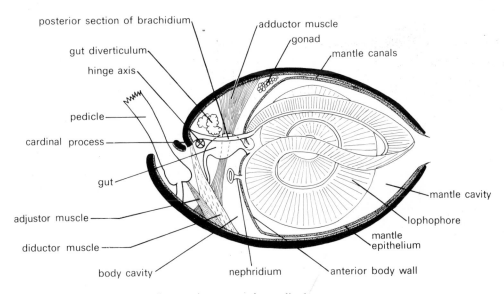

posterior section of brachidium
adductor muscle
gonad
gut diverticulum
mantle canals
hinge axis
pedicle
cardinal process
gut
adjustor muscle
diductor muscle
mantle cavity
lophophore
mantle epithelium
body cavity
nephridium
anterior body wall

Figure 7.2 *Magellania flavescens:* median section, somewhat stylised.

umbo. It grows in a logarithmic spiral (p. 140), as does the shell of a bivalve, and at each growth increment new shell material is accreted round the growing edge. The valves meet along a line of junction or **commissure.** In conventional orientation the umbones are located at the **posterior** end of the brachiopod, and since the **hinge line** is in front of them the valves gape at the **anterior** end.

Each valve is externally sculptured with faint **ribs** radiating from the umbo. These become stronger anteriorly so that the anterior part of the commissure is crenulated where the ribs meet it. The **growth lines** are subconcentric round the umbo and likewise become rather crenulated anteriorly where the ribs are stronger. Each of these growth lines represents a record of the former position of the edge of the shell, and as they are especially prominent where growth has ceased for a short time the shell retains a permanent record of its own ontogeny.

The pedicle valve is the larger of the two and is inturned posteriorly so that the most convex part of it can be seen from the upper surface. This part of the shell is the **palintrope.** The umbo is perforated by a round **pedicle foramen** through which the pedicle emerges. The latter is a stalk of horny tissue, cemented at its distal end to rocks or pebbles, and by it the brachiopod is raised above the bottom. Between the umbo and the inner edge of the valve is a triangular cavity, the **delthyrium,** closed by a pair of **deltidial plates.** They are marked by straight parallel growth lines, transverse to the brachiopod's plane of symmetry.

Though the commissures of the two valves fit exactly nearly all the way round the shell, the smaller brachial valve has its umbo tucked just into the pedicle valve anterior to the delthyrium. Through this region runs the transverse **hinge axis** operating on internal teeth (pedicle valve) and sockets (brachial valve).

Internal anatomy. In Figure 7.2 the internal organs are displayed in a section cut through the plane of symmetry. These organs relate to structures made of calcium carbonate which can be seen in the separated valves. The shell is secreted by a **cellular epithelium** which secretes calcareous material mainly at the valve margin. This epithelium underlies every part of the shell. It is two layers thick in the anterior part of the brachiopod, where the inner layer forms the **mantle** enclosing a large **mantle cavity.** Towards the posterior end of the shell the mantles of the two valves abruptly leave the internal valve surface and join, forming a single sheet of tissue. the **anterior body wall,** which crosses between the valves and separates the mantle cavity from the posterior **body cavity.** The only organ contained within the mantle cavity is the lophophore, which itself is invested with a continuation of the mantle epithelium. This is the main food-gathering and respiratory mechanism of the brachiopod. In *Magellania* the lophophore has a hydrostatically supported fluid-filled canal (the **brachial canal**) as its axis. This is supported by a long loop-shaped calcareous ribbon (**brachidium**) which is symmetrical about the median plane and attached at two points to the inside of the brachial valve at the rear of the mantle cavity. The brachidium may remain attached to the shell after death when the investing tissue has rotted away, but more often it is not preserved. From the strong brachidium-supported axis of the lophophore there spring a large number of slender parallel filaments. These are sticky and lined with cilia. The beating of these cilia generates the currents that bring in and exhale water. Inhalant currents come in laterally, and all the filaments are so arranged as to provide an effective net for trapping food particles; they are strained off whilst the filtered water is exhaled anteriorly in a single stream. The caught food particles are passed down the filament in a mucus belt to a food groove running along the brachial axis so to the mouth.

The mouth is situated in the anterior body wall; it leads to a small gut with an oesophagus, stomach and blind-ended intestine, but no anus. There are also digestive glands or diverticuliae associated with the gut. Excreta are voided through the mouth into the mantle cavity and disposed of by the exhalant current. There are no other perforations in the anterior body wall other than the paired funnel-shaped **nephridia** ('kidneys'), which remove nitrogenous waste taken up by wandering cells (**coelomocytes**) and also act as passages for the escape of gametes from the gonads. Sometimes the latter become so swollen with eggs or sperm as the breeding season approaches that they can expand into the **mantle canals.** These canals are tubular branching extensions of the body cavity (coelom) which run between the mantle and the inner epithelium. Within them coelomic fluid circulates, mainly being used for respiration. In dead specimens of *Magellania*, divested of soft tissue, the scars of these canals are clearly seen on the inside of the shell.

The only other organs found in the body cavity are the pedicle and related structures and the muscles for opening and closing the shell. The pedicle is a cylindrical stalk having a thick external cuticle, within which is a thin epithelium and a central core of connective tissue. Though the shell is fixed to the pedicle it can be turned in any direction by two sets of **adjustor muscles** near its base. These allow the upright brachiopod to swing into or away from the current.

Brachiopods open and close the shell using two sets of paired muscles. The **adductor muscles** which close the shell are analogous, though not homologous, to the adductors of bivalves. They join the two valves somewhat obliquely, and their points of attachment are both anterior to the hinge axis, so that when they contract the shell must close. By contrast the **diductor muscles** which open the shell have a quite different line of action with their bases on opposite sides of the hinge. They are attached to the pedicle valve just outside the adductors, but they are fixed to the brachial valve by a calcareous boss, the **cardinal process,** on the posterior side of the hinge axis. If the diductor muscles contract whilst the adductors correspondingly relax the cardinal process swings downwards, and as the rest of the shell lies anterior to the hinge the valves gape open by a few degrees. Feeding may then begin.

These two sets of muscles, and the adjustors also, leave well-defined **muscle scars** inside the valves, visible in isolated valves of dead individuals and sometimes in fossil valves also. Of the scars in the pedicle valve the larger outer pair are diductor scars, the smaller inner pair the adductor scars. There are likewise two main pairs of muscle scars in the brachial valve, but these all belong to the adductor muscles since the latter divide in two.

Separated valves from which the soft tissues have been removed thus show the following characters. Brachial (dorsal) valve: sockets, cardinal process, adductor muscle scars, brachidium, mantle canals; there may also be a **median septum** or partition in the plane of symmetry. Pedicle (ventral) valve: palintrope, teeth, pedicle foramen, delthyrium closed by deltidial plates, adductor and diductor muscle scars, adjustor scars and mantle canals. The delthyrium, incidentally, is the ancestral site of the pedicle foramen. In brachiopods that retain the more 'primitive system' the delthyrium is open and through this the pedicle emerges. In the early ontogeny of *Magellania* the pedicle is first located in an open delthyrium, but the foramen gradually migrates to its

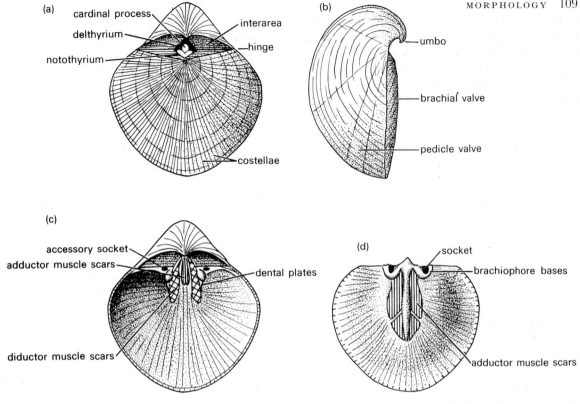

Figure 7.3 *Visbyella visbyensis* (Sil.): (a) complete shell, upper surface; (b) lateral view; (c) internal view of pedicle valve; (d) brachial valve. (All x 5)

adult size as the brachiopod grows, and the delthyrium closes off behind it by the growth of the deltidial plates.

Magellania is an example of a brachiopod with a **non-strophic shell.** The hinge axis passes through the teeth and sockets which are its only fulcra. The shell edges adjacent to this are curved and not coincident with the hinge line. Most living brachiopods (e.g. the Terebratulida, Terebratellida and Rhynchonellida) are of this type. Many groups of Palaeozoic brachiopods (e.g. the Orthida and Spiriferida), however, have **strophic shells.** In these the hinge line is straight, often extends the full width of the shell and is coincident with the hinge axis.

VISBYELLA

Visbyella (Order Orthida) is a Silurian brachiopod with a strophic shell (Fig. 7.3). It belongs to the earliest of all articulate brachiopod stocks to have

evolved, the Orthida, first known from the Cambrian. The two valves are dissimilar in size and convexity. The large pedicle valve is very deep, the brachial only slightly plano-convex, but both have the same surface sculpture of fine ribs or **costellae** radiating from the umbo and curving outwards laterally to the commissure. A pronounced umbo characterises the pedicle valve, anterior to which is the large open delthyrium without deltidial plates. This is flanked by two triangular **interareas,** together forming a flat slightly sloping shelf whose anterior edge is the straight hinge line. These interareas have closely spaced growth lines parallel with the hinge. If the valves are separated the large crenulate teeth are visible, situated anterior to the hinge line where the delthyrium reaches its greatest width. Joining the teeth to the floor of the pedicle valve are the vertical **dental plates,** which enclose a deep **umbonal cavity** continuous with the delthyrium. Flooring this cavity and extending some way anteriorly are the recessed muscle scars; the

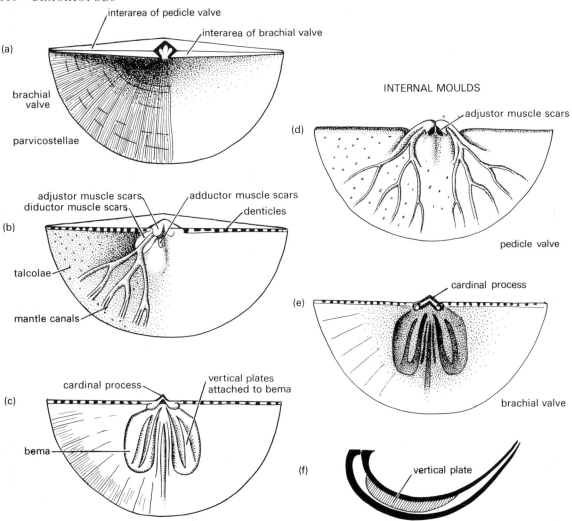

Figure 7.4 *Eoplectodonta penkillensis* (Sil.): (a) complete shell, upper surface; (b) internal view of pedicle valve; (c) brachial valve; (d) internal mould of pedicle valve; (e) brachial valve as found in rocks; (f) vertical median section showing disposition of vertical plates.

Figure 7.5 *Eoplectodonta penkillensis* (Sil.), Pentland Hills, near Edinburgh. Pedicle valve with mantle canals and adductor muscle scars (x 8). Internal mould.

diductor scars are outside those of the adductors and the whole muscle field is heart-shaped. The adjustor scars are not recessed and are barely visible.

In the brachial valve there are interareas though they are narrower than those of the dorsal valve. Between them there is a small triangular opening, the **notothyrium,** which lies directly opposite the delthyrium so that the **pedicle opening** is diamond-shaped and is formed by the delthyrium and notothyrium together. A bilobed or trilobed **cardinal process** projects posteriorly and almost fills the notothyrium when the valves are closed. The adductor scars are very large. Supports for the lophophore are represented only by a pair of divergent **brachiophore bases,** which are simply a pair of oval knobs into which the sockets are recessed; no other parts of the lophophore are known. In both valves there are mantle canals of distinctive form. The costellae are more pronounced towards the commissure so that they interlock along a crenulated edge.

EOPLECTODONTA

Eoplectodonta (Order Strophomenida) (Figs 7.4, 7.5) is a Silurian genus in which, as in many brachiopods of this order, the two valves are concavo-convex. They fit inside one another, the brachial valve having a concave exterior, the pedicle valve being deeper and convex externally. The shell is strophic, and the hinge extends the full width of the semicircular shell, being bordered by very narrow interareas. There is a large cardinal process which blocks the pedicle opening when the valves are closed. Externally the valves have a sculpture of very thin ribs, with somewhat more pronounced single ribs spaced at intervals and dividing the shell into radial segments. There are no teeth. These have been lost in the evolutionary history of strophomenides, but their function has been taken over by many small **denticles** which run along the hinge line, those of the two valves interlocking. The ventral muscle field has a bilobed diductor scar within which the smaller adductor scars are located. In the brachial valve there is a large V-shaped cardinal process which projects posteriorly carrying the point of attachment of the diductor muscles well to the rear of the hinge. This is necessary in view of the very narrow body space inside the shell and the restricted line of action of the muscles. A series of **vertical plates** supported on a raised platform

(bema) hang down from the roof of the brachial valve, almost touching the floor of the pedicle valve when the shell is closed (Fig. 7.4f). Presumably this structure was in some way connected with the lophophore which may have adhered to it. Lateral to this lie the adductor scars. An interesting feature of the inside of the shell are the numerous small projections **(taleolae),** which are calcite rods obliquely set in the shell and radially arranged parallel with the costellae. These are rod-like units lodged within the shell itself and projecting internally, and they are found only in the Strophomenida.

Since the delthyrium and notothyrium are entirely sealed by the large cardinal process there cannot have been a functional pedicle to this strange thin shell, and individuals must have lain flat upon the sea floor.

Preservation, study and classification of articulate brachiopods

The brachiopod species described above, one living and two fossil, belong to successful and important groups and give some indication of the variety of form and function in Class Articulata. Yet they by no means show the full structural range even though most brachiopods do not depart too radically from the kind of morphology shown in these. Nevertheless there are some peculiar genera amongst the Brachiopoda, especially some Permian strophomenides which are so modified, at least externally, that they are hardly recognisable as brachiopods.

The majority of brachiopods can be assigned to their correct order and usually referred to their family on the basis of external morphology alone, but other features – the lophophore, muscle scars, hinge, dentition, cardinal processes, etc. – may have to be used in identifying a brachiopod at genus and species levels. Unfortunately the brachidium is found only in some of the many fossil groups. Amongst Palaeozoic orders the brachidium is present only in the Spiriferida and Terebratulida and possibly some Pentamerida (Fig. 7.6). In these the brachidium may be preserved by mineral incrustation, if the shell is not infilled, or buried in matrix where it is. In the latter case the structure can be revealed by dissection or thin sectioning.

Many articulate brachiopods are preserved with their shell material intact. Since the shell is calcitic it

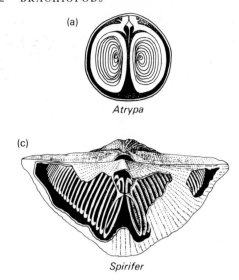

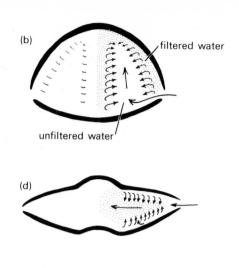

Figure 7.6 Spiral brachidia: form and function. *Atrypa reticularis:* (a) interior of brachial valve from below (x 1·25); (b) vertical section showing spiral brachidia with unfiltered and filtered water. *Spirifer striata:* (c) silicified specimen with shell partially removed showing spiral brachidia (x 1 approx.); (d) vertical section showing laterally directed spires with filtering system the reverse of that in *Atrypa*. ((a) and (c) after Davidson, 1851; (b) and (d) after Rudwick, 1970.)

is not subject to the potential diagenetic transformation of an aragonitic shell. The external morphology of such shells is often very well preserved and, provided that they can be freed cleanly from the matrix, may readily be examined. Since it is very often the internal morphology that is really diagnostic, isolated valves are of particular value. If separated valves do not occur in the fauna it may be necessary to undertake serial grinding in order to reveal internal structure. A parallel grinding machine that can grind to fixed distances is used for this, but each successive ground face must be photographed or drawn before regrinding. Though this is time-consuming such serial sections are often the only source of information on internal morphology. A complete internal reconstruction can be prepared from them by building a wax model or, more rapidly, by computer modelling.

If the brachiopod is preserved in a silty matrix but still retains its shell, the internal structures may be studied after removal of the shell by acid or by calcining, i.e. by burning off the outer shell so that the mould in matrix is visible.

Many Palaeozoic brachiopods and some more recent ones may be preserved in siltstone or mudstone from which all the calcareous material, including the shell, has been leached away. The resultant internal moulds often appear very different from moulds of the external surface. Hence in the *Eoplectodonta* specimens illustrated in Figures 7.4d, e and 7.5 the umbonal cavity of the pedicle valve seems very pronounced, the mantle canal grooves are represented by ridges and the muscle fields are in negative relief. In the brachial valve of *Visbyella* (Fig. 7.3d) deep indentations represent the brachiophore bases and the cardinal process. Much of the relief represents, of course, variations in thickness of the shell. The internal structures, such as the muscle scars, are often strikingly well preserved and can be studied with minimal preparation. It is in fact very often easier to work with and identify brachiopods from internal and external moulds than to use specimens with the calcitic shell preserved.

Furthermore, latex replication allows 'positives' to be taken which show all the features of shell morphology just as they were in life. Here a thin film of rubber solution is poured onto a shell surface and allowed to dry; then a latex block is built up on top of this from successive thicker layers and when fully dry may be stripped off. It is like producing a photographic print from a negative, and the result is normally just as clarifying.

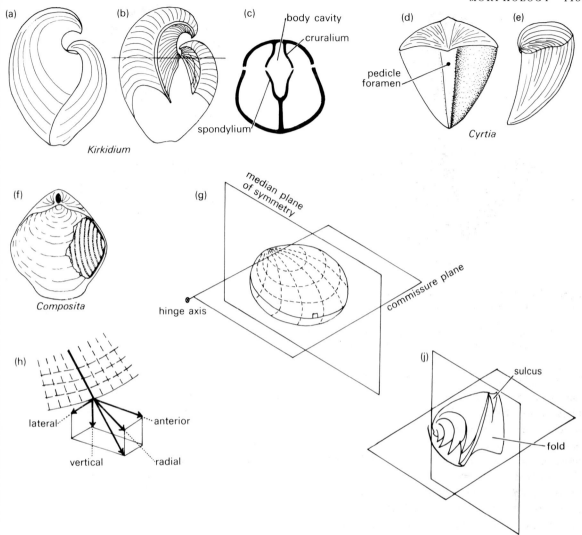

Figure 7.7 Brachiopod growth and form. *Kirkidium knightii* (Sil.) (a) lateral view (x 0·5); (b) with external shell removed exposing spondylium and cruralium; (c) in section with body cavity enclosed by spondylium and cruralium, *Cyrtia exporrecta* (Sil.): (d), (e) in frontal and lateral views, showing greatly expanded pedicle valve interarea (x 1·25); (f) *Composita* (Carb.) spiriferide homeomorphic on terebrat-ulides (x 1·5); (g) radial and concentric growth elements of a rectimarginate shell, showing radial sectors, concentric zones, median plane of symmetry, commissure plane and hinge axis; (h) growth vectors operating at any point on the edge of a shell; (j) antero-lateral view of a rhynchonellide shell showing serial deflections, fold and sulcus ((g)–(j)) modified from Rudwick, 1959).

Major features of brachiopod morphology

Form of brachiopod shells

Brachiopod shells, as noted before, grow by accreting new material from the mantle at the valve edges. The ultimate form of any developing brachiopod shell is a product of the relative growth rates of the different parts of the valve edges (Rudwick 1959). At any point of the valve margin, growth may be resolved into radial, anterior, lateral and vertical components. If all of these keep pace throughout development the result will be a **rectimarginate** shell, i.e. one with a planar commissure (Fig. 7.7g, h). On the other hand, if the vertical component grows more rapidly than the others there will be a localised growth anomaly, resulting in a vertical deflection of the commissure. Whether the deflections are median, paired or serially arranged (Fig. 7.7j) depends on their locality, whilst their amplitude and whether they are sharp or gradual is a product of the rapidity of change of relative growth.

Relatively few shell shapes have actually been adopted by brachiopods, presumably for good functional reasons. Furthermore, many of the shell shapes that are found in nature, even if they depart markedly from the biconvex norm, tend to appear over and over again in unrelated stocks through time. In such **homeomorphy,** the result of convergent evolution, the descendants of a defined stock come to have shell shapes (or indeed other characters) closely reminiscent of those of other and separate groups.

Composita (Order Spiriferida) (Fig. 7.7f) is in external appearance strikingly like many terebratulides of the same and different ages. Both are non-strophic, though the majority of spiriferides are strophic. The only real external difference is in the style of perforation of the umbo, though the internal morphology is radically dissimilar of course. This is but one example of a ubiquitous phenomenon. There are **isochronous** and **heterochronous homeomorphs** (i.e. present at the same time or at different times respectively), and they may occur either within the same taxon or in different taxa.

Some brachiopods, such as the rhynchonellides and spiriferides, have pronounced serial zigzag deflections all along the anterior part of the commissure, with a concave central **sulcus** on the brachial valve fitting into a pronounced **fold** in the pedicle valve (Fig. 7.7j). Such structures probably resulted from differentiation of growth rates at localised sites not only in the vertical components but in the radial ones too. These sharp and serially arranged zigzags gave a relatively much greater effective length of commissure over which food particles could be taken in, for no extra gape. When the shell gaped slightly small particles could be drawn in whilst large and harmful sand grains were excluded. Only at the sharp angle of the zigzag could larger particles enter, and these angles are normally protected by **setae** (which also act as early warning sensors) or occasionally by spines. The fold and sulcus system of rhynchonellides seems to be instrumental in separating lateral inhalant currents from a median exhalant stream. Spiriferides also have a median fold and sulcus, but they are relatively smaller and rarely ornamented by zigzags and probably mark the exhalant system.

The costellae or ribs may sometimes be prolonged into hollow or solid spines. One function of such spines appears to have been for anchorage. The large *Productus* has many spines on the base of the pedicle valve which seem to have had this function, and the thin-shelled strophomenide *Chonetes* which has spines along the hinge margin could conceivably have used them for fixing itself upright on the sea floor with the spines held vertically. But in the rhynchonellide *Acanthothiris* the spines of the pedicle and brachial valves when first formed at the edge of the mantle were hollow and have been interpreted as sensory tubes for extending the sensitive tip of the mantle well away from the body: another kind of early warning sensor (Rudwick 1965b). These spines, initially hollow, became filled up with calcite as they migrated away from the mantle edges and eventually took up an anchoring function. External ornamentation in some silicified brachiopods is shown in Figure 7.8.

Microstructure of brachiopod shells (Fig. 7.9)

In all articulate brachiopods the shell is multilayered, and the various layers can usually be distinguished in fossil brachiopods as well as in living ones. Shell structure and development has been extensively studied in the Recent rhynchonellide *Notosaria nigricans* (Fig. 7.9a) which serves as a standard model (Williams 1968). There are three shell layers: an outer non-calcareous **periostracum,** a middle calcareous **primary layer,** and an inner **secondary layer** of calcareous and inorganic material.

Within the periostracum there are three main proteinaceous layers which underlie an outer gelatinous sheath. This sheath protects the growing edge of the shell and is the first element to be formed, but being gelatinous it is soon rubbed off and does not extend far beyond the edge of the shell.

The primary shell layer beneath the periostracum is of rather structureless crystalline calcite, whilst the very distinctive secondary layer consists of elongated calcitic rods (fibres) with rounded ends, inclined at about 10° to the shell surface and having a regular system of stacking and a characteristic trapezoidal

Figure 7.8 Silicified brachiopods from the Visean (Lower Carb.) of County Fermanagh, Ireland, showing details of surface sculpture and morphology. *Tylothyris laminosus* (McCoy): adult shell of spiriferacean viewed (a) posteriorly showing high ventral interarea with delthyrium almost completely covered by deltidium; (b) dorsally. (c) *Cleothyridina fimbriata* (Phillips): an athyrid spire-bearer showing part of the spire; Typical spinose lamellae externally. (d) *Productidina margaritacea* (Phillips): juvenile specimen of a productoid; Pedicle valve viewed posteriorly. (e, g) *Dasyalosia panicula* (Brunton): adult shells showing two directional spines on pedicle valve (e), and well developed interarea (g). (f) *Overtonia fimbriata* (J. de C. Sowerby): young pedicle valve with characteristic interdigitating ridged and spine pattern. (a, b x 1·5, c, f x 2, d x 7, e, g x 4). (Photographs by courtesy of Dr. Howard Brunton)

cross-section. Whilst the primary layer is of constant thickness, the secondary layer keeps on growing throughout the life of the brachiopod and so is thickest nearest the umbones.

A section through the growing edge of the shell shows how these several layers are secreted. The mantle, which secretes the shell, is infolded into a groove under the shell edge; here there is a **generative zone** from which all the shell-secreting cells are produced. As each new cell is formed it moves towards the edge of the shell, producing the various layers in turn from its outer surface. First the shell secretes the gelatinous sheath, then the three layers of the periostracum in turn. Thus by the time the cell has reached the outer edge of the shell it has already produced four different kinds of material in succession. It then begins to grow calcite rhombs from its upper surface, embedded in a protein cement; these then grow together and amalgamate as the crystalline primary layer. The cell then swivels round an axis of rotation, so that its distal surface now faces outwards, and becomes permanently fixed in its final place as

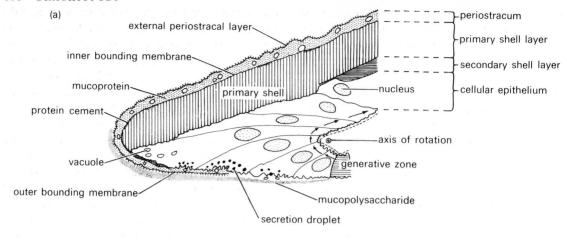

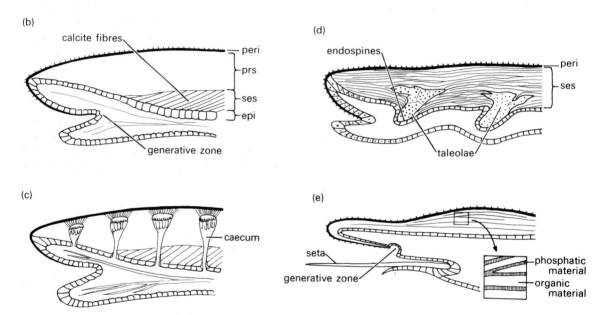

Figure 7.9 Shell structure in brachiopods: (a) standard secretory regime in *Notosaria nigricans* (for explanation see text); (b) impunctate shell (e.g. rhynchonellide); (c) endopunctate shell (e.g. terebratulide) with caecae possessing core cells freely suspended in the cavity – the secondary shell fibres are shown as cut in section; (d) pseudopunctate shell (e.g. strophomenide) with no primary layer but with taleolae prolonged internally as endospines; (e) shell of *Lingula* with primary shell constructed of alternate layers of phosphatic and organic material – a marginal seta is shown (Mainly based on Williams and Rowell in '*Treatise*' Part (H))

the growing edge of the shell moves away from it, all the while accreting new material in conveyor belt fashion. When the secretion of primary cell material is complete, the secretory function of the cell changes for the sixth and last time and it produces a single long calcitic fibre: one of many identical elements in the secondary layer. In this layer the regular pattern of the many long inclined fibres naturally reflects the ordered arrangement of the individual cells that produce them. Since new material is continually added to the lower end of the fibre by the surface of each cell, it is not surprising to find the secondary layer thickest at the umbones, which are the oldest part of the shell.

The modern terebratulide *Waltonia* produces its shell according to the same 'standard secretory regime' as *Notosaria*, but the periostracum is much thicker and of a more complex labyrinth-like structure.

Most articulate brachiopods, living and fossil, probably secreted their shells in much the same way as *Notosaria*. Differences in shell structure of some groups, however, indicate that the secretory programme must have been modified. In the 'advanced' Strophomenida (Fig. 7.9d) the shell below the periostracum is only a single layer of laminar calcite traversed by inclined calcitic rods (**taleolae**). But in the ancestral strophomenide group, Superfamily Plectambonitacea (Suborder Strophomenidina), to which *Eoplectodonta* belongs, there is a secondary layer of inclined calcitic fibres like that of *Notosaria*. Hence the unusual laminated shell of the later strophomenides must be equivalent to the standard primary layer, and in the evolution of this order the secondary layer has been lost, perhaps by neoteny.

Recent Terebratulida and Thecideidina have an endoskeleton of calcareous spicules, secreted within the living tissue and assisting in its support. Such spicules are found in some Cretaceous terebratulides.

Endopunctation and pseudopunctation in brachiopod shells

In articulate brachiopods such as *Notosaria* the shell is **impunctate,** i.e. has no perforations or cavities within the shell structure (Fig. 7.9b). Other kinds of articulates (e.g. terebratulides) have **endopunctate** shells (Fig. 7.9c) in which the shell structure is penetrated from the inside by large regularly arranged elongated cavities known as **punctae,** normal to the shell surface. These contain tubular outgrowths of the mantle, the **caecae.** Caecae are formed at the edge of the shell during the standard secretory regime by small knots of cells which behave independently of the rest of the mantle. As the growing edge of the shell moves away the caecae are permanently locked in position, but since each caecum retains its contact with the mantle throughout the deposition of the primary and secondary layers the lower end may become very long.

The caecum nearly reaches the outer shell surface and is connected to it by a 'brush' of tiny tubes filled with mucopolysaccharide. Below the brush are core cells filled with glycogen and proteins; these hang down freely into an empty cavity below, so that the only connection of the caecum with the mantle is by the flattened peripheral cells that line the wall of the punctae. Caecae are primarily storage chambers; they are also respiratory, and inhibit boring organisms. In addition, the periostracal brush can release an organic 'glue' (for the repair of injury) if the periostracum is accidentally ruptured.

Impunctate and endopunctate shells in articulate brachiopods seem to cut across established systematic boundaries, and it is probable that endopunctation evolved more than once. Thus most orthides are impunctate, apart from one superfamily, the Enteletacea. All atrypides and pentamerides are impunctate. Most spiriferides and rhynchonellides are impunctate also, but there is one mainly punctate spiriferide superfamily, the Spiriferinacea, and one monogeneric punctate rhynchonellide superfamily, the Rhynchoporacea. All terebratulides are punctate.

In strophomenide shells (Fig. 7.9d) the thin and often irregularly formed taleolae give a spurious impression of punctation, especially to weathered specimens. Hence the term **pseudopunctate** is used to distinguish them, perhaps unfortunately, for in other respects their microstructure is clear enough.

Hinge and articulation of brachiopods

Strophic and non-strophic shells are distinguished by their hinge structure. There are no apparent intermediates between them; indeed it would be hard to imagine a brachiopod of intermediate kind. The various structures associated with the hinge region are dealt with in turn.

Pedicle opening. The delthyrium and notothyrium are more closely seen on strophic shells since the hinge is straight. Primitively, as in orthides, enteletides and pentamerides, they together form a diamond-shaped opening for the emergence of the pedicle. But this is not their sole function, for in brachiopods with high interareas they also allow a space for the diductor muscles when these are contracting. Since the diductor muscles are attached to the cardinal process their line of action is brought close to the hinge, and if there were no opening the muscles would catch on the hinge line and fray.

Where there is no pedicle the delthyrium and

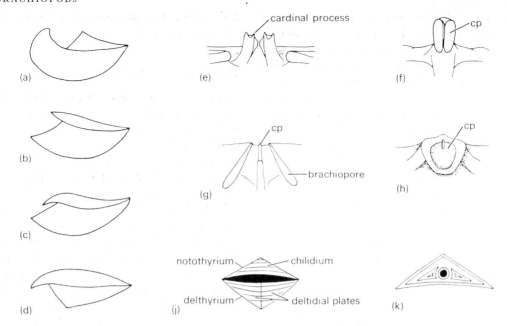

Figure 7.10 (a–d) Inclination of interareas: (a) brachial valve hypercline, pedicle valve anacline; (b) brachial valve anacline, pedicle valve apsacline; (c) brachial valve apsacline, pedicle valve apsacline; (d) brachial valve apsacline, brachial valve procline. (e)–(h) Cardinal processes: (e) *Strophomena;* (f) *Pustula;* (g) *Hesperorthis;* (h) *Leptellina;* (j) pedicle foramen with closing structures. (k) Triangular stegidial plates closing delthyrium. (Redrawn from Williams and Rowell in '*Treatise*' Part (H))ᵗ

notothyrium may be protected by various kinds of plates. In the pedicle valve there may be a single plate (the **deltidium**) (Fig. 7.10j) or, as in *Magellania*, a pair of plates (**deltidial plates**), united by a median suture and arched so that they do not interfere with the line of action of the muscles. The notothyrium may likewise be plated by a pair of **chilidial plates** or by a single **chilidium.** Rarely in spiriferides there may be a **stegidium** (Fig. 7.10k). This special type of delthyrial covering is a series of concentric triangular accretions formed during growth which successively close the delthyrial cavity from the outside in.

Certain strophomenides have only a very tiny hole in the umbo for the emergence of the pedicle. Likewise in the spiriferide *Cyrtia* (Fig. 7.7d, e), which has a much expanded pedicle valve interarea, the pedicle opening resides in the centre of the interarea. Any pedicle emerging from such an opening could not have enabled the brachiopod to be held upright on the sea floor but could certainly have acted as a tether to hold it so that the shell was not swept away by currents.

Non-strophic shells such as the large Silurian *Kirkidium* (Fig. 7.7a–c) have a large long pedicle valve with a strongly curved umbo but an open delthyrium some distance anterior to the umbo. In juvenile shells of *Magellania* and other terebratulides the pedicle opening is located in a normal delthyrium. As the shell grows this opening migrates posteriorly towards the umbo by resorption of the shell. Meanwhile deltidial plates form anteriorly and eventually join in the midline, closing off the pedicle opening entirely from the delthyrium; this can then be distinguished as a pedicle foramen.

Interareas Interareas proper are found only in strophic shells. Their attitude relative to the hinge line is very important in taxonomy, and the nomenclature of some of the various attitudes is illustrated in Figure 7.10a–d. Of these **apsacline** and **anacline** respectively are the most common. Some shells have the interareas of the pedicle valve attenuated into a pronounced beak-like **rostrum,** which in *Uncites* is curiously twisted to one side.

Cardinalia. The structures at the posterior end of the brachial valve, collectively known as **cardinalia,** are highly differentiated and serve various functions. Of these the medially placed cardinal process is often the most prominent (Fig. 7.10e–h). In its simplest form, as found in the more primitive articulates, it is no more than a pair of undifferentiated muscle bases behind the notothyrium (Fig. 7.10h). A more elaborate kind of structure is a **median ridge;** this primitively separated the two areas of muscle attachment, but in more advanced forms it actually became the site of diductor attachment through the inward migration of the muscle bases (Fig. 7.10g).

A further development of this kind of structure led to a shaft with a head **(myophore)** (Fig. 7.10f), which increases the area of muscle attachment and may be variously crenulated, forked, multilobate or comb-like.

Other cardinalia include **socket walls** and **brachiophores** (the basal parts of the lophophore).

Teeth, sockets and accessory structures. Teeth are always found in the pedicle value, sockets in the brachial. This system evolved only once apparently and then remained relatively stable. The teeth consist of knobs of secondary calcite with smooth or crenulated surfaces. In many brachiopods they may be supported by a pair of vertical dental plates which join them to the floor of the pedicle valve.

The teeth may be supplemented or even replaced (e.g. *Eoplectodonta*) in many strophomenides by denticles which grow along the margins of the hinge line. Members of Superfamily Productacea (Suborder Productidina) have neither teeth nor denticles.

Sockets, forming part of the cardinalia, have often large and thick interior walls, often as large as the teeth themselves. In such cases the interior socket wall may even indent a **secondary socket** close to the tooth on the pedicle valve. Socket walls can be swollen and, as in *Visbyella,* closely associated with the brachiophores marking the lophophore bases (Fig. 7.3d).

Teeth and sockets are constructed of secondary shell. As the shell grows the teeth move away from the hinge line, taking up successive though closely spaced positions and leaving a 'track', which is the cumu-lative effect of the position that these structures had throughout ontogeny.

Brachiopod muscle attachments

Adductor, diductor and adjustor muscles are normally attached directly to the insides of the two valves, forming scar patterns which are genus- or species-specific.

Sometimes, however, all the muscles may be raised off the floor by **muscle platforms,** such as are clearly displayed in the pentameraceans (Fig. 7.7a–c). In these the pedicle valve has a vertical wall divided in two distally so that it appears Y-shaped in cross-section. This structure, the **spondylium,** lies directly opposite a **cruralium:** a pair of near-vertical plates which are outgrowths of the brachial processes, hanging down from the roof of the pedicle valve. Together these structures enclose a narrow vertical cavity, open anteriorly about a third of the way from the umbo, in which the muscles and probably all the other organs of the body cavity were enclosed as well.

Muscle platforms such as these are not uncommon in brachiopods, even in the inarticulates, such as *Trimerella* (Fig. 7.12f), and seem to have arisen several times during evolution. Rudwick (1970) has noted that it is not necessary to have a contractile muscle running the full height of the shell in order to open and close it; a shorter muscle could do the job more economically and with greater control. Muscle platforms tend to occur in those shells which are highly biconvex; they allow for the attachment of a short but optimally effective muscle system. Not all strongly biconvex shells have this system; the surviving rhynchonellides and terebratulides have part of the muscle strand replaced by a non-contractile **tendon,** which is another way of operating with equal efficiency.

The brachiopod lophophore (Fig. 7.11)

The brachiopod lophophore, whose structure in *Magellania* has been described, is functionally similar in all brachiopods.

In the brachiopods, though the filamentous structure of the lophophore is constant there are several distinct ways in which the lophophore itself is arranged within the mantle cavity. Furthermore, though some brachiopods have a calcareous support

(a)

great brachial canal

food groove

filaments

(b)

(d) original trocholophe

terminal generative zone

(e) original trocholophe

terminal generative zone

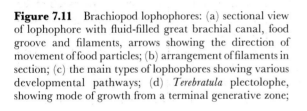

(c)

trocho-
mouth

zygo-

schizo-

early spiro-

plecto-

ptycho-

spiro-

(f)

Amphigenia

(g)

Gryphus

Figure 7.11 Brachiopod lophophores: (a) sectional view of lophophore with fluid-filled great brachial canal, food groove and filaments, arrows showing the direction of movement of food particles; (b) arrangement of filaments in section; (c) the main types of lophophores showing various developmental pathways; (d) *Terebratula* plectolophe, showing mode of growth from a terminal generative zone; (e) *Rhynchonella* spirolophe, same; (f) *Amphigenia* (Dev.), primitive loop structure of early terebratulide (Centronellidina); (g) *Gryphus* (Rec.) short loop of Terebratulidina. ((a)–(c) Redrawn from Rudwick, 1970; (d)–(e) redrawn from Williams and Wright, 1961, *Palaeontology* **4,** 149–76; (f)–(g) from '*Treatise*' Part (H))

for the lophophore (spiriferides, atrypides, terebratulides), others do not, and in the orthides, as in most strophomenides, pentamerides, and fossil inarticulates, it is hard to tell what the original lophophore structure must have been. Living rhynchonellides likewise have no calcified lophophore support, and in these as in living inarticulates the lophophore was held up purely by the hydrostatic pressure of fluid in the great brachial canal (which early in life becomes closed off from the coelom). In many strophomenides there are traces of possible lophophore structures impressed upon the inside of the brachial valve, whilst the arrangement of the vertical plates in *Eoplectodonta* and related genera suggests the form that the lophophore must have had; other possible lophophore structures give little information on the original form.

The **lophophore support** is represented in rhynchonellides and some pentamerides only by a pair of short struts (**crura**) projecting from the cardinalia. In groups where the brachidium is a calcareous ribbon (terebratulides, atrypides, spiriferides) it always grows (or grew) by accretion on the (anterior) growing edge and by resorption on the (posterior) trailing edge as the shell enlarges. Invariably new filaments are added at the growing ends of the lophophore alone, never intercalated within the filament series.

Changes in the form of the brachidium during ontogeny provide a basis for the classification of lophophores into different types (Fig. 7.11c). All lophophores pass through the same early stages in development. The first stage, whether supported or not, is a **trocholophe.** Here the lophophore is merely a pair of curving 'horns' projecting horizontally on either side of the mouth and forming an incomplete ring. Only one genus of micromorphic adult brachiopods has this kind of structure, and since the trocholophe forms the initial developmental stage of all brachiopods this particular case is probably a neotenous development. The next stage in development is the **schizolophe,** where the two original horns have become larger but are bent back and run parallel, directed towards the mouth. All brachiopod lophophores go through this stage too; a few micromorphic terebratulides have schizolophes in the adult. From here there are three possible developmental pathways:

(a) one, starting with a **zygolophe,** which leads to a **plectolophe:** the long or short loop of many terebratulides;
(b) one leading to the multilobed **ptycholophe,** as found in the modern *Lacazella*, a thecidean, and in some fossil strophomenides;
(c) one leading to the twin spiral system of the **spirolophe,** perhaps the commonest and most widespread functional system, possessed by Recent rhynchonellides and inarticulates, as well as by fossil spiriferides and atrypides and possibly other fossil groups.

In both the plectolophe of terebratulides and the spirolophe of many groups the twin ribbons may be joined by a stout rod, the **jugum,** near the posterior end.

The mode of operation of spirolophes is well known in some modern rhynchonellide brachiopods, and some inferences can be made about how it functioned in fossils. As in all brachiopods, the ciliated filaments borne on the brachia create currents which normally enter the shell laterally (Fig. 7.6). When the water passes into the shell it first of all lies outside the lophophore in an **inhalant chamber.** It then has to pass through the cilia from the outside to the space enclosed within the spires, i.e. into an **exhalant chamber.** As it does so food particles are strained off.

The water from the exhalant chamber comes out medially, and the marked deflections of the anterior margin of the rhynchonellides appear to separate inhalant from exhalant currents.

Notosaria is a Recent rhynchonellide whose twin conical spires point downwards. These spires form the inhalant chamber and contain unfiltered water which it passes through the filaments to an exhalant chamber outside the spires. In Family Atrypidae (Suborder Atrypidina) (Fig. 7.6a, b) the morphology of the lophophore is striking similar with the sole difference that the spires point upwards rather than downwards, and there is a calcified brachidium. It seems reasonable to assume that current systems functioned in a similar manner and that the unfiltered water lay inside the spires whilst the exhalant chamber was outside them.

Plectolophes, as developed in modern terebratulides, are highly efficient structures of complex three-dimensional form. Each arm develops as a curving horn (the trocholophe stage), then recurves sharply as a bar running parallel with the original arm and lying above it (the zygolophe stage). In the fully developed plectolophe this bar turns inwards to terminate in a vertical spiral coil. The two spiral median coils developed from the paired arms lie side by side and with their filaments occupy much of the space in the centre of the shell; between them is inhalant water.

Members of Superfamily Terebratellacea (Suborder Terebratellidina) have plectolophes supported along most of their length by a long calcareous loop in which the two ends have joined up, though the median coil is supported by hydrostatic pressure alone. But in terebratulaceans only the basal parts are supported by a much shorter loop; the rest remains erect through hydrostatic pressure, though some additional support is given by the calcareous spicules within the lophophore itself.

Ptycholophes are generally rather flat structures and do not form the complex three-dimensional coils found in the other two lophophore types. Presumably they are less efficient filtering systems. They are uncommon today, occurring only in Family Megathyrididae (Superfamily Terebratellacea) and thecideans, but they may have been the normal lophophore type in strophomenids, as is witnessed by the brachial ridge system of *Eoplectodonta*. Most fortunately, intact calcified ptycholophes have

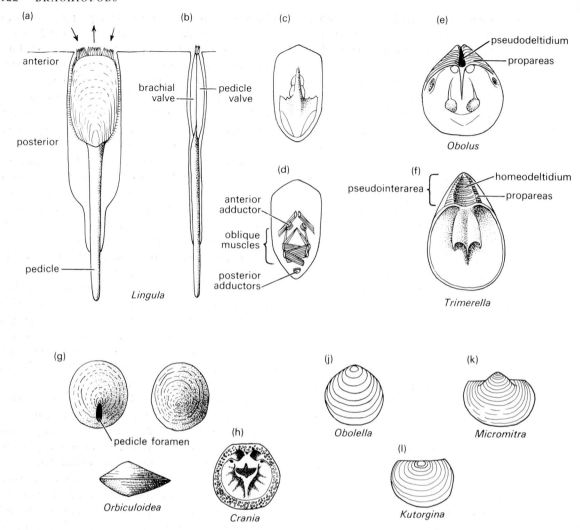

Figure 7.12 Inarticulate brachiopods: *Lingula* (Rec.): (a)–(c) in life position in burrow showing inhalent and exhalant currents; (c) interior of brachial valve; (d) internal muscles in brachial valve; ((a)–(d) x 1); (e) *Obolus*, pedicle valve interior (x 2); (f) *Trimerella*, interior of pedicle valve (x 0·4); (g) *Orbiculoidea* showing pedicle and brachial valves and lateral view (x 3); (h) *Crania* pedicle valve interior (x 2); (j) *Obolella* (x 4); (k) *Micromitra* (x 4); (l) *Kutorgina* (x 1·5). ((a)–(d) Based on Rudwick, 1970; (e)–(l) based mainly on photographs by Rowell in '*Treatise*' Part (H))'

recently been found (Grant 1972) in small Permian productaceans (Fig. 7.14f).

Class Inarticulata

LINGULA

Lingula (Order Lingulida) is a phosphatic-shelled inarticulate brachiopod which has persisted since the Cambrian with relatively little change: a type example of bradytelic evolution. Modern *Lingula* lives successfully in a brackish to intertidal environment (Craig 1951). The early invasion of *Lingula* into such a 'difficult' environment for which there were few competitors is probably one of the primary reasons for its virtual evolutionary stasis. Nevertheless there is reason to believe that some fossil *Lingula* species lived in environments other than the intertidal, so the use of

Lingula as an environmental indicator is limited. Some differences in *Lingula*-bearing 'palaeocommunities' through time have been documented by West (1976).

Modern *Lingula* (Fig. 7.12a–d) has a pair of almost identical, bilaterally symmetrical valves. They are gently convex, and from between them projects an elongated pedicle. The animal lives in a vertical position in a burrow, the pedicle extending deep into the sand with the upper edges of the two valves just below the surface. By analogy with other brachiopods the pedicle end is described as posterior, whilst the slightly larger valve is called the pedicle valve and the smaller the brachial valve. Horny setae round the exposed edges of the shell mark off three apertures leading to the inside; the outer two for incurrent water, the central for excurrent water.

The shell is lined by a mantle, and the anterior body wall, the lophophore and other internal organs are as in the Articulata. The lophophore is a spirolophe with the two spires directed inwards, each being an 'inhalant spire' carrying unfiltered water which is strained and exhaled medially. The gut, however, has an anus discharging posteriorly, and there are four nephridia. The circulatory system is simple, and there is a nervous system with a ganglion near the mouth, as in the articulates. One of the greatest differences between *Lingula* and articulate brachiopods lies in the musculature for opening and closing the shell (Fig. 7.12c, d). Since there is neither a hinge line nor teeth and sockets the shell cannot open by leverage, and so the muscular system is unlike that of articulates. The two valves close by adductor muscles: a single posterior muscle and a pair of anterior ones. When these relax the valves move apart slightly through the elasticity of the muscles, but otherwise the valves can only rotate, shear or slide against one another through the action of various **oblique muscles** which are large and well developed. The pedicle has a leathery external cuticle within which is an epithelium, embryologically part of the mantle. Inside it there are muscles enabling the pedicle to contract, so that when the end is firmly fixed in the sand the brachiopod can be withdrawn inside its burrow.

Taxonomy and evolution of the Inarticulata

There are four (possibly five) orders of Inarticulata of which only two are numerically important and long-ranged: Lingulida and Acrotretida. The others are small Lower Palaeozoic groups whose relationships with other brachiopods are unclear. All these had already appeared by the Lower Cambrian.

Order Lingulida (L. Cam.–Rec.) consists of brachiopods with shells of calcium phosphate or more rarely calcium carbonate, the pedicle, where present, emerging posteriorly between the valves.

Superfamily Lingulacea contains *Lingula* and its relatives in which the shells are elongated ovals with growth lines, concentric round a marginal umbo. Family Obolidae also belongs to this superfamily; they are suboval or rounded shells (Fig. 7.12e) in which the pedicle valve often has an internal flange divided into two **propareas,** which resemble the interareas of articulate brachiopods and have a triangular **pseudodeltidium** in between them. The musculature of *Obolus* can be reconstructed with reference to the muscle attachments, and though there are clearly defined adductor and oblique muscles some evidence suggests that the leverage system could have functioned more like that of an articulate than that of *Lingula*.

The other superfamily, Trimerellacea (Ord.–Sil.) (Fig. 7.12f), consists of large brachiopods with thick biconvex shells, probably aragonitic, having in the pedicle valve a large **pseudointerarea** and strongly defined propareas with a **homeodeltidium** in between. A peculiar feature is the common presence of muscle platforms supported by a central buttress and often turned down at the edges to enclose paired chambers. This development of muscle platforms is, of course, quite independent of that in articulates.

Inarticulates of Order Acrotretida (L. Cam.–Rec.) often have circular valves of differing convexity. Some possess high conical pedicle valves and flat disc-shaped brachial valves. They are usually phosphatic though calcareous in Superfamily Craniacea (Suborder Craniidina). The pedicle, where present, emerges through an opening which perforates the pedicle valve subcentrally. In the Craniacea the pedicle is absent, and the pedicle valve cements itself to an appropriate substratum. Living acrotretide genera, e.g. the shallow water *Discinisca*, and fossils such as *Acrotreta* and *Orbiculoidea* may be distinguished from other brachiopods by their subcentral umbo with subconcentric growth lines and by the pedicle opening.

Order Obolellida (L. Cam. – M. Cam.) (e.g. *Obolella*) comprises calcareous shelled inarticulates, elongated oval in form, which have a very variable position for the pedicle opening and usually a pseudointerarea.

Inarticulates of Order Paterinida (L. Cam. – M. Cam.), such as *Paterina* and *Micromitra,* have phosphatic shells with pseudointerareas. Their muscle scar system, however, is quite unlike that of any other inarticulates.

Finally, there is Order Kutorginida (L. Cam. – M. Cam.). These are small calcareous shelled brachiopods lacking teeth but with a hinge line. In most respects they do in fact resemble articulates, and opinion is divided as to which class they should be referred.

Shell structure in inarticulates

As in articulates, the shell is secreted by cells produced in a generative zone in the mantle, under the valve rim. In *Lingula* (Fig. 7.9e) the periostracum is underlain by a primary shell with many alternating stratified layers of organic and phosphatic material. They are all subparallel with the shell margin, and rather than being continuous layers they are elongated lenses whose mode of formation is incompletely understood. The phosphatic layers are thicker where they overlie the body cavity. There is no secondary layer; nor are there punctae.

The craniaceans (e.g. *Crania*) (Fig. 7.12h) are unusual inarticulates in having a shell of protein and calcium carbonate. They survive today as the most successful of calcareous shelled inarticulates. In the brachial valve the shell is triple-layered. Externally there is an impersistent mucopolysaccharide sheath with the periostracum below. Below this is the primary shell made up of needle-like calcite crystallites, and finally there is a secondary layer of calcitic laminae sheathed by protein sheets. But the secondary layer is found only in the brachial valve; it is absent in the pedicle valve, which cements itself to the surface by its outer mucopolysaccharide film. Craniaceans are punctate with branching caecae in both valves, used for storing polysaccharides and protein. Where the muscles are inserted the calcite of the primary layer forms blades normal to the surface.

ONTOGENY

The sexes are separate in brachiopods. Fertilisation of the eggs takes place in the sea water after the copious genital products have been shed via the nephridia through the mantle cavity and out into the sea. Breeding habits are poorly known, being only well documented in *Lingula.* Apparently some species have a single annual breeding time; others produce multiple broods during the year.

In a few Recent brachiopods the development of the fertilised eggs takes place in special invaginated **brood pouches** in the shell of the female, and this seems to have been the case in certain fossil brachiopods as well.

The development of the zygote and larval stages is quite different in articulate and inarticulate brachiopods, providing good embryological evidence that the two classes are natural divisions. In *Lingula,* as in all inarticulates, the free-swimming early larva has two ciliated segments: one becomes the body, the other the mantle. From the mantle segment there develops a mound-like ridge, which starts to grow lophophore filaments whilst the larva is still free-swimming. At the same time the larval shell (**protegulum**) develops as a single plate (though this is unusual in any brachiopod), later splitting into two separate plates for the brachial and pedicle valves respectively. The pedicle too grows from the mantle segment, and when it is large enough the brachiopod settles upon it, usually by the time some ten or so mantle filaments have been produced. All other structures (e.g. muscles, gut and nephridia) originate from the body segment.

The rhynchonellide *Notosaria* (Fig. 7.1g) shows the very different character of articulate development. The larva has three segments; the upper is the globular body segment or anterior lobe, the middle the presumptive mantle, and the lower the pedicle segment. Initially the mantle segment hangs down freely over the pedicle segment as a cylindrical 'skirt'. This then inverts itself to cover the body segment, leaving the pedicle exposed. Shell secretion then begins on the now reversed outer surface of the mantle, but the brachial and pedicle valve protegula are separate from the beginning. After a very brief larval life (shorter than in most inarticulates) the young brachiopod settles on the pedicle and completes its development.

CLASSIFICATION (Fig. 7.13)

Brachiopod taxonomy has emerged from a state of disorder and confusion only in recent years. Fossil brachiopods are not easy to classify, and the difficulties are compounded by homeomorphy. But perhaps more importantly, confusion reigned because the taxonomic scheme in common use, originally proposed by Beecher, was for want of a better system retained long after the principles upon which it was based had been shown to be biologically unsound.

The basis of a new classification of brachiopods was laid by Muir-Wood (1955) who made it clear that the concept of working down from 'orders' predetermined by a few key characters had to be abandoned in favour of a scheme of building up from the generic level to higher units, thus recognising but avoiding the problems arising from homeomorphy.

The normally adopted classification, given below and illustrated with time ranges in Figure 7.13, is that of Williams and Rowell in the *Treatise*. Though it is still at the stage of being a 'working classification' it seems to represent evolutionary relationships more effectively than its predecessors. It may change somewhat with time, and not all of it is agreed upon. Rudwick (1970), for instance, has used a slightly modified version, but the *Treatise* system, simplified and abbreviated, is followed here for the sake of convenience.

PHYLUM BRACHIOPODA (Cam.–Rec.): Solitary marine bivalved coelomates, inequivalved but bilaterally symmetrical normal to the commissure plane. Shell chitinophosphatic or calcareous; pedicle attached, cemented or free. Valves formed by mantle, enclosing body and mantle cavities, the latter with a filamentous lophophore.

CLASS 1. INARTICULATA (L. Cam.–Rec.): Brachiopods with chitinophosphatic or calcareous valves lacking teeth or sockets; valves attached by muscles and body wall only. Muscular pedicle formed as outgrowth of mantle; lophophore unsupported. Gut with true anus.

ORDER 1. LINGULIDA (Cam. –Rec.): Shell phosphatic, rarely calcareous, biconvex; umbones usually located terminally. Pedicle growing out between valves or absent.

SUPERFAMILY 1. LINGULACEA (L. Cam.–Rec.): Phosphatic shells with subequal valves, e.g. *Lingula, Lingulella*.

SUPERFAMILY 2. TRIMERELLACEA (M. Ord. – U. Sil.): Large thick calcareous shells; muscle platforms in both valves, e.g. *Trimerella, Dinobolus*.

ORDER 2. ACROTRETIDA (L. Cam.–Rec.): Shell phosphatic or calcareous, circular or subcircular. Pedicle opening through pedicle valve alone or absent.

SUBORDER 1. ACROTRETIDINA (L. Cam.–Rec.): Phosphatic shells with pedicle, e.g. *Acrotreta, Siphonotreta, Orbiculoidea*.

SUBORDER 2. CRANIIDINA (L. Ord.–Rec.): Shell calcareous, lacking pedicle, free or cemented. e.g. *Crania, Valdiviathyris*.

ORDER 3. PATERINIDA (L. Cam. – M. Ord.): Shell phosphatic, round or oval; pseudo-interareas in both valves with delthyrium and notothyrium, e.g. *Paterina, Micromitra*.

ORDER 4. OBOLELLIDA (L. Cam. – M. Cam.): Calcareous biconvex shells, subcircular or elongate oval; pseudointerarea in pedicle valve; brachial valve beak marginal, e.g. *Obolella*.

ORDER 5. KUTORGINIDA (L. Cam. – M. Cam.): (Uncertain whether this order should be placed with the Inarticulata or Articulata.) Shell calcareous with cardinal area in both valves; delthyrium and notothyrium present, but no teeth, socket, or cardinal process, e.g. *Kutorgina*.

CLASS 2 ARTICULATA (L. Cam.–Rec.): Shell calcareous, endopunctate, impunctate or pseudopunctate. Teeth and sockets present, though sometimes lost secondarily. Crura usually present, sometimes prolonged as a brachidium. Valves opened by diductors, closed by adductors. Pedicle arises from larval rudiment. Gut without anus.

ORDER 1. ORTHIDA (L. Cam. – U. Perm.): Articulata with unequally biconvex strophic shells. Delthyrium and notothyrium usually open; cardinal process normally present. Shell impunctate, rarely punctate. Ventral muscle field small; muscle platform rare. Brachidia absent, but lophophore probably schizolophous or spirolophous. Suborders include:

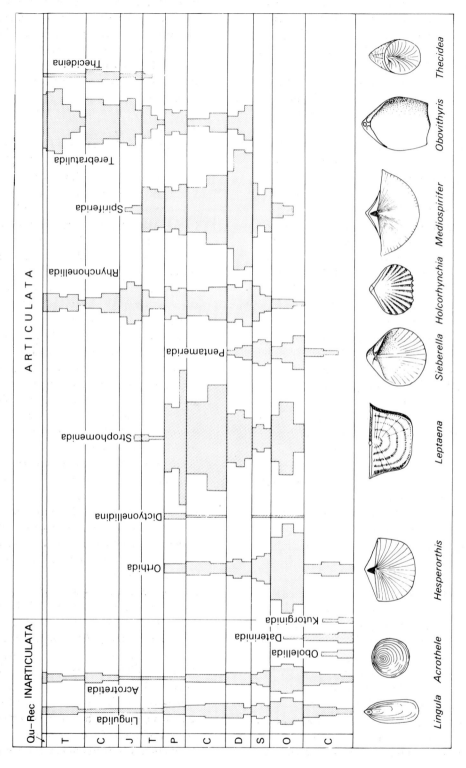

Figure 7.13　Time ranges and abundance of brachiopod orders, with representative genera illustrated. (Based on '*Treatise*' Part (H))

SUBORDER 1. ORTHIDINA (L. Cam.–Perm.): Normally impunctate, finely ribbed shells (other than in SUPERFAMILY ENTELETACEA in which the shells are punctate). e.g. *Billingsella, Orthis, Heterorthis, Valcourea, Visbyella, Schizophoria, Dalmanella.*

SUBORDER 2. CLITAMBONITIDINA (Ord.): Wide-hinged impunctate pseudopunctate shells, having pronounced pseudodeltidium. e.g. *Clitambonites.*

SUBORDER 3. TRIPLESIIDINA (L. Ord. – U. Sil.): Biconvex impunctate shells with long forked cardinal process, e.g. *Triplesia.*

ORDER UNCERTAIN

SUBORDER DICTYONELLIDINA (M.Ord.–Perm.): Small group of unusual brachiopods in which umbo of pedicle valve is marked with peculiar triangular umbonal plate, e.g. *Eichwaldia.*

ORDER 2. STROPHOMENIDA (L. Ord. – L. Jur.): The largest of all brachiopod orders, containing the most unusual forms. Shells strophic pseudopunctate, plano-convex to concavo-convex or rarely biconvex, with long hinge lines. Pedicle foramen usually closed. Shells sometimes cemented, with or without tubular spines. Cardinal process often bilobed. Suborders include:

SUBORDER 1. STROPHOMENIDINA (Ord.–Trias.): The largest group, with simple or reduced teeth, often supplemented or replaced by denticles, e.g. *Eoplectodonta, Plectambonites, Strophomena, Stropheodonta, Rafinesquina, Davidsonia.*

SUBORDER 2. CHONETIDINA (L. Sil – L. Jur.): Small to large plano-convex shells, with prominent spines along the pedicle valve interarea, e.g. *Strophochonetes, Chonetes, Cadomella.*

SUBORDER 3. PRODUCTIDINA (L. Dev. – U. Perm.): Shell usually plano-convex, usually with deep mantle body cavity, occasionally conical and bizarre in form. Valves often geniculate, usually prolonged into a 'trail'; delthyrium and notothyrium frequently closed. Shell usually with tubular spines, e.g. *Productus, Gigantoproductus, Gemmellaroia, Waagenoconcha, Richthofenia, Dictyoclostus, Eomarginifera.*

SUBORDER 4. OLDHAMIDINA (U. Carb. – U. Trias.): Irregularly shaped brachiopods with conical pedicle valve and highly lobed brachial valve recessed within it and subparallel with its axis; no interareas, hinge or pedicle opening, e.g. *Lyttonia.*

ORDER 3. PENTAMERIDA (M. Cam. – U. Dev.): Biconvex, impunctate, usually non-strophic shells with spondylium in pedicle valve and sometimes a pair of vertical plates (cruralium) opposite the spondylium, the whole enclosing the muscle cavity. Suborders include:

SUBORDER 1. SYNTROPHIIDINA (M. Cam. – L. Dev.): Pentamerides with prominent fold and sulcus and open delthyrium; cruralium rare, e.g. *Porambonites.*

SUBORDER 2. PENTAMERIDINA (M. Ord. – U. Dev.): Large strongly biconvex shells, usually lacking fold and sulcus; cruralium well developed, e.g. *Pentamerus, Stricklandia, Sieberella.*

ORDER 4. RHYNCHONELLIDA (M. Ord.–Rec.): Shell normally non-strophic, usually with beak and functional pedicle; delthyrium partially closed; crura in brachial view. Usually with coarse ribs meeting along a zigzag commissure, and often with a pronounced fold and sulcus. Spirolophous. Muscle platforms developed only in SUPERFAMILY STENOCISMATACEA. Impunctate except for monogeneric SUPERFAMILY RHYNCHOPORACEA, e.g. *Rhynchonella, Pugnax, Camarotoechia, Gibbirhynchia, Uncinulus, Tetrarhynchia, Wilsonia.*

ORDER 5. SPIRIFERIDA (M. Ord–Jur.): Shells biconvex, strophic or non-strophic, with spiral brachidium, with or without jugum, punctate or impunctate, delthyrium open or closed. Suborders include:

SUBORDER 1. ATRYPIDINA (M. Ord. – U. Dev.): Impunctate biconvex spiriferides with short hinge line and narrow or absent interareas. Open delthyrium or umbonal pedicle foramen. Spiralia may be dorsally or laterally directed, e.g. *Atrypa, Catazyga, Glassia, Dayia.*

SUBORDER 2. RETZIACEA (U. Sil.–Perm.): Rhynchonelliform shells with laterally directed spiralia, e.g. *Rhynchospirina.*

SUBORDER 3. ATHYRIDIDINA (U. Ord.–Jur.):

Smooth impunctate shells with narrow hinge line; beak often truncated by umbonal foramen; laterally directed spiralia joined by jugum of peculiar and complex form, e.g. *Athyris, Composita, Meristella, Tetractinella*.

SUBORDER 4. SPIRIFERIDINA (L. Sil. – L. Jur.): Spiriferides with a long hinge line; typically ribbed; interareas well developed, especially that of the pedicle valve. Delthyrium open or constricted. Spiralia laterally directed. Mostly impunctate, but SUPERFAMILY SPIRIFERINACEA mainly punctate. e.g. *Spirifer, Cyrtia, Liriplica, Reticularia, Punctospirifer*.

ORDER 6. TEREBRATULIDA (L. Dev.–Rec.): Biconvex articulates with short non-strophic hinge, and with functional pedicle emerging through umbonal foramen. Delthyrium generally closed by delthyrial plates. Shell punctate. Brachidium loop-like, usually plectolophous. Suborders include:

SUBORDER 1. CENTRONELLIDINA (L. Dev.–Perm.): Archaic terebratulides with primitive loop of oval form, e.g. *Centronella, Amphigenia, Stringocephalus*.

SUBORDER 2. TEREBRATULIDINA (L. Dev.–Rec.): Short loop; median septum usually absent; internal spicules developed, e.g. *Dielasma, Terebratula, Pygope, Plectothyris*.

SUBORDER 3. TEREBRATELLIDINA (Trias.–Rec.): Long loop with cardinalia and median septum, e.g. *Magellania, Zeilleria, Digonella, Terebratella*.

ORDER UNCERTAIN

SUBORDER THECIDEIDINA (Trias.–Rec.): Small thick-shelled articulates, usually cemented and without pedicle, capable of opening very wide. Ptycholophe recessed deep within granular interior. Variously considered as of strophomenide, spiriferide or terebratulide affinities, e.g. *Thecidea, Lacazella*.

EVOLUTIONARY HISTORY

The earliest known brachiopods are of Lower Cambrian age. Some supposed Precambrian 'brachiopods' of teardrop shape with concentric rings have turned out to be inorganic, resulting from small scale mudflows. Brachiopods are the earliest lophophorates in the fossil record and on their earliest appearance are quite diverse. The Lower Cambrian brachiopod fauna contains representatives of all inarticulates, as well as kutorginides and the articulate Superfamily Billingsellacea (Suborder Orthidina). Calcite was used in the latter two and in the obolellides; all other groups had phosphatic shells. Whilst the Lower Cambrian brachiopod fauna was limited, the later Cambrian brachiopods were substantially more diverse and included the earliest pentamerides as well as some orthides. This late Cambrian expansion was a precursor to the great Ordovician articulate radiations, which were paralleled on a minor scale by a comparable inarticulate radiation. In the earliest Ordovician the orthides and pentamerides expanded and diversified, and there arose the earliest strophomenides (plectambonitaceans), also rhynchonellides and spiriferides, whilst the last of the minor inarticulate groups vanished. The orthides were undoubtedly the root stock of the articulates, and the strophomenides and spiriferides were evidently derived from them. But rhynchonellides could have originated through pentamerides, themselves of orthide ancestry.

The Silurian saw some reduction of the diversity of inarticulates, orthides and strophomenides, though rhynchonellides and spiriferides continued to expand. In the uppermost Silurian the last of the important orders, the Terebratulida, was added, possibly derived from an atrypide ancestor. During the Devonian the spiriferides reached their peak, and there seems to have been a general brachiopod expansion. But by the late Devonian there was a general decline and widespread extinction, and the pentamerides died out. In the later Palaeozoic there was a new burst of evolution, particularly in the Productidina (a suborder of the Strophomenida) which include many large and peculiar though highly functional genera. These first appeared in the Lower Devonian and extended through the Permian. Their great success was partially because they successfully colonised a habitat previously closed to brachiopods: the quasi-infaunal or almost buried (and hence protected) mode of life. Amongst these are included *Gigantoproductus*, the largest of all brachiopods. The Permian was a time of great evolutionary plasticity in strophomenides, culminating in the origin of some very peculiar Permian groups – Superfamily

Lyttoniacea (Suborder Oldhamidina) and Super-family Richthofeniidae (Suborder Productidina) amongst others – which colonised reef environments and which, in spite of their aberrant and quite unbrachiopod-like appearance, were highly adapted to particular environments. Some of these, interpreted functionally by Rudwick and Cowen (1968), involved very promising adaptations, such as rhythmic flow feeding supplanting ciliary pumping.

But with the severe extinctions of late Permian time, in which all fossil groups were affected, even these new devices did not help the strophomenides to survive. Only a very few (such as the family Thecospiridae) made the transition to the Mesozoic. The rest became extinct, as did the majority of the Palaeozoic groups. By the end of the Middle Jurassic the last spiriferides had gone, leaving only rhynchonellides and terebratulides, which together with the small suborder Thecideidina make up the groups that have survived until Recent times.

Thecideidines (e.g. *Lacazella*) are a small group of calcareous-shelled ptycholophous articulates whose affinities are disputed. On the basis of shell structure they seem to be allied to spiriferides, but the lophophore structure suggests a strophomenide ancestry. They remain an interesting enigma.

Inarticulates seem to have changed little since their decline at the end of the Ordovician, though their habitats have become less diverse if the known presence of deep water lingulides in the Silurian is a reliable guide.

The character of brachiopod faunas thus seems to have changed greatly throughout the Phanerozoic, and even a superficial analysis of any fauna will normally enable it to be referred to a particular system.

Competition with bivalves, especially those with siphons, may have been an important factor in limiting the post-Palaeozoic expansion of the brachiopods. Yet as Rudwick (1970) has shown, this influence has been less direct than it may seem at first sight, for the infaunal habitats so successfully exploited by the siphonate bivalves are not those in which brachiopods were ever very successful, other than the quasi-infaunal Productidina. Perhaps it was only when the latter were extinct that it was possible for the bivalves to exploit the infaunal habitat to a higher degree than before, and brachiopods have been of only limited importance since.

ECOLOGY AND DISTRIBUTION

Ecology of individual species

Pedicle attachment

Modern brachiopods are normally attached to rocks and dead shells on the sea floor, or sometimes in soft muds, by means of the pedicle. Whether or not extinct pedicle-bearing brachiopods could anchor in soft substrates has not been determined. Some of them probably could not, since there is a tendency for brachiopods found *in situ* to grow in nests or clusters whose early members anchored themselves to a chance dead shell and later by their own dead shells provided a settling place for later generations. In such nests at least three annual broods have been found (Hallam 1962), containing brachiopods of the same or different species. The nests came to a sudden end, probably by being overwhelmed by sediment which preserved them.

Spiriferides, which sometimes grew in similar clusters with only a small basal attachment area, are sometimes found to be asymmetrical and distorted, probably through crowding by adjacent individuals (Ager & Riggs 1964).

Functional pedicles seem to have been present in most rhynchonellides and terebratulides, most orthides and spiriferides, some pentamerides and a few strophomenides.

Free-lying

Most brachiopods with a closed pedicle opening (e.g. many strophomenides, some spiriferides, and orthides) must have lain on the sea floor. Where as in some cases there is a very small pedicle opening (e.g. *Cyrtia*, Fig. 7.7d, e), a slender cord-like pedicle emerging from it could have had a 'tethering' function but would not have been supportive. Free-lying brachiopods are unknown in Recent faunas but seem to have been common in the Palaeozoic. Any sediment collecting round the commissure could have easily been cleared by the rapid clapping of the valves, and in some cases such clapping might, as in the bivalve *Pecten*, have allowed a limited amount of swimming movement. Brachiopods that were free-lying in the adult had a functional pedicle during their early development.

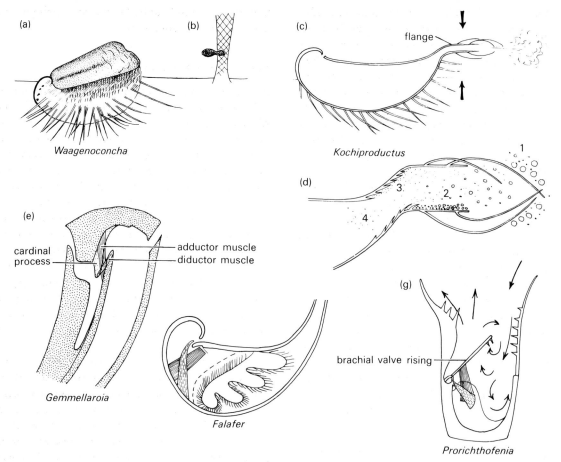

Figure 7.14 Ecology and functional morphology of brachiopods: (a) *Waagenoconcha*, a Permian spiny productide in life attitude (x 0·8); (b) juvenile *Waagenoconcha* attached by spines to a bryozoan ((a) and (b) from Grant, 1966). (c) *Kochiproductus*, a Carboniferous spiny productide of quasi-infaunal habit shown cleaning sediment from flange by clapping valves together (x 0·4); (d) cross section through flange of *Kochiproductus* showing various sediment baffles: 1 – spines excluding large particles; 2 – flange acting as settling table for intermediate particles; 3 – taleolae (endospines) trapping small particles; 4 – only food particles get through ((c) and (d) redrawn from Sheills, 1968); (e) *Gemmellaroia* (Permian) showing how brachial valve acts as a tight plug (x 3) (based on Rudwick and Cowen 1968); (f) *Falafer*, a Permian productide showing ptycholophe (x 10) (based on Grant, 1972); (g) *Prorichtho-fenia* (= *Cyclacantharia*), Rudwick's conception of currents generating by a rising brachial valve flushing out the interior (x 2) (redrawn from Rudwick 1961).

Cementation

The modern calcareous inarticulate *Crania* cements its pedicle valve to rocks, the commissure of the brachial valve being shaped to fit it. Cementation is known in some fossil articulates though it never seems to have been particularly common. It is best known in strophomenides, where a scar or cicatrix often marks the umbo of the pedicle valve suggesting attachment in the juvenile stages, though the adult broke away and was free-lying.

Quasi-infaunal

Though the infaunal niche so well exploited by *Lingula* was never fully colonised by articulates, many productides and perhaps some other strophomenides seem to have spent their adult lives partially buried in sediment and with a sediment cover over the concave brachial valve. Adult productides (Fig. 7.14a–d) were anchored by the strong spines that projected from the pedicle valve. These spines, formed at the margin of the valve, were originally hollow, later becoming filled with calcite as the mantle withdrew. Since the margins of both valves are sharply flexed upwards and drawn out into a narrow crescentic flange or **trail,** only this thin crescent would show above the general level of the sediment, and the brachiopod would thus be well camouflaged. *Waagenoconcha* (Fig. 7.14a, b) is a productide with delicate spines on the brachial valve which probably prevented settled sediment from being winnowed away. During the ontogeny of *Waagenoconcha* the settling larva attached itself to cylindrical bryozoans in the same environment, which it embraced with one or more pairs of curving cardinal spines and to which it remained attached for some time. When the shell became too large and heavy it broke off and, landing on the sea floor, grew long stout spines which it used for anchorage in the mud, eventually becoming buried as far as the trail (Grant 1966).

In the case of *Kochiproductus* (Fig. 7.14c, d) the trail forms a wide flat horizontal flange which must have lain flat on the sea floor. Experiments on models with a flow tank (Sheills 1968) showed that the wide flange acted as a settling table for extraneous and inedible particles, whilst smaller particles that passed through the first baffle were trapped by taleolae extended like a row of stakes at the entrance to the mantle cavity.

Brachiopod assemblages and 'community' ecology

Fossil brachiopods tend to be found in recurrent assemblages, often composed of several species. In these assemblages other invertebrates also occur, but the brachiopods are usually dominant. These assemblages have been greatly used in recent years in attempts to understand the structure of ancient (mainly epifaunal) communities and the controls operating upon them.

The recurrent assemblages characterise particular environments and in life must have related to such parameters as depth, temperature, salinity and substrate. Most palaeoecologists working with recurrent assemblages have tried not only to define them in terms of their composition and the relative abundance of their faunal elements, but also to determine, as far as possible, the nature of the controls.

So much recent work has been done on this most vital of palaeoecological fields that the bulk of literature is considerable. Only a few examples are here selected for discussion.

Silurian palaeocommunities

An early classic study of brachiopod-dominated palaeocommunities was that of Ziegler (1965) and Ziegler *et al.* (1968), based on recurrent assemblages of Upper Llandovery age in the Welsh Borderland (Fig. 7.15). During the earlier Llandovery a shoreline ran eastwards from Haverfordwest and, curving northwards near the town of Llandovery, extended due northwards. On the shelf sea between the shore and the graptolite shale area to the west lived a rich fauna. By the time the shoreline had retreated to the east of the Malvern Hills in the Upper Llandovery the shelf was much broader and shelly faunas were differentiated over it. Five distinct though gradational communities dominated by brachiopods (though with subsidiary bivalves, gastropods, corals and trilobites) lay in distinct concentric belts parallel with the shore.

Nearest the shore lay a *Lingula* community, with abundant small rhynchonellides (probably *Stegerhynchus* or *Rostricellula*) and *Lingula*. The succeeding *Eocoelia* community had similar rhynchonellides (*Eocoelia*) but a more diverse range of other brachiopods. In the further-out *Pentamerus* community the eponymous genus dominated all others, often reaching a large size; specimens are sometimes found in life position. The large *Costistricklandia* was the most abundant brachiopod in the penultimate community, though smaller brachiopods were also important. And in the outer shelf *Clorinda* community there were at least a dozen genera of relatively small brachiopods, so that it was the most diverse community of all. Beyond this lay an area of graptolitic muds accumulating in a deeper water environment. Similar palaeocommunities have been found in rocks of the same age in other parts of the world, and in some cases they seem to be equivalent to those defined by Ziegler *et al.*

The examples of Ziegler *et al.* show a relatively straightforward pattern of distribution in narrow bands which were largely independent of substrate and parallel with the shore. This seems to suggest that the distribution pattern is in some way related to depth of water. Depth alone, however, has been generally regarded as a rather elusive factor whose

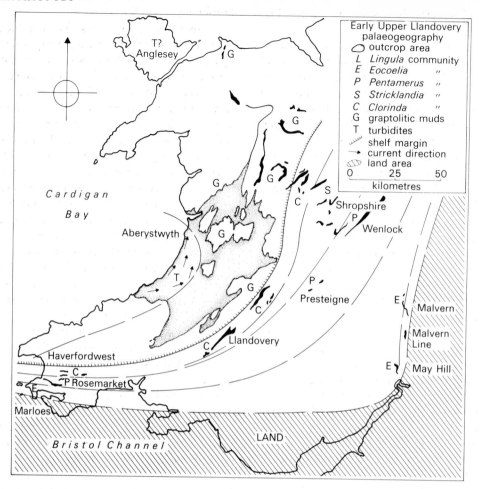

Figure 7.15 Palaeogeography and brachiopod communities in Wales in the early Upper Llandovery.

(Redrawn from Ziegler, Cocks and Bambach, 1968)

influence on sediments or animals is indirect. It is not depth as such that is a primary control, but certain important factors that vary with depth, including pressure, water turbulence, salinity, substrate and food supply. Temperature, however, which decreases in deeper water, according to Boucot (1975) may be the most important of all these variables.

Where other Silurian brachiopod palaeocommunities have been described, however, such as the Ludlovian assemblages of Lawson (1975), it is less easy to assess the controls of distribution. Lawson, suggesting that substrate may be more important than depth in some cases and noting that brachiopods belonging to different associations may have varying tolerance ranges, rejected attempts to impose

the simple pattern exhibited by the Upper Llandovery palaeocommunities on other kinds of assemblage.

Devonian brachiopod assemblages

A primary task for palaeoecology is to try to integrate the existing bulk of data on Palaeozoic assemblages and to establish how they have evolved in fluctuating environments through time. This work, following that of Bretsky (1969) (see Bibliography following Ch. 1) and Boucot (1971), promises well, but the major hazard encountered is in determining the environmental equivalence of disjunct associations which replaced each other in time but which may have few characters in common with their precursors.

The major work of Boucot (1975) has shed much light on this topic. He undertook a large scale study of Siluro-Devonian brachiopod distribution, primarily to assess the size of breeding populations (in terms of the areas they occupied) and so to determine how rates of evolution and extinction varied in populations of different sizes. He was able to confirm the predictions of population dynamicists that evolution would be rapid in smaller populations and postulated a complex series of biological and physical controls over extinction events.

In establishing a 'standard community framework' for this study Boucot redefined all known Ordovician to Devonian palaeocommunities in terms of **benthic assemblages (BA).** A benthic assemblage is 'a group of communities occurring repeatedly in the same position relative to a shoreline'. According to Boucot, these were probably controlled by water temperature, decreasing with depth. Boucot thus regarded the *Lingula* through *Clorinda* 'communities' of Ziegler *et al.* (p. 131) as BAs 1–5, zoned according to distance from shoreline. He also defined a sixth BA lying seawards of BA5. He was able to relate all known Ordovician to Devonian assemblages to a system of six BAs, though the components might vary laterally depending on water turbulence. Separate quiet and rough water assemblages could be distinguished in most of these BAs.

The composition of the BAs changed through time, often abruptly. Where there were reefs, with their largely endemic and rapidly evolving populations, they were usually associated with BAs 2–3. Assigning a given brachiopod association to a particular BA may be difficult where there has been an abrupt faunal replacement so that there is no continuity with former associations. However, where neighbouring associations have not changed then the assemblage in question may be 'bracketed' between them and so referred to its correct BA. If an unknown association lies between BA3 and BA5 it is probably BA4, even though it is disjunct.

Boucot's model of benthic associations is so far only a provisional step towards the integration of 'palaeo-community' data and the interpretation of this in environmental and evolutionary terms, but it provides a necessary framework for future developments of community studies in Palaeozoic brachiopods.

Permian reef associations

In western Texas, South-East Asia and Sicily there are immense reefs of Permian age composed mainly of algae, bryozoans and sponges. The Capitan reef complex of western Texas is the best known of these. It marks a late stage in reef development, forming a giant barrier reef round the margins of deep water basins which it encircled; these had only a narrow outlet to the sea on the west. Behind the reef lay large salt flats: evaporating pans in which dolomite, gypsum and anhydrite were deposited in sequence away from the reef as the water became progressively more hypersaline. These were replenished by sea water washing over the top of the reef. These reefs are not especially fossiliferous and contain only localised developments of brachiopods, in pockets of low diversity. But prior to the development of the great algal reefs numerous smaller reefs or bioherms grew in somewhat deeper water. These patch reefs, best exposed in the Glass Mountains, grew up to 80 m high though were usually smaller and are abundantly fossiliferous, containing some of the most highly modified brachiopods ever to have lived (Grant 1971).

Most of the Permian biohermal brachiopods are of relatively normal form; the majority were pedicle-attached and able to exploit several different habitats within and peripheral to the bioherm. Other kinds, especially some of the productides, were definitely 'antireefal' and lived in the flat sediment between the reefs. But others again were confined to the reef, often being of conical form with long tubular spines forming an interlacing meshwork which contributed to the reef fabric. These rather coral-like brachiopods were all modified strophomenides in which the conical pedicle valve was cemented in the early stages.

Some bioherms were composed largely of Richthofeniidae (Suborder Productidina) (e.g. *Cyclacantharia*) (Fig. 7.14g). These have long pointed pedicle valves supported by tubular spines. The brachial valve is reduced to a thin flat horizontal plate recessed well within the pedicle valve; it could open to 80° or so and has a protective grille of spines or meshwork above the opened valve. Rudwick (1961) made a working model of this and showed that the brachiopod could have generated strong currents, bringing in food and flushing out the mantle cavity simply by rapidly flapping the brachial valve using the adductor and diductor muscles alternately.

Though this interpretation has been criticised by Grant (1972) – on the grounds that the known structure of the large ptycholophe in other Permian productides (Fig. 7.14f) would make this unlikely, and that the brachiopod could have functioned perfectly well without such a valve-flapping mechanism – the possibility still remains that the prorichtofeniides could have used flapping valves to supplement or replace ciliary pumping as a means of collecting food.

In *Gemellaroia* (Fig. 7.14e), known from Sicily, the brachial valve is modified as a thick tight-fitting plug which, like a cork in a bottle, could only have opened by moving vertically upwards. Unlikely though it may seem, the two sets of muscles, arranged vertically and at nearly 180° to each other, would allow this possibility. *Gemellaroia* may have normally been exposed at low tide, and such a hermetic seal could have been an adaptation that prevented dessication.

Whilst the functional morphology of some of these peculiar forms may be open to question, surely Rudwick is correct in saying that these adaptations 'bear witness to a degree of plasticity unparalleled at any other time in brachiopod evolution'.

Mesozoic brachiopod associations

In many instances there seems to be a broad and general correlation of brachiopod shell form with environments. It was recognised long ago, for example, that Lower Palaeozoic orthides with coarse ribs preferred arenaceous environments, and there is a general tendency in the Mesozoic for similar anatomical features to occur in unrelated stocks in similar environments. Thus any assemblage from very shallow waters generally includes rhynchonellides with coarse ribs and large pedicle openings supporting stout pedicles, as well as terebratulides with sharp commissure folds. A quite different assemblage was to be found in 'perireefal' environments, with large asymmetrical rhynchonellides and terebratulides with elongated beaks. Some peculiar brachiopods, e.g. the perforate *Pygope,* were confined to calm deep seas, whilst small thin-shelled rhynchonellides were evidently epiplanktonic.

Such differences, though oversimplified and hard to quantify, could prove to be a useful tool in brachiopod palaeoecology, especially if the differences can be interpreted in functional terms.

FAUNAL PROVINCES

The global distribution of brachiopods is most fully documented in the Lower Palaeozoic, which is in any case that part of the geological column where they are of greatest stratigraphical value and where they show an instructive parallel with contemporaneous trilobites. Williams (1973, 1976), working on Ordovician brachiopods, and Boucot (1975), working on Siluro-Devonian faunas, have summarised much of the information available, though there are other views.

Williams used cluster analysis as a method of elucidating faunal distribution in Ordovician brachiopods. This demonstrated that there were 'several' generic assemblages of statistically related clusters largely independent of one another in time and space. Such assemblages represented the fossilised remains of original palaeocommunities, inhabiting different environments but collectively making up distinct provinces.

Provinciality in the Ordovician was well marked from the Arenig to about the mid Caradoc, thereafter decreasing until by the mid Ashgill the fauna was essentially cosmopolitan. During the earlier Ordovician there were two distinct 'realms': the American and the European. Within each of these realms there were a number of distinct and stable 'provinces' between which there was only limited exchange. In the American realm there were two provinces at a maximum; in the European there were four or, at one transitory period, five. These provinces were probably partially controlled by temperature since the presence of much carbonate sediment, associated particularly with the provinces of the American realm and with the 'Baltic' province of the European realm, would seem to indicate warm water.

The really important break in generic composition is found along the line separating the American and European realms, which follows the line of the Appalachian, southern Scottish and Scandinavian mountain belts. The distinctness of the two realms was recognised before this fold belt was seen in plate-tectonic terms, but the issue was more clearly resolved when it was later appreciated that the main realms (just as with contemporaneous *Selenopeltis* and *Bathyurid* provinces in trilobites) had been separated by the wide Iapetus ocean which had slowly closed and allowed the faunas of the two sides to merge prior to the eventual collision and welding of the plates that bore them.

By the later Ordovician the world fauna of brachiopods was more-or-less ubiquitous, but during the last Ordovician stage (the Hirnantian) there was a period of widespread glaciation centred on a south pole located in North Africa. Possibly due to this very cold period the number of brachiopod genera was greatly reduced.

Silurian brachiopod faunas recovering from this decimation became cosmopolitan, though two new provinces arose: one characterised by *Clarkeia* which became established in South America, another with *Atrypella* which arose in the USSR, Greenland and western North America. During the Devonian there was further provincial differentiation, resulting eventually in five faunal provinces which have been fully documented by Boucot (1975).

The striking parallels between trilobite and brachiopod distribution in the Lower Palaeozoic shows how palaeobiogeography can be a powerful tool in timetabling major global events.

STRATIGRAPHICAL USE

Many brachiopod genera and species are relatively long-ranged and thus of limited value in correlation. This is particularly so with Mesozoic brachiopods, which though sometimes useful on a local scale are of less stratigraphical applicability over wider areas.

In the Palaeozoic, however, brachiopods have proved of much greater value; together with trilobites they are the primary stratigraphically useful fossils in the shallow water facies of the Ordovician.

Brachiopod assemblages are, of course, very much controlled by facies and, on a global scale, by realms and provinces, but correlations at the stage level based upon concurrent brachiopod ranges within provinces can be very good. Thus stratigraphy and structure in the Caradocian rocks of the Girvan region of Ayrshire, Scotland, have been correlated by reference to the standard unbroken brachiopod/trilobite-bearing series of Virginia, Alabama and Tennessee, both areas being part of the Scoto-Appalachian province of the American realm.

So successful has this stratigraphical work been that a case can be presented for using 'hybrid' stratigraphical tables in areas such as the British Isles, where the incursion of 'American' forms at a particular period in time (through plate suturing) allows the recognition of zones and stages originally defined in the United States.

There is now a good correlation between the stages of the Ordovician and Silurian shelly facies and concurrent graptolite zones; indeed there are seven Caradocian stages in England and Wales for only two graptolite zones, indicative of a fine refinement of correlation. The American shelly facies stages are less finely drawn but still good. However, such correlations are still more firmly based in some parts of the world than in others.

In the Carboniferous, brachiopods have been of some stratigraphical value when combined with corals in the zones erected for the Lower Carboniferous in England by Vaughan (1905) and Garwood 1913) (see Bibliography following Ch. 5). Though still valid, these zones are now supplemented by stratigraphical work using microfossils, which will probably give a more precise correlation.

BIBLIOGRAPHY

Books, treatises, symposia

Moore, R. D. (ed.) 1965. '*Treatise*' Part (H). Brachiopods. (2 volumes)

Rudwick, M. J. S. 1970 *Living and fossil brachiopods*. London: Hutchinson. (Invaluable treatment, with a functional bias)

Individual references

Ager, D. and E. A. Riggs 1964. The internal anatomy, growth and asymmetry of a Devonian spiriferid. *J.*

Paleont. **33,** 749–60. (Crowding results in asymmetry)

Boucot, A. J. 1971. Practical taxonomy, zoogeography, paleoecology, paleogeography, and stratigraphy for Silurian–Devonian brachiopods. *Proc. North Amer. Paleont. Convention, Chicago, 1969* (F), 566–611. (Early study; base for Boucot 1975)

Boucot, A. J. 1975. *Evolution and extinction rate controls.* Amsterdam. Elsevier. (Develops concept of benthic assemblages)

Craig, G. Y. 1951. A comparative study of the ecology and palaeoecology of *Lingula*. *Trans Edin. Geol. Soc.* **15,** 110–20. (Fossil *Lingula* in life position)

Davidson, T. 1851. *A monograph of the British fossil Brachiopoda*. Palaeont. Soc. monogr. (Standard, superbly illustrated reference in five volumes: taxonomy)

Grant, R. E. 1966a. A Permian productoid brachiopod: life history. *Science* **152**, 660–2. (Change in life habits during ontogeny)

Grant, R. E. 1966b. Spine arrangement and life habits of the productoid brachiopod *Waagenoconcha*. *J. Paleont*. **40**, 1063–9. (As above)

Grant, R. E. 1971. Brachiopods in the Permian reef environment of west Texas. *Proc. North Amer. Paleont. Convention, Chicago, 1969* (J), 1444–81. (Bioherms with unusual specialised faunas)

Grant, R. E. 1972. The lophophore and feeding mechanisms of the Productidina (Brachiopoda). *J. Paleont*. **46**, 213–48. (First description of productid lophophores and functional interpretation)

Hallam, A. 1962. Brachiopod life assemblages from the Marlstone Rock bed of Leicestershire. *Palaeontology* **4**, 653–9. (Successive annual broods)

Lawson, J. D. 1975. Ludlow benthonic assemblages. *Palaeontology* **18**, 509–25. (Distinction of four brachiopod-dominated assemblages and their controls)

Muir-Wood, H. 1955. *A history of the classification of the phylum Brachiopoda. Brit. Mus. Nat. Hist.* 1–124. (First attempt to erect a modern classification)

Rudwick, M. J. S. 1959. The growth and form of brachiopod shells. *Geol. Mag*. **96**, 1–24. (Growth vectors in rectimarginate and plicated shells)

Rudwick, M. J. S. 1961. The feeding mechanism of the Permian brachiopod *Prorichthofenia*. *Palaeontology* **3**, 450–71. (Working model shows operation of flapping valves)

Rudwick, M. J. S. 1964. The function of zig-zag deflections in brachiopods. *Palaeontology* **7**, 135–71. (Zigzags interpreted as protective)

Rudwick, M. J. S. 1965a. Adaptive homeomorphy in the brachiopods *Tetractinella* Bittner and *Cheirothyris* Rollier. *Paläont. Z.* **39**, 134–46. (An extreme case of homeomorphy interpreted functionally)

Rudwick, M. J. S. 1965b. Sensory spines in the Jurassic brachiopod *Acanthothiris*. *Palaeontology* **8**, 604–17. (Spines as early warning sensors)

Rudwick, M. J. S. and R. Cowen 1968. The functional morphology of some aberrant strophomenid brachiopods from the Permian of Sicily. *Bull. Soc. Paleont. Ital*. **6**, 113–76. (Richthofeniids and others replace ciliary pumping by valve movement to generate currents)

Sheills, K. A. C. 1968. *Kochiproductus coronus* n. sp. from the Scottish Visean, and a possible mechanical advantage of its flange structure. *Trans Roy. Soc. Edin*. **67**, 477–507 (Use of flow table in determining particle settling on flange)

West, R. R. 1976. Comparison of seven lingulid communities. In *Structure and classification of palaeocommunities* R. W. Scott and R. R. West (eds), 171–92. Pennsylvania Dowden, Hutchinson & Ross. (Lingulids were adapted to different environments)

Williams, A. 1968. A history of skeletal secretion in brachiopods. *Lethaia* **1**, 268–87. (Excellent general review)

Williams, A. 1973. Distribution of brachiopod assemblages in relation to Ordovician palaeogeography. In *Organisms and continents through time*, N. F. Hughes (ed.), 241–69. Spec. pap. in palaeont., no. 12. (Methodology, statistics; development and changes in faunal provinces)

Williams, A. 1976. Plate tectonics and biofacies evolution as factors in Ordovician correlation. In *The Ordovician System*, M. G. Bassett (ed.), 29–66. Proceedings of a Palaeont. Ass. symposium, Birmingham, September 1974. Cardiff. Wales: Univ. of Wales Press. (Faunal provinces, and stratigraphical problems raised by plate closure and faunal migration)

Ziegler, A. M. 1965. Silurian marine communities and their environmental significance. *Nature* **207**, 270–2. (Assemblage distribution, with maps)

Ziegler, A. M., L. R. M. Cocks and R. K. Bambach 1968. The composition and structure of Lower Silurian marine communities. *Lethaia* **1**, 1–27. (Classic well-illustrated work)

8 Molluscs

The Mollusca are one of the most diverse of all invertebrate phyla and include a whole range of animals, living and fossil, which at first sight seem to be so different as to be unrelated. There are the curious plated chitons (amphineurans), the slugs and snails (gastropods), tooth-shells (scaphopods), bivalves, their possible progenitors (the extinct rostroconchs) and finally the most complex and successful of all molluscs: the cephalopods, which include the modern squids, cuttlefish, octopuses and pearly nautilus and the fossil ammonoids and belemnites. Molluscs are mainly a marine phylum, and only a few bivalves and gastropods have been successful in fresh water. One group of gastropods only, the pulmonates, was ever able to make the transition from sea to land. The diversity of different kinds of molluscs is extreme, yet they are all united by common ground plan, and the same basic structures are to be found in all molluscs however much they have become differentiated.

FUNDAMENTAL ORGANISATION

It is probably simplest to consider the fundamental organisation of molluscs with reference to a hypothetical **archimollusc** in which all the basic features of molluscan structure are present though not particularly specialised in any one direction (Fig. 8.1). This hypothetical molluscan ancestor has close similarities to the chitons and an even closer resemblance to the modern limpet-like primitive mollusc *Neopilina* (Fig. 8.2), which was unknown until 1953 when many specimens were dredged from the deep sea during the Danish 'Galathea' Expedition. *Neopilina* is a member of the molluscan Class Monoplacophora, which has fossil representatives known from Cambrian rocks alone. The discovery of this 'living fossil' has been of the greatest interest, especially since, contrary to expectations, it shows traces of segmentation.

The structure of the archimollusc is quite simple. It has a cap-like **shell** secreted by a layer of tissue known as the **mantle.** Below this is the body with a mouth at one end and anus at the other, the latter discharging into a posterior space, the **mantle cavity,** in which also reside the gills.

In other molluscs this structure is modified in different ways. Most molluscs have an external shell of calcium carbonate, though this has been lost in various gastropods, particularly the marine nudibranchs and the terrestrial slugs. In squids and cuttlefish, as well as in the extinct belemnites, the shell has become modified as an internal skeleton, and in the octopuses it has been entirely lost. Where the shell is present it is always secreted by the mantle, which directly underlies the shell. Throughout growth calcium carbonate is added to the edge of the shell from the mantle. In most molluscs this mantle consists of only a few cell thicknesses, but in the cephalopods it is greatly thickened and supplied with powerful muscles. The mantle cavity is a constant feature of all molluscs and (in different molluscan classes) has been put to various uses. The gills, whose primary function is respiratory exchange, are present in all molluscs except the land gastropods (pulmonates), where the internal surface of the mantle is highly vascular and is modified as a lung. In the bivalves the gills are not only respiratory but also adapted for the gathering of suspended food particles from the water. The mantle cavity has an outlet to the sea through which waste material from the anus discharges, as well as respired water. This outlet has also undergone a differential modification, most markedly in the cephalopods where water, taken into the mantle cavity and squirted out as a jet through a funnel, is the primary means of propulsion.

Above the mantle cavity in all molluscs is the **visceral mass,** which contains the gut, digestive glands and kidneys and the nervous, circulatory and muscular systems. These too in structure,

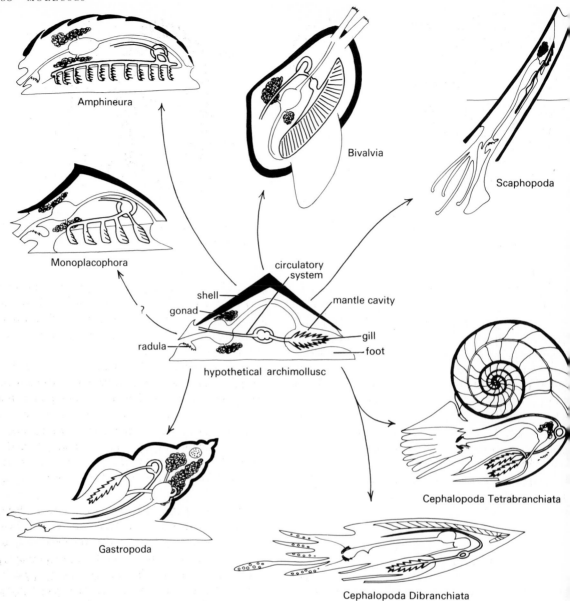

Figure 8.1 Morphology and relationships of molluscan classes with reference to a 'hypothetical archimollusc'

arrangement and physiology have most features in common.

Such then is the fundamental plan of molluscan organisation, but why should molluscs have diverged from it in so many directions? It seems that much of the evolutionary differentiation of molluscs is bound up with their mode of nutrition, and it is interesting to see how their structural differentiation is closely connected with how they feed. The monoplacophorans, chitons and gastropods are all slow-moving molluscs which creep along on a muscular 'foot' They are all provided with a **radula,** a belt of serially arranged teeth within the mouth, which fits them to be herbivores, carnivores or scavengers. Only the gastropods have a well-developed head with sense organs which enable them actively to hunt. The

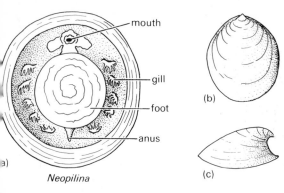

Figure 8.2 *Neopilina* (Monoplacophora): (a) ventral view, showing central foot and serially arranged gills (x 2 approx.) (based on Lemche in '*Treatise*' Part (I)); (b)–(c) dorsal and lateral views (x 1).

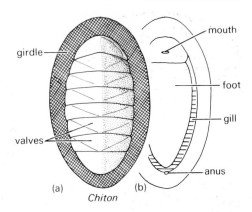

Figure 8.3 *Chiton* (Amphineura): (a) dorsal view; (b) ventral view (x 2).

bivalves are mostly suspension feeders and less commonly deposit feeders. They have no head, no jaws and normally only limited capacity for movement. The cephalopods, on the other hand, are fast-moving hunters; here the radula is used mainly in swallowing, but the cutting up of food is done by the powerful jaws. The head with its highly developed sense organs and well-organised brain, the animal's power of movement, its buoyancy devices and the tentacular apparatus all evolved in accordance with this habit of catching active prey. It is the mode of feeding that has been at the root of molluscan differentiation; this should be remembered when considering the structural plan of the various molluscan groups.

CLASSIFICATION

An outline classification of the main divisions within Phylum Mollusca is as follows:

CLASS 1. MONOPLACOPHORA (Camb.–Rec.): (Fig. 8.2) A group of primitive marine molluscs with univalved limpet-like shells (though unrelated to limpets proper, which are gastropods). They have paired serially-repeated muscles, gills, nephridia (excretory organs) as well as other internal organs. The foot is circular and central and ringed by the mantle cavity in which lie the gills. Monoplacophorans are the only molluscan class with a true internal segmentation, suggesting a zoological relationship with a segmented ancestor, possibly an annelid worm. Fossils such as *Pilina* and *Tryblidium* are known only from the Palaeozoic. *Neopilina*, *Vema* and a few other genera occur today mainly in the deep sea.

CLASS 2. AMPHINEURA (U. Camb.–Rec.): (Fig. 8.3) Marine molluscs having a bilaterally symmetrical shell with seven or eight calcareous plates, but otherwise resembling the archetypal molluscan plan in having an anterior mouth with radula, a posterior mantle cavity with anus and gills, and a ventral foot. Amphineurans are rare fossils but occur scattered throughout the Phanerozoic, e.g. *Chiton*.

CLASS 3. SCAPHOPODA (Ord.–Rec.): (Fig. 8.4a) Marine molluscs with small tapering curving shells open at both ends. The anterior wider end with the mouth is permanently embedded in sediment; the animals feed on small organisms using specially adapted tentacles. The anus is at the upper end and the gills are much reduced. Recent species are more abundant than fossil ones and occur dominantly on the continental slope or shelf. The one detailed study so far made of the ecology of a fossil scaphopod, the Ordovician *Plagioglypta*, suggests similar habits to modern forms, e.g. *Dentalium*.

CLASS 4. BIVALVIA (LAMELLIBRANCHIA or PELECYPODA) (Ord.–Rec.): Bivalved molluscs or 'clams' with no definite head, but having the soft parts enclosed between paired but unequilateral calcareous shells united by a toothed dorsal **hinge.** The valves can be shut by strong internal

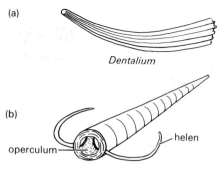

Figure 8.4   (a) *Dentalium*, a scaphopod shell (x 0·75); (b) hyolithid skeleton (x 3). (Based on Runnegar, B. *et al.*, 1975, *Lethaia* **8,** 181–91)

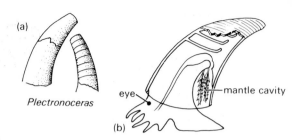

Figure 8.5 *Plectronoceras*, an early cephalopod: (a specimens from the late Cambrian of Shantung, China (× 3); (b) inferred restoration of soft parts with shell partially removed to show mantle cavity with gills. (After Yochelson Flower and Webers, 1973)

musculature, and opened by the outward pressure of a springy **ligament** along the hinge when the muscles relax. The gills are large and modified for filter feeding, and the mantle cavity is connected to the outer environment by **siphons.**

CLASS 5. ROSTROCONCHIA (L. Cam.–Perm.): (Fig. 8.14) Extinct molluscs of bivalve-like appearance but with one or more of the shell layers continuous across the dorsal margin so that a dorsal commissure is lacking. The juvenile shells are univalved and coiled.

CLASS 6. GASTROPODA (Cam.–Rec.): Snails of all kinds – marine, land and fresh water – which creep along on a flattened 'foot'. A true head is present with eyes and other sense organs, and the single univalved shell is coiled, **planispirally** or more often **helically.** The internal organs are twisted by a 180° **torsion** so that the mantle cavity faces anteriorily. Some secondarily shell-less groups that have lost their torsion are known.

CLASS 7. CEPHALOPODA (U. Cam.–Rec.): The most advanced of all molluscs, having external or internal **chambered** shells with the chambers linked by a **siphuncle** and giving buoyancy. They have a properly defined head with elaborate sense organs, and move by jet propulsion of water from the mantle cavity. The modern *Nautilus*, squids and octopuses are here included, as well as ammonites, belemnites and some chambered Cambrian molluscs (Fig. 8.5), extinct relatives of *Nautilus*.

Of these various classes only the bivalves, gastropods and cephalopods are common and important fossils and will be treated in some detail. Some attention i also given to the Rostroconchia, whilst the other three classes are rare fossils and will not be discussed further.

SOME ASPECTS OF SHELL MORPHOLOGY AND GROWTH

Since the shell is the only part of the mollusc normally to be fossilised certain aspects of shell shape should be first considered.

Coiled shell morphology (Fig. 8.6)

Whilst some molluscan shells are simply straight tubes or cones, very many others, whether bivalved or univalved, are coiled. This coiling is most evident in gastropods and cephalopods, but even the individual valves of a bivalve shell are coiled; the 'open' side of the valve which in life encloses the viscera is analogous to the **aperture** of other coiled shells. Leaving aside for the moment the highly modified shells of belemnites and squids, it is not hard to see that most coiled molluscan shells are simply hollow cones rolled up on themselves to a greater or lesser extent. In such rolled-up cones, which grow at the apertural end only, there are very interesting mathematical properties, for the coiling, represented by a line traced along the edge of the shell from the first-formed part (**protoconch**) to the aperture invariably has the form of a **logarithmic** or **equiangular spiral.** D'Arcy Thompson (1917,

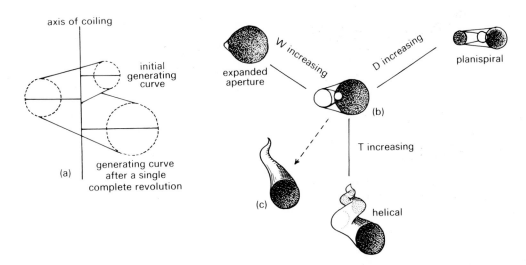

Figure 8.6 Theoretical morphology of the coiled shell (based on Raup 1966): (a) schematic diagram of part of a helically coiled gastropod shell; (b) computer-simulated shell shapes derived by varying various geometrical parameters D, T, and W; (c) a theoretical, but non-functional, shell shape not adopted by any molluscan group.

1961) (see Bibliography following Ch. 1) may be quoted here: 'it is peculiarly characteristic of the spiral shell that it does not alter as it grows; each increment is similar to its predecessor, and the whole, after each spurt of growth is just like what it was before'. Hence the spiral shell can accrete new material at the one end only without changing its shape. In an ideal shell (such as that of *Nautilus*) of radius r, the equation of this spiral shape is $r = a^{\theta}$, where a is a constant and θ is the whole angle through which the spiral has been traced.

The usefulness of this way of growing is undoubtedly the reason for its adoption by so many living organisms – in the shells of brachiopods and molluscs, in the foraminiferides, in the horns of mammals, and even in the eyes of trilobites – where the ability to grow at one end only without changing form has been important.

Nevertheless, in nature coiled shells often only approximate to the ideal mathematical form, and often it is the departures from the hypothetical that are of the greatest functional interest. In ammonites (fossil cephalopods) there may be several minor changes in the spiral angle throughout growth, and not all parts of the ammonites necessarily obey the rules of the logarithmic spiral form. The spacing of

the ribs on the shell is nothing to do with the growth of the spiral and may alter significantly at various periods throughout growth, whilst the aperture is often peculiarly contracted.

Following D'Arcy Thompson's early studies, the theoretical morphology of the coiled shell has attracted considerable attention. The best-known recent studies are those of Raup (1966), who used a computer-based graphical method to produce various kinds of hypothetical coiled shapes and was then able to see how many of these possible types had in fact been adopted in nature. He was able to generate a large number of ideal shapes using only four parameters in the programme; the shell shapes were projected onto an oscilloscope screen. Taking the example of a helically coiled gastropod (Fig. 8.6a), and allowing that its shell is no more than a hollow cone growing at the apertural end and coiling about a vertical axis as it grows, these parameters are:

(a) The shape of the tube in section, otherwise known as the shape of the generating curve (effectively equivalent to the shape of the aperture). In the example illustrated here it is circular, but the apertural shapes of gastropods usually depart from this form.

(b) The rate of whorl expansion *(W)* after one revolution. In the diagram $W=2$, since the diameter of the tube after a single revolution is twice what it was one whorl before.

(c) The position and orientation of the generating curve with respect to the axis *(D)*. In our example the circular tube is separated by a constant distance from the coiling axis, equal to half its own diameter; this is *D*.

(d) The rate of whorl translation along the axis *(T)*, i.e. the relative distance between successive revolutions along the axis as compared with away from the axis.

In some of Raup's models the parameters were kept constant; in others one or more of them were made to change as the model was being generated (Fig. 8.6b). Thus a gastropod with an increasing *W* would have a concave lateral surface. In general the coiled shells of any one molluscan group cannot readily be confused with those of others. They evolved to different functional ends. Thus gastropods are usually helically coiled and tend to have a low *W*, but *T* is very variable and may be extremely high, giving very long thin high-spired shells. In bivalves *T* is low and *W* very high, whilst in brachiopods *W* is again high but $T=0$. Both these shells have very expanded apertures. Cephalopods are normally planispiral; *W* is never normally high and $T=0$. Because of the difference in *W* between cephalopods and brachiopods there is no overlap in shell shape between them, though some planispiral gastropods are approximately equivalent in shape to coiled cephalopods. What makes such differences in the form of many planispiral and helical shells is, however, the shape of the generating curve, i.e. the tube in cross-section. In ammonites, for instance, the shape of **cadicone** shells (Fig. 8.21a) contrasts markedly with that of a compressed **playcone;** yet it is only really this parameter which is radically different.

Of all the possible shell shapes that Raup was able to generate only relatively few have been found biologically useful, and similar shell shapes have been adopted time and time again within the groups that bear them. The other types (e.g. Fig. 8.6c) have not been able to be put to useful functional purposes and so are rarely, if ever, found in any living or fossil group.

Whilst the discussion of theoretical morphology is extremely useful, it has to be remembered that the shells so produced are only models. External features and surface sculpture cannot easily be generated simply, and what has been done in this field is only intended to show how, by isolation of certain important parameters, much information about growth, form and evolution may be gained.

Septation of the shell

There is a fundamental distinction between those shells of molluscs which are divided by internal partitions (**septa**) and those which are not. Cephalopods always have septate shells, whereas those of bivalves and most gastropods are devoid of septa, as are the shells of the Amphineura and Scaphopoda. But there are some gastropods, living and fossil, which do have septa in the upper part of the shell, and they are not uncommon. There is, however, a great difference between gastropod and cephalopod shells, in that cephalopods always have a tubular siphuncle running through all the septa and connecting the chambers; this structure is intimately connected with the buoyancy of cephalopods both recent and fossil. Gastropods do not have such a structure, and their viscera normally fill up the whole of the inside of the shell, except in the septate forms.

Some high-spired Cambrian monoplacophoran shells have been described which have septa (Yochelson *et al.* 1973); these closely resemble the shells of the earliest known cephalopods of the genus *Plectronoceras* (Fig. 8.4), which comes from the Upper Cambrian of China. *Plectronoceras,* however, is 'siphunculate' whereas the monoplacophorans are not. It is quite possible that the earliest cephalopods were derived from septate monoplacophorans, the first forms being bottom-crawling as in the ancestors; the later ones, rapidly evolving means of buoyancy (effected by the siphuncle) and propulsion, were able to exploit the vacant nektonic niche, which was hitherto closed to molluscs. But the fossil record gives no further evidence of how this remarkable transition could have taken place; nor is there evidence from comparative anatomy and embryology. As is commonly and frustratingly the case in attempting to unravel phylogenies, the intriguing intermediate forms were rapidly superseded and have left no trace of their existence.

PRINCIPAL FOSSIL GROUPS

Class Bivalvia

All bivalves have a shell consisting of a pair of calcareous valves between which the soft parts of the body are enclosed. Unlike the brachiopods, which are also bivalved, they are very abundant and diverse today, and most shorelines are littered with their dead shells. The majority of bivalves are marine; most are benthic and live infaunally or epifaunally. Some genera have successfully colonised fresh water habitats. They are first found in rocks of Lower Ordovician age but are of relatively limited abundance in the Palaeozoic except locally. In the Mesozoic they became much more common, as did the gastropods, and from the early Tertiary onwards they have come to dominate the hard-shelled shallow marine fauna. The name Bivalvia was originally given by Linnaeus in 1758, adopted from the usage of Bonnani in 1681, but for rather complex historical reasons the later terms Pelecypoda and Lamellibranchia have been more commonly in use. There is much to be said, however, for suppressing the two latter names both for taxonomic correctness and to avoid confusion.

CERASTODERMA (Fig. 8.7)

The modern cockle *Cerastoderma edule* (Subclass Heterodonta, Order Veneroida) is an infaunal bivalve which lives in the intertidal zone in European and other waters. It is still sometimes referred to in the literature as *Cardium edule*, though since its morphology departs too greatly from that of the type species of *Cardium* originally described by Linnaeus the genus *Cerastoderma* has later had to be erected for it.

C. edule inhabits a burrow a few centimetres deep, in which it lives with its two valves joined dorsally by a hinge, with the line of closure (the commissure) being vertical. The two valves are normally slightly open, allowing the muscular axe-shaped foot to protrude between them anteriorly and the siphons connecting the animal with the surface to project backwards and upwards.

Shell morphology and orientation The two valves on either side of the commissure are virtually mirror images of each other; they are said to be **equivalves.** An intact *Cerastoderma* when examined from either end appears heart-shaped, with the two valves symmetrical about the vertical commissure and their **umbones** close together and facing each other. These umbones, as with brachiopods, are the early-formed part of the shell. From the side the valves do not appear symmetrical but are inequilateral and somewhat lopsided. From the umbones radiate twenty-two to twenty-eight strong ribs, crossed by concentric **growth lines** which are records of the former positions of the edge of the shell. The umbones are set slightly towards one end of the shell, and it is this which is important in determining the orientation of the shell and which valve is which. In standard orientation for *Cerastoderma* the umbones face anteriorly and are closest to the anterior end; this is typical of most though not all bivalves. If an intact *Cerastoderma* with both valves present is held in the hand with the commissure vertical, the hinge horizontal and the umbones directed away from the observer, the **right** and **left valves** are immediately distinguished. In this orientation the **ligament,** which holds the valves together and is instrumental in opening the shell, is nearest to the observer and thus posterior.

If either valve is examined from the inside (Fig. 8.7b) the following internal features are observed:

(a) A flattened vertical area (the **hinge plate**) bearing **teeth** between which are **sockets** corresponding to the teeth on the other valve; these act as guides ensuring that the two valves go exactly back into place when they close, making a secure and tight fit. Dentition in bivalves is of various kinds; in *Cerastoderma* the teeth are of **heterodont** type and fall into three groups. Thus directly below the umbo (Fig. 8.7e, f) are the large **cardinal teeth,** two in the left valve and three in the right; the two sets of elongated and obliquely set **lateral teeth** lie some distance away. Bivalve dentition is very important in classification and identification and is considered in detail later.

(b) The ligament: a rubbery material connecting the two valves and holding them open. It lies in an elongated **pit** posterior to the umbo. The ligament is rarely preserved when the shell has been dead some time.

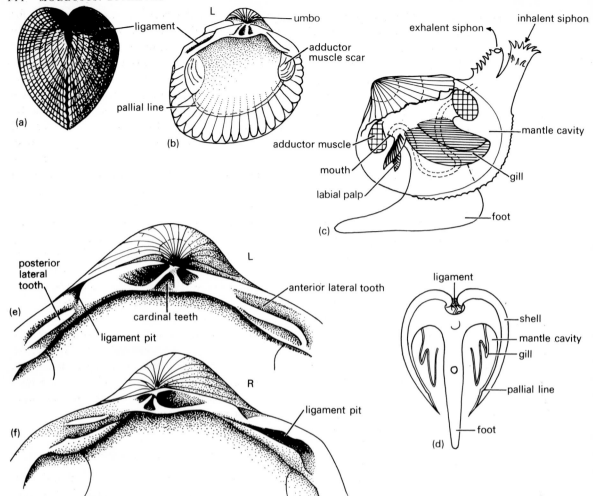

Figure 8.7 *Cerastoderma edule*, a recent infaunal heter-odont bivalve: (a) posterior view of intact shell; (b) interior of left valve; (c) right lateral view of living animal in life position with shell partially removed to show soft parts; (d) vertical section of living animal; (e)–(f) cardinal region of left and right valves respectively.

(c) Two large ovoid smooth areas towards the ends of the shell. These are the **scars** or attachment sites for the **adductor muscles,** which run between the valves and when contracted keep the shell closed.

(d) The **pallial line** which joins the two adductor scars and runs parallel with the edge of the shell and some distance within it. This marks the periphery of the innermost of the calcareous layers of which the shell is constructed. Though the pallial line of *Cerastoderma* is entire, many bivalves have a deep indentation (the **pallial sinus**) towards the rear of the shell. This is present only in genera with retractable siphons and allows a pocket in which these can be tucked away in case of danger (Fig. 8.12). The pallial line has nothing to do with the muscles as such, though it apparently connects them.

Internal anatomy. The shell (Fig. 8.7c, d) is formed by the mantle, which is analogous though not homologous with the mantle of brachiopods. There are three layers in the shell, all of which are secreted by different parts of the mantle. At its periphery the mantle forms three folds, only the outermost of which is secretory and shell-forming. The middle fold is sensory, the inner one muscular. The outer part of the shell is a thin layer (the **periostracum**) made of dark-coloured tanned protein. This is distinct in modern shells but rarely preserved in fossils. It is formed by the outermost mantle fold at the shell edge and is the only shell layer to cross the dorsal margin. Below this are two much thicker layers of crystalline calcium carbonate, formed by deposition in a proteinaceous (**conchiolin**) matrix. The inner layer stops short at the pallial line.

Within the shell the upper region is occupied by the visceral mass and the lower by the mantle cavity. The mouth is set anteriorly and the anus at the posterior end of the mantle cavity. The large central foot is a muscular organ used in digging. The animal can alternately contract and expand it and thus can use it both to dig its way into the substratum and to move horizontally within it. On either side of the foot the long filamentous gills hang down into the mantle cavity. Near the mouth two extra gill-like structures are found: the **labial palps.** These together with the gills are used in feeding.

Since *Cerastoderma* lives as an infaunal suspension feeder protected within the sediment, it has to be able to maintain connection with the surface waters for feeding and respiratory exchange. This is facilitated by two large siphons which project upwards from the posterior end of the animal. Water is drawn into the shell through the inhalant siphon by the ciliary pumping action of the **gill filaments.** These vertical filaments form a comb-like structure lined with innumerable cilia, which beat in successive and co-ordinated waves of movement and collectively set up one-way pressures, thus generating quite strong inward currents. The gills and labial palps not only generate the currents but also trap the food particles, by secreting a sticky mucus. The food particles adhere to this and are conveyed to the ventral edge of the gill or palp by the cilia. They then move in a line along this edge until they come to the mouth. Waste water containing carbon dioxide, exchanged by the gills, is taken away through the exhalant siphon together with excreta from the anus, which is placed very close to the exhalant siphon.

The only other important structures within the shell are those connected with its opening and closure. *Cerastoderma*, like other bivalves, has many natural enemies, especially birds, gastropods and starfish, and its only defence against them is to withdraw the foot inside the shell and to keep the valves tightly shut for as long as possible. This closure is effected by the strong horizontal adductor muscles, which are both about the same size (the **dimyarian** condition) and attached to facing points on the internal walls of the opposing valves. When the adductors relax the valves open automatically to let the foot out, forced apart by the ligament which acts like a compressed spring and whose effect, acting alone, would keep the valves permanently open. Recently dead shells in which the muscles have decayed or been eaten may still retain the ligament, and the two valves still joined together are always found in the open position.

Bivalves thus open and close their valves by an antagonistic muscle/ligament system, which is very effective but requires continual expenditure of energy when the shell is closed. The shell closure system may be contrasted with that of brachiopods, which operates on two sets of antagonistic muscles.

Though there is no head in bivalves there are well-developed circulatory, excretory and nervous systems, adequate for all their physiological needs. *Cerastoderma* is very well adapted for life as an infaunal suspension feeder. It has a high filtering rate and, having colonised the rather 'difficult' intertidal environment, is highly successful.

Range of form and structure in bivalves

Even though certain features of bivalve organisation, such as the absence of a head, have limited their evolutionary potential, the range in their form and the adaptations that bivalves have undergone show a remarkable degree of inherent or actualised evolutionary plasticity. There are some features in the hard and soft parts that have remained fairly stable since their origin. They have not altered much or have undergone only minor evolutionary changes. Such characters are very useful in classification for defining the higher taxonomic categories; they include shell microstructure, gill morphology and dentition, though the latter has on the whole been more variable than the others. Other characters, however, including the overall form of the shell, the musculature and the presence and relative development of siphons and associated structures, are more directly related to the bivalves specific adaptations to particular modes of life and are thus of more value at lower taxonomic levels.

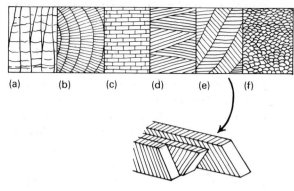

Figure 8.8   Bivalve shell layer morphology as seen in thin section: (a) simple prismatic; (b) compound prismatic; (c) sheet nacreous; (d) foliated; (e) crossed-lamellar, with inset showing disposition of stacked argonite lamellae; (f) homogeneous. (Based on Taylor, J. D., & M. Layman, 1972, *Palaeontology*, **15**, 73–87)

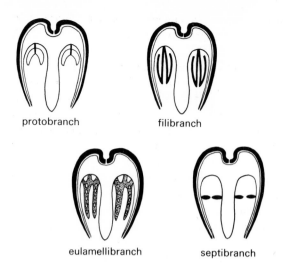

protobranch   filibranch

eulamellibranch   septibranch

Figure 8.9   Bivalve gill morphology, four basic types shown by transverse sections. Shell black, foot projecting centrally. (Redrawn from Moore, Lalicker and Fisher 1953, *Invertebrate fossils*. New York: McGraw Hill)

Shell microstructure and mineralogy (Fig. 8.8)

The calcareous shells of bivalves are multilayered and consist of two intermixed phases (Wilbur 1961): (a) an organic matrix, and (b) crystalline calcium carbonate in the form of calcite or aragonite. Some bivalve shells such as those of oysters are entirely calcitic; others are entirely aragonitic, but perhaps the majority have different layers composed of calcite and aragonite. Vaterite, another form of calcium carbonate, is reported from injured shells which the mantle has been able to regenerate.

The two phases tend to be found in bivalves in a number of recurrent patterns, occurring in discrete shell layers. These have been used in the unravelling of phylogenies. Six primary types have been differentiated (Taylor *et al.* 1969, Bathurst 1975):

(a) **Simple prismatic** structure with columnar polygonal calcite or aragonite prisms.
(b) **Composite prismatic** structure with tiny radiating acicular crystals.
(c) **Nacreous** structure in which tabular sheets of aragonite are found resembling a brick wall when cut in section. These are usually found in middle and inner shell layers.
(d) **Foliated** structure of lath-like calcitic crystallites arranged in sheets.
(e) **Crossed-lamellar** structure which is normally aragonitic. Here the shell is made of closely

spaced lamellae within each of which are found thin stacked plates of aragonite, those of adjacent lamellae being inclined in opposite directions to one another. In some cases intergrowths of blocks of crystals are found (**complex crossed-lamellar** structure) with four principal orientations.
(f) **Homogeneous** structure with small granular anhedral crystals.

Of all these, nacreous structure seems to be phylogenetically the oldest; it is also the strongest, which raises the unsolved question of why the other types evolved if they are less strong? Underlying the areas of muscle attachment there is a specialised region of irregularly prismatic crystals: the **myostracum.** In *Cerastoderma* the calcareous part of the shell is two-layered, the outer layer being of crossed-lamellar form, the inner being complex crossed-lamellar. A few bivalves, such as the Superfamily Lucinacea (Subclass Heterodonta), have three calcareous layers. There is some evidence that the type of shell structure actually present in bivalves may partially reflect their mode of life. Thus crossed-lamellar, complex crossed-lamellar and composite prismatic shell layers have the highest hardness values but do not have particularly high compressive or tensile strengths. They are commonest in burrowing bivalves, in which a good resistance to abrasion, but not necessarily resistance to bending stresses, would be useful.

The mineralogical differences in the shell may be under ecological control; for instance, in the modern Superfamily Mytilacea (Subclass Pteriomorpha) tropical species have entirely aragonitic shells whereas the percentage of calcite increases progressively towards cooler waters.

Whilst shell structure and mineralogy are very useful in taxonomy, they are rarely preserved unchanged in older fossils. Tertiary fossils may retain their original aragonite, but there are no known molluscs with aragonite older than the Carboniferous.

Gill morphology (Fig. 8.9)

The structure of the gills is never preserved in fossils, but it is most valuable in taxonomy. The gills hang down into the mantle cavity on either side of the foot. In Order Nuculoida (Ord.–Rec.) the gills are **protobranch,** being small and leaf-like rather like those of amphineurans and cephalopods. Their unmodified appearance and the fact that they occur in a primitive group suggests that they are not far removed from the ancestral type. **Filibranch** gills form lamellar sheets of individual filaments in a W-shape, as do **eulamellibranch** gills (occurring in *Cerastoderma*), but in the latter there are cross-partitions joining the filaments and making water-filled cavities between them. The vast majority of bivalves have gills belonging to the latter two types. **Septibranch** gills, which are confined to a single superfamily of rock borers, the Poromyacea (Subclass Anomalodesmata), run transversely across the mantle cavity, almost enclosing an inner chamber which maintains only a small connection with the outer cavity.

Dentition (Fig. 8.10)

There are several kinds of tooth and hinge structure in bivalves of which the following can be clearly recognised:

(a) In **taxodont** dentition the teeth are numerous and subparallel or radially arranged, and in modern taxodont bivalves such as *Glycimeris* and *Arca* they are all rather similar. *Nucula* and its relatives have taxodont teeth, as do many of the Ordovician bivalves, though in some of these there is quite considerable differentiation of the teeth. In the Ordovician **palaeotaxodont** genus *Praeleda*, for instance, there is a posterior section with low ridge-like teeth and an anterior part with much larger ridge-like teeth, presumably allowing the protrusion of a large foot below them which they protected from above whilst the shell was open.

(b) **Dysodont** dentition consists of small simple teeth near the edge of the valve, as in *Mytilus*, *Pecten* and *Ostrea*.

(c) **Isodont** teeth are very large and located on either side of a central ligament pit, as in *Spondylus*. Such teeth are characteristic of one superfamily only, the Anomiacea (Subclass Pteriomorpha), though they also occur in *Pecten*.

(d) **Schizodont** teeth are confined to Superfamily Trigoniacea (Subclass Palaeoheterodonta, Order Trigonoida); they are very large and have many parallel grooves normal to the axis of the tooth. In *Trigonia* the left valve has three teeth, the right two.

(e) Most Tertiary and Recent bivalves have **heterodont** teeth, such as are exemplified by *Cerastoderma*. Normally there are two or three cardinal teeth below the umbo, as well as the elongated lateral teeth anterior and posterior to these. Heterodont dentition can be traced back to the Ordovician, and it has been suggested that both the palaeotaxodont and **palaeoheterodont** teeth in the bivalves of that time were derived from an ancestor with multiple-ridged teeth.

(f) **Pachydont** dentition occurs only in the peculiar hippuritid (rudistid) bivalves, which cement themselves to the substratum by a very large left valve of coral-like form. The teeth are very large, heavy and blunt. Evidently this peculiar dentition is directly connected with the unusual mode of life.

(g) **Desmodont** dentition is a common form of hinge structure in which the teeth are very reduced or absent, but accessory ridges lying along the hinge margin take their place, and often, as in *Mya*, a large projecting internal process (the **chondrophore**) carries the

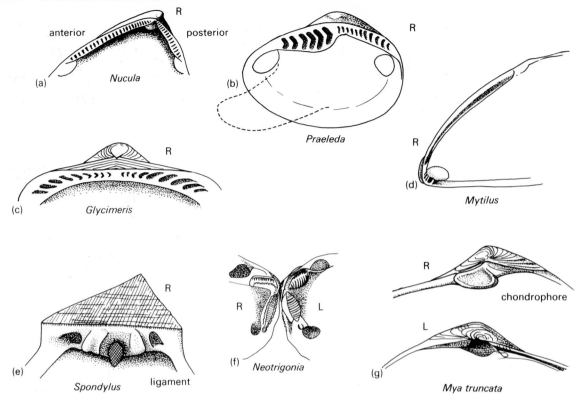

Figure 8.10 Bivalve hinge-lines and dentition (not to scale): (a) *Nucula* (Tert.–Rec.), right valve (note that the umbones face posteriorly), taxodont hinge; (b) *Praeleda* (Ord.), right valve, modified taxodont hinge, foot position inferred; (c) *Glycimeris* (Tert.–Rec.), right valve, taxodont hinge; (d) *Mytilus* (Rec.), right valve, dysodont hinge; (e) *Spondylus* (Cret.), right valve, isodont hinge; (f) *Neotrigonia* (Rec.), schizodont hinge; (g) *Mya truncata* (Rec.), desmodont hinge with chondrophore. ((b) Redrawn from Bradshaw, M. 1970, *Palaeontology* **13,** 623–45; (e) from Woods, 1946, *Palaeontology* (Cambridge: CUP).)

ligament. All bivalves possessing desmodont hinges are infaunal suspension feeders.

The great variety of dental structure in bivalves contrasts strongly with the situation in brachiopods, where it is almost constant.

Muscles and ligaments

The ligament is a variable structure in bivalves and may have two parts: one external and the other internal. The latter may reside in a pit between the teeth (e.g. in isodont shells) or may be supported by a chondrophore (e.g. *Mya*). Either component may be absent.

Cerastoderma is an example of an **isomyarian** bivalve, in which the two adductor muscle scars are more or less the same size. *Mytilus* (Fig. 8.11e) and other byssally attached genera, however, have a greatly reduced anterior adductor (**anisomyarian** condition), and in the swimming scallop *Pecten* (Fig. 8.11j) the anterior adductor has vanished altogether and there remains only the large **monomyarian** posterior adductor.

Other shell structures

Most bivalves have the umbones anterior to the midline (**prosogyral**), but there are some **opisthogyral** genera with posterior umbones. Some bivalve shells, whether proso- or opisthogyral, have depressed areas (the **lunule** and **escutcheon**) placed in front of and behind the hinge respectively (Fig. 8.11k). Their function may have something to do with burrowing,

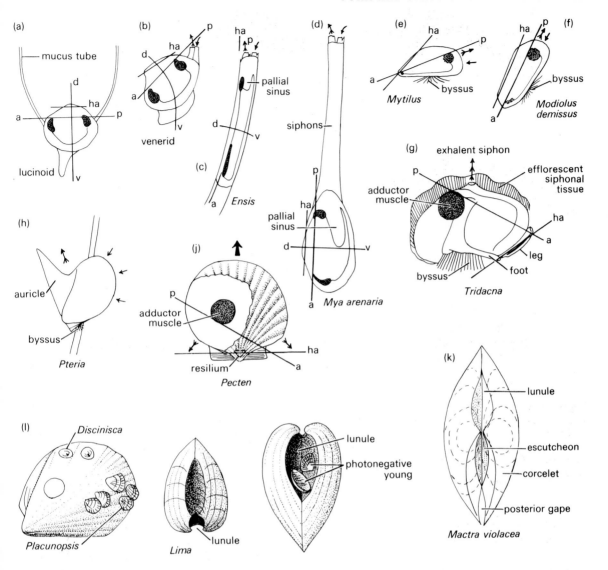

Figure 8.11 Bivalve morphology and mode of life, showing disposition of: ha – hinge axis; d–v – dorso-ventral axis; a–p – antero-posterior axis; and shell modification. Small arrows represent incurrents (plain) and excurrents (feathered). (a) Lucinoid in burrow, with mucus tube to surface (x 0·5); (b) venerid in shallow burrow (x 0·5); (c) razor-shell, *Ensis*, in long shallow burrow (x 0·3); (d) *Mya arenaria* in deep burrow with long siphons and pallial sinus (x 0·25); (e) *Mytilus*, a surface dweller attached by byssus (x 0·4); (f) *Modiolus demissus* in half-buried and inclined life attitude (x 0·4); (g) *Tridacna*, with exhalant siphon and efflorescent siphonal tissue containing algal symbionts (x 0·1); (h) *Pteria*, showing possible sheltering function of auricle (x 0·3); (j) *Pecten*, a swimming bivalve, in which the anterior adductor is reduced completely (x 0·4) – Broad arrows represent direction of shell movement when water is ejected near hinge ((a)–(j) Based on Stanley, 1970). (k) *Mactra violacea* in dorsal view, showing corcelet, escutcheon and lunule, and posterior gape (x 1·2); (l) *Lima lineata* (Trias.) (left) in life position with epizoic brachiopods *(Discinisca)* above, and oysters *(Placunopsis)* below (x 0·5) (based on Seilacher, A. 1954, *Neues Jahrb. Geol. Pal.* (Monatshefte) 163–83). (Right) Photonegative young are sometimes found sheltering in the (ventral) lunule (x 0·5). (Based on Jefferies 1960)

but their presence and shape depend on how the shells grow. Nevertheless the Triassic *Lima lineata* (Subclass Pteriomorpha, Superfamily Limacea) apparently found a use for its lunule, as has been shown in some interesting ecological studies. This *Lima* (Fig. 8.111), like its modern species, apparently rested in life with its lunule on the sea floor, as is apparent from the orientation of epizoic oysters and other small organisms on the shells and from the position of the shells themselves when fossilised. Juvenile specimens of *Lima* are often found clustered at the anterior end of the lunule. They have been interpreted (Jefferies 1960) as photonegative young which entered the dark space between the umbones of the adult and lived there for a while. There are no large juveniles in the lunules, however, and presumably they must have moved out when they had outgrown their dark shelter. Many modern *Lima* species build nests of byssus in dark cavities, and it seems that they must have inherited this preference for the dark from their ancestors, at least as far back as the Triassic.

The orientation of bivalves and the distinction of left from right valves is facilitated if the following considerations are borne in mind:

(a) most bivalves are prosogyral, though *Nucula*, *Lima* and *Donax* are exceptions to this rule;
(b) the pallial sinus and external ligament are always posterior;
(c) though the adductors may be the same size, if one is reduced or absent it is always the anterior one.

Classification of bivalves

Bivalves have always been found hard to classify. The problems are not acute at lower taxonomic levels, for species, genera and families seem on the whole to fall into clearly defined natural groupings. These are based upon shell form and structure, the presence or absence of a pallial sinus, dentition and other characters. Some of the family groups have remained very conservative over long periods of time, and a number of genera have persisted with relatively little change since the Palaeozoic.

But it is not so easy to define higher taxonomic categories of a kind that can be related in a phylogenetically meaningful way. There are too few morphological clues, and most of the more useful stable characters are in the soft parts or shell microstructure. This causes difficulties with the fossil forms, where the soft parts have vanished and where the aragonitic part of the shell is normally recrystallised or dissolved. Furthermore, parallel evolution, which has confused many lines of descent, again raises taxonomic difficulties.

A recent classification, given below and based on that in the '*Treatise*', uses shell microstructure, dentition and (to some extent) hinge structure, gill type, stomach anatomy and the nature of the labial palps as the stable characters, and in this classification six subclasses are defined. Such characters as shell shape are so closely related to life habits that they are 'unstable' and useful only in classification at lower taxonomic levels.

SUBCLASS 1. PALAEOTAXODONTA (Ord.–Rec.): Includes only the one ORDER NUCULOIDA. Small, protobranch, taxodont, infaunal, labial palp feeders with aragonitic shells; e.g. *Nucula*, *Ctenodonta*.

SUBCLASS 2. CRYPTODONTA (Ord.–Rec.): A largely toothless (dysodont), infaunal, aragonitic-shelled, mainly Palaeozoic group with *Solemya* (ORDER SOLEMYOIDA) as the only living representative. *Edmondia* and *Sanguinolites* are representative of the only other Order, the Palaeozoic PRAECARDIODA.

SUBCLASS 3. PTERIOMORPHIA (Ord.–Rec.): A rather heterogeneous group of normally byssate bivalves with variable musculature and dentition. Shells may be calcitic, aragonitic or both.
ORDER 1. ARCOIDA: Isomyarian filibranchs with crossed-lamellar shells, taxodont; e.g. *Arca*, *Glycimeris*.
ORDER 2. MYTILOIDA: Anisomyarian filibranchs and eulamellibranchs, prismatic/nacreous shells byssate, dysodont; e.g. *Mytilus*, *Pinna*, *Lithophaga*.
ORDER 3. PTERIODA: Anisomyarian or monomyarian filibranchs or eulamellibranchs, shell structure varied, byssate or cemented. Includes all scallops, oysters, and pearl clams, e.g. *Pecten*, *Pteria*, *Gervillea*, *Inoceramus*, *Lima*, *Ostrea*, *Exogyra*.

SUBCLASS 4. PALAEOHETERODONTA (Ord.–Rec.): A dominantly Palaeozoic aragonitic-shelled group including the following orders:
ORDER 1. MODIOMORPHOIDA: The Palaeozoic precursors of most later bivalves with heterodont teeth (actinodonts), e.g. *Modiolopsis*, *Redonia*.

ORDER 2. UNIONOIDA: Heterodont non-marine genera with a long time range, e.g. *Unio*.

ORDER 3. TRIGONOIDA: Bivalves with large trigonal shells and well developed schizodont teeth. Common in the Mesozoic when they were amongst the most numerous bivalves; now represented by *Neotrigonia*, the one living genus; e.g. *Trigonia*.

SUBCLASS 5. HETERODONTA (Trias.–Rec.): Heterodont eulamellibranchs to which most modern genera belong, nearly all with aragonitic crossed-lamellar shells, adapted to varied modes of life and especially to infaunal siphon feeding. The hinge structures may degenerate to a desmodont condition.

ORDER 1. VENEROIDA: Active heterodonts with true heterodont teeth, e.g. *Lucina, Thetis, Cardita, Crassatella, Cerastoderma, Venus, Mactra, Tellina*.

ORDER 2. MYOIDA: Thin-shelled burrowers and borers, very inequivalve, hinge degenerate, siphons well developed, e.g. *Mya, Corbula, Pholas, Teredo*.

ORDER 3. HIPPURITOIDA: Large, often coralloid, cemented extinct bivalves with pachydont dentition, e.g. *Diceras, Hippurites, Radiolites*.

SUBCLASS 6. ANOMALODESMATA (Trias.–Rec.): Burrowing or boring forms, very modified, with aragonitic shells and desmodont dentition. One order only, the PHOLADOMYOIDA, e.g. *Pholas, Pholadomya, Lithophaga*.

Evolutionary history of bivalves

The bivalve *Fordilla* has recently been recorded from the Lower Cambrian but only in the Arenig did bivalves become common, when they underwent an initial great burst of adaptive radiation, which was rapid and spectacular. It resulted in the establishment, during the Ordovician, of several superfamilies having taxodont, dysodont or heterodont hinges and belonging to four feeding types. First there were the earliest labial palp feeders of the Order Nuculoida, which have paired flexible extensions of the palps which project from the shell and collect food particles directly from the sediment. Secondly, there were the shallow burrowing types e.g. the Astartidae (Order Veneroida) with no real siphons. Then there were epifaunal, byssally attached bivalves, the pteriomorph Order Mytiloida, amongst others, and finally there

arose the infaunal mucus-tube feeders, the superfamily Lucinacea (Subclass Heterodonta) which were the only Palaeozoic deep-burrowers. All these feeding types survive to the present, as do the superfamilies in which they arose. Apparently the earliest types were infaunal, but epifaunal genera began to diversify later in the Ordovician.

In one genus of Ordovician bivalves, *Babinka*, the shell has a rather curious conformation which has attracted considerable attention (McAlester 1962). This genus is an early lucinoid known only from the Llanvirnian of Czechoslovakia. *Babinka* has an isomyarian shell of standard form, but between the large adductor scars are two chains of smaller muscle scars like strings of beads. The upper chain has eight scars, the lower very many tiny ones. These have been interpreted as foot (pedal) and gill (ctenidial) muscle scars and bear a remarkable similarity to those of the monoplacophoran *Neopilina*. *Babinka*, being a lucinoid, was probably infaunal and a mucus-tube feeder, but it does seem that it retains the muscle pattern as an inheritance from an ancestral mollusc, something quite like a monoplacophoran. The evidence is slender but is in accordance with deductions about molluscan phylogeny from other sources.

It is worth noting that the presence of the muscular foot gave an inherent advantage to bivalves over the articulate brachiopods, which were the rival hard-shelled suspension feeders. Only by means of the foot could bivalves colonise infaunal environments successfully, and this the Articulata were never really able to do. Indeed the second great expansion of the bivalves during the early Mesozoic and continuing throughout the Caenozoic was directly due to the fact that they could burrow. But burrowing ability alone was not enough, for what gave the impetus for the full exploitation of the infaunal habitat was the fusion of the posterior edges of the mantle to form true siphons (Stanley 1968). In the early Mesozoic a large number of new heterodont or desmodont superfamilies arose of which by far the majority were siphonate, using their siphons for feeding whilst remaining protected deep below the sediment surface.

Many of these live in the intertidal zone where they either occupy deep permanent burrows (e.g. *Mya, Macoma*) or, like the more shallow-burrowing *Cerastoderma*, can burrow in again quickly if washed out and thus re-establish themselves again. Quite possibly the intertidal zone was not greatly colonised by bivalves during the Palaeozoic since they did not have the requisite structural potential. Thus the mantle fusion and siphon formation, originating at the very end of the Palaeozoic, appear to have given the bivalve groups that inherited this structure a great new evolutionary potential. Probably the desmodont hinge, which was derived from a heterodont predecessor, is a subsidiary modification also associated with the infaunal siphon-feeding habit. The expansion of siphon feeders into new, previously unexploited habitats (intertidal, deep burrowing, boring, etc.), dependent upon one new key character alone, has been described as an ideal model of adaptive radiation.

Bivalve functional morphology and ecology

Shell form and mode of life in bivalves is illustrated in Figure 8.11. Within the bivalves the structure of the hinge, dentition, mineralogy, and shell structure and composition do not appear to be characters of much adaptive significance. They are important in classification for defining major taxonomic groups but have little to say about the adaptation of the different sorts of bivalves to their environment.

On the other hand, the shape and general morphology of bivalve shells directly reflects their mode of life. Indeed our current understanding of the ways in which modern bivalves are adapted to particular modes of life enables reasonable inferences to be made as to how extinct bivalves lived.

Modern bivalves can be grouped into several morphoecological categories (Stanley 1970):

(a) infaunal shallow burrowing;
(b) infaunal deep burrowing;
(c) epifaunal, attached by **byssus** threads to the substratum;
(d) epifaunal, cemented to the rock;
(e) free-lying;
(f) swimming;
(g) boring and cavity-dwelling.

Examination of the dead shells of most bivalves and indeed fossil ones also will allow the correct category to be inferred. In the burrowing species, shell form is also related to whether or not the animal was a slow or rapid burrower and to the nature of the sediment into which it burrowed. Most bivalves, like *Cerastoderma*, are suspension feeders. There are a few genera, however, members of the Nuculacea and the veneroid Superfamily Tellinacea, which are deposit feeders. Nuculaceans extrude specialised extensions of their labial palps into the sediment and so collect organic particles from it, whilst tellinaceans use their long inhalant siphons to suck up food particles from the sediment surface rather like a vacuum cleaner. Other than these, the bivalves discussed in the following sections are all suspension feeders.

Burrowing bivalves

Bivalves that burrow in soft substrata, such as our type example *Cerastoderma*, have a well-defined sequence of movements which enable them to penetrate the sediment. First the foot probes downwards and swells with blood from the circulatory system; then the siphons close. This is followed by a rapid adductive movement of the two valves which dilates the foot further and squeezes water out from between them. When the foot is subsequently retracted the shell sinks down into the sediment. Then the muscles relax prior to the onset of the next cycle. The anterior adductor usually contracts first, followed by the posterior one; hence a rocking movement is imparted to the shell, which can be up to 45° in some of the more discoidal shells but is normally less. Whilst this process of 'digging in' is going on the siphons extend to keep contact with the surface.

Most shallow burrowing genera (e.g. *Lucina*, *Donax*, *Venus*) have equivalved shells with the two adductor muscle scars about the same size and deep pallial sinuses. The anterior–posterior line in these (joining the dorsal tips of the adductor scars) is approximately parallel with the hinge line. The anterior and posterior sections of the commissure may be permanently parted; the shell may therefore gape at either or both ends for the foot and the siphons to come out. Pedal and siphonal gapes are more characteristic of the deeper-burrowing genera. Some shallow burrowers (e.g. *Tellina*, *Divaricella*) have a curious external sculpture of ridges on the outside of the shell. *Divaricella* has an unusual W-shaped pattern of fine ridges (Fig. 8.12e). When it burrows its rocking movement of some 45° is aided by the grip given to the shell by these ridges as it 'saws' its way down into the

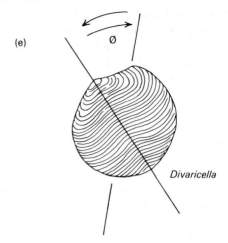

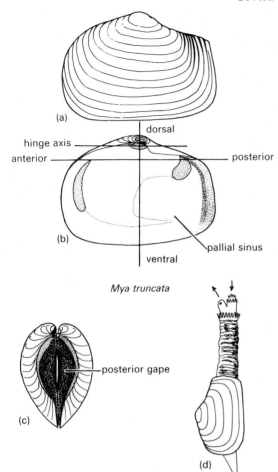

Figure 8.12 *Mya truncata*, an infaunal bivalve adapted for intermediate-depth burrowing: (a) left valve, external view (x 0·6); (b) right valve, internal view; showing the three axes (dorso-ventral, antero-posterior and hinge axis); (c) posterior view; (d) in life position; (e) *Divaricella*, a bivalve which burrows using its surface sculpture as a saw, by rocking through the angle θ x 1. (Redrawn from Stanley, 1970).

posterior gapes. The siphons are fused together, and though they may be collapsed as an escape reaction when the blood supporting them is drained they cannot be withdrawn inside the shell. Razor-shells such as *Ensis* (Fig. 8.11c) and *Solen* have very long, almost tube-like shells, with reduced teeth at the anterior end alone and permanent anterior and posterior gapes. They occupy tubular burrows down which they can move for protection when threatened.

Not all deep burrowers are of such modified form. Some lucinoid bivalves such as *Phacoides* (Fig. 8.11a) have conventional-looking shells of near-circular form and a nearly horizontal hinge axis. There is a long posterior (exhalant) siphon, but the inhalant current is drawn in through a long mucus tube connecting with the sediment surface. This habit is probably ancient for lucinoids are found as far back as the Ordovician.

Byssally attached bivalves

Many bivalves secrete threads of the protein collagen, with which they attach themselves to the sea floor. The common mussel, *Mytilus* (Fig. 8.11e), typifies this habit. The threads, known as byssus, are secreted from a gland located at the base of the foot. They form

fine sand in which it lives. The oblique ridges of the more elongate *Tellina* assist the burrowing function in much the same way. Such patterns as these are unusual, and the majority of burrowers are smooth-shelled and streamlined.

Shallow burrowers do not normally have very elongated shells. There are, however, a number of well-known deep-burrowing genera, which have shells of very drawn-out form adapted for life in deep excavations which are virtually permanent. Representatives of many bivalve families have independently evolved to this mode of life, and their shells have become modified in very similar ways. The large *Mya*, for instance (Figs 8.10g, 8.11d, 8.12) is a sluggish bivalve which burrows in firm sand or mud. It has a long elliptical desmodont shell, very thin, with a curiously modified ligament (chondrophore), much reduced teeth, and pronounced anterior and

sticky secretions which harden rapidly after their formation. *Mytilus* and the majority of other byssally attached forms are attached in an upright position, with the sagittal plane vertical. The shell is elongated, often with a flattened ventral surface offering support and stability to the shell. There is a **byssal notch** marking the base of the shell, from which the byssus emerges. Usually too the anterior part of the shell is much reduced, and the anterior adductor is greatly diminished in size so that the anterior–posterior line is oblique to the hinge axis, though there is no clear agreement why this should be so.

Mytilus is epifaunal, but many of its modern relatives live partially or completely buried. The more cylindrical shell of *Modiolus* (Fig. 8.11f) is adapted for life in salt marsh conditions, where it lives with only the posterior part of the shell exposed. *Tridacna* (Fig. 8.11g) is a genus of enormous thick-shelled clams which inhabit tropical reefs and are likewise byssally attached. Its siphons are directed straight up and are enormously expanded, whilst the byssus is mid-ventral with the hinge just posterior to it. Within the expanded siphon tissue are innumerable algae (zooxanthellae) living in a symbiotic relationship with this clam as they do with corals.

Though most byssally attached bivalves live with the commissure plane vertical, there are some in which this is not so. *Pteria* (Fig. 8.11g), for instance, has one of the two **'ears'** along the hinge line greatly enlarged, extending the hinge like a wing. In the specialised habit of *Pteria*, which lives attached to alcyonarian stems, the function of the wing seems to be that exhalant currents are removed as far as possible from the inhalant region and are not recycled. In benthic 'eared' bivalves the extension of the shell prevents it from being overturned by currents.

Cemented and free-lying forms

The best known of those bivalves which attach themselves by cementation to the substratum are the oysters (Order Pterioda, Suborder Ostreina) whose morphology and evolution have been fully documented in the *Treatise* by Stenzel (1971). These are perhaps the most successful of all bivalves, having a very efficient feeding mechanism which integrates the activities of palps and gills in a way that no other bivalves have been able to do. Oysters are abundant in ancient and modern sediments and are normally preserved in their natural life position. Many hardgrounds in the Mesozoic and Caenozoic are marked by the presence of many oysters which attached themselves to what was then a hard, recently submerged substratum. When they settle, oyster larvae attach themselves to the sea floor by their left valve, which becomes cemented to the rock. Since the two valves must close exactly along the commissure any irregularity in the left valve is reflected in the right valve also. Thus if the larva settles down on a dead ammonite shell, both the left and right valves will have an impression of the ammonite. Oysters have a single large adductor to close the valves and are often of somewhat arcuate form, with the gills lying horizontally and the anterior–posterior axis at some 60° to the hinge axis. Though in most genera the commissure is more or less flat, some fossil and Recent genera (e.g. the Cretaceous *Arctostrea* (Fig. 8.13e) and the modern *Ostrea frons*) have zigzag commissures like those of many brachiopods, and these presumably fulfilled a similar function (p. 114). Oysters may build up substantial biostromes, and as they often lie in belts parallel with shorelines they have been used successfully in determining former shore positions.

Other cemented forms include various species of the very spiny isodont genus *Spondylus,* which live in coral reefs. The Cretaceous *S. spinosus* (Fig. 8.13a–c) has spines arranged at right angles to the shell margins; these apparently acted as snowshoes, preventing this free-living bivalve from sinking into the soft ooze in which it lived.

An extinct group of cemented bivalves, the hippuritoids or rudistids (Subclass Heterodonta), became very highly modified so that they are hardly recognisable as bivalves at all. In *Hippuritella* and *Radiolites* (Fig. 8.13g, h) from the Upper Cretaceous the valves are very unequal in size. The right valve, which may be up to 20 cm high, is conical; the left valve is nearly flat and sits on top of the right valve like a lid. The right valve has very thick walls of two layers, a relatively small body cavity, and a single gigantic tooth which articulates with an equally large pair of teeth hanging down from the lower surface of the left valve. In external morphology there are some similarities to the bizarre Permian strophomenide brachiopod genera *Gemellaroia* and *Cyclacantharia*, and in a remote sense even to solitary corals, but there is certainly little functional similarity to the latter.

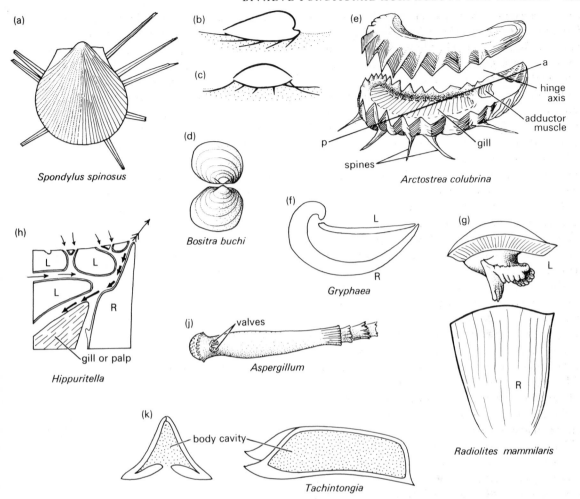

Figure 8.13 Anatomy of modified bivalves: *Spondylus spinosus* (Cret.): (a) dorsal view, (b), (c) showing how spines are used as 'snowshoes' (x 0·75) (based on Carter, R. M., 1968, *J. Paleont.* **46,** 325–40); (d) *Bositra buchi*, a possible Jurassic nektoplanktonic bivalve, with valves opened in inferred life position (based on Jefferies and Minton, 1965); (e) *Arctostrea colubrina*, a Cretaceous oyster, lying cemented by its right valve, showing position of gill, spines, adductor muscles, hinge axis, and antero-posterior axis (x 0·6) (based on Carter 'R. M., 1968, *Palaeontology* **11,** 458–85); (f) *Gryphaea* (Jur.) median section (x 0·6); (g) *Radiolites* *mammillaris* showing coralloid right valve (ventral) and smaller left valve (dorsal) (x 0·6); (h) *Hippuritella*, section through intact specimen near shell edge showing possible directions of inhalent (plain arrows) and exhalent (feathered arrows) currents. Broad arrows show inferred movement of food particles (redrawn from Skelton, 1976); (j) *Aspergillum*, a highly modified sand-dwelling bivalve, with reduced valves (x 0·75); (k) *Tachintongia* (Perm.) showing body cavity (x 0·15). (Redrawn from Runnegar and Gobbett, 1975)

Some rudistids seem to have been able to suck in water through the perforated left valve, and detailed functional analyses have recently been made showing probable current directions (Skelton 1976) (Fig. 8.13h). Most rudistids were small, only a few centimetres high, but there are some immense species up to 50 cm high. They are found in Cretaceous limestones in France and elsewhere, sometimes forming extensive clusters which may even form small reefs, though only some species actually contribute towards reef formation.

Amongst other free-lying genera is the large oyster *Gryphaea* (Fig. 8.13f) whose evolutionary development has been much debated. *Gryphaea* has a very thick convex left valve, necessary for stability, for if an individual was overturned the commissure would be blocked, with fatal results.

Amongst the strangest of all bivalves is the free-lying *Tanchintongia* (Fig. 8.13k), an enormous genus from the Permian of Malaysia (Runnegar & Gobbett 1975). It is some 40 cm long, with very thick valves having lateral flanges upon which individuals were able to rest. The permanent gape shows that it must have lived below the tide level. Such genera as this illustrate the extremes of evolutionary differentiation possible in bivalves.

Swimming forms

The large scallop *Pecten* (Fig. 8.11j) normally lies free on the sea floor, but it can swim by vigorous and repeated clapping of the valves together so as to expel water in successive jets on both sides of the 'ears'. Such activity is exhausting for the bivalve and cannot be sustained for very long, but it is normally used only to escape from predators.

In *Pecten* the two valves are unequal in size, the ventral being the more convex, but they are nearly equilateral. On either side of the umbo the hinge is prolonged into two 'ears' of nearly equal sizes. The ligament is set centrally and internally and emplaced in a small triangular pit. A single large adductor muscle occupies much of the central space. Evidently this kind of shell was derived from a byssally attached ancestor; the two kinds of shell have many features in common, though that of *Pecten* is extended by an increased umbonal angle, assisting its capacity as a hydrofoil.

Lima is a byssally attached genus, but it can also swim as an escape reaction. Individuals can release their byssus and swim by valve clapping with the commissure held vertically. The mantle is here prolonged into 'tentacles' around the commissure; these row like oars whilst the animal is swimming, which adds to the speed of movement.

Probably *Pecten* and *Lima* inherited their natatory ability from a common ancestor, and there is evidence that as far back as the Carboniferous some bivalves could swim. Fossil Pectinacea are not uncommon. Some of them were like *Pecten* in morphology and habit (e.g. the Carboniferous *Pterinopecten*); others such as the Devonian to Jurassic posidonian genus *Bositra* (Fig. 8.13d) and its relatives may have been specialised for a nektoplanktonic (entirely free-swimming) mode of life. *Posidonia* is characteristically present in black shales (ecologically equivalent to the earlier graptolitic facies) in which the only other fossils are ammonites or goniatites, but it may also occur in shallow water limestones deposited as lime muds fine enough to preserve the shells (the parallel with graptolite preservation is again noteworthy). All posidonians are thin-shelled with only two shell layers. They have gapes on either side of the hinge as in their living pectinacean relatives, but they never at any time in their ontogeny had a byssal notch, and they probably never went through an attached phase at any time in their life history. Specimens are almost always preserved with both valves open, though in modern limaceans and other pterioids this is rarely so. Experiments by Jefferies and Minton (1965) showed that the valves would be preserved in the open position only if their normal opening angle exceeded 60°, which may indeed have been their normal angle of opening in between swimming contractions. Further experiments and calculations have indicated that such a swimming bivalve would not have sunk rapidly, especially if the drag effects were increased by a fringe of stiff tentacles around the commissure, as in the modern *Lima*.

Posidonia therefore may well have exploited an ecological niche, that of the permanently swimming nektoplankton, which no later bivalve has been able to invade since the extinction of this genus in the Cretaceous.

Boring and nestling bivalves

Certain bivalves are adapted for life in hard substrates. The stone- and wood-boring genera *Lithophaga* (Order Mytiloida) and *Teredo* (Order Myoida, Suborder Pholadina) have elongated shells of cylindrical form, and like modified deep burrowers they live with their long axis vertical and have very extended siphons. The shells of these bivalves are very thin and must be very resistant to abrasion, for it is with the edges of the shells that they excavate their burrows. Frequently the shell edges are provided with stout spines, used as scraping tools in excavation, which is effected by rocking movements of the shell about the long axis. Not all highly attenuated bivalves are borers. Some live in firm sediments, e.g. *Aspergillum* (Order Anomalodesmata) (Fig. 8.13j), and are well adapted for this mode of life; evidently this represents the primitive condition from which the rock and wood borers later evolved.

Nestling bivalves cannot bore but are photo-negative, like *Lima*, and occupy pre-existing cavities. Some are byssally attached, and their shells grow to fit the cavity even if this is of irregular shape.

Ecology and palaeoecology of bivalves

Whilst shell shape and other factors have been shown as useful in interpreting the mode of life of extinct bivalves, other kinds of ecological study are more directly concerned with organism/environment relationship. The following example is just one that shows how biological and geological criteria can be used together to give information about past environments with facies-controlled faunas.

The Great Estuarine Series (M. Jur.) of the Inner Hebrides of Scotland consists of a shale and sandstone sequence in which bivalves, along with some gastropods and ostracods, are the most dominant fossils (Hudson 1963). Individual bivalves are very common but belong to only a few species, and within the sequence there are hardly any normal marine fossils except at the very top. Such association is normally indicative of a 'difficult' environment, which few species have been able to colonise and in which the species that have done so are successful. The beds were deposited in very shallow water as shown by mudcracks at certain horizons suggesting periodic desiccation. Successive lithologies within the sequence contain bivalve faunas. Hudson found remarkable analogies with the present shallow lagoonal bays of the Texas coast, from the point of view both of environment and of the bivalve genera living there. Several of these modern genera *(Mytilus, Ostrea, Unio)* have direct counterparts in the Great Estuarine Series; furthermore, in Texas the bivalve assemblages in faunal content, which can be closely compared with the Jurassic ones, are salinity-controlled. The Texan *Crassostrea* species and mytilids live in water of rather reduced salinity, and Hudson inferred that analogous Jurassic associations likewise lived in hyposaline water.

By matching the analogues Hudson showed clearly that the overlapping assemblages were controlled by salinity variations through geological time: from fresh water (dominated by the bivalve *Unio* and the small *Viviparus*); through brackish water (the most 'difficult' environment to colonise for physiological reasons), where only the euryhaline *Neomiodon* was present; thence to brackish marine, where oysters and mytilids thrived; and finally to fully saline marine environments.

More recent work on oxygen and carbon isotopes has shown the essential correctness of this picture, though a few modifications were needed, notably that *Unio* and *Neomiodon* of the Jurassic apparently lived in normal marine conditions.

Stratigraphical use of bivalves

Bivalves are on the whole far too long-ranged in time to be of much zonal value. But they have been used in a broader stratigraphical sense, as in Lyell's division of the Tertiary, which has abundant bivalves and gastropods, into four series based upon the relative percentages of the molluscan faunas present therein now living.

The one circumstance under which bivalves have been used successfully as stratigraphical indicators is in the British Carboniferous Coal Measures, where non-marine bivalves are abundant at certain horizons. These genera (*Carbonicola*, *Naiadites* and *Anthraconaia* amongst others) are not unlike the modern fresh-water *Unio*, but they differ in certain morphological characters. The species are not easy to distinguish and are rather long-ranged. But even so, six or seven zones have been defined using concurrent ranges of different species of non-marine bivalves, and the zones thus defined have been corroborated by plant and spore fossils.

Class Rostroconchia

In recent years a small group of Palaeozoic molluscs has been recognised as being of unique phylogenetic interest and has been separated out from other molluscs as Class Rostroconchia (Pojeta & Runnegar 1976).

Rostroconchs look superficially like bivalves and were probably fairly similar to them internally, e.g., in the possession of a protrusible foot, signified by a marked anterior gape in the shell. Where they differ is in the morphology of the hinge line, for they do not possess a functional hinge at all. These molluscs began their growth by producing a small limpet-like,

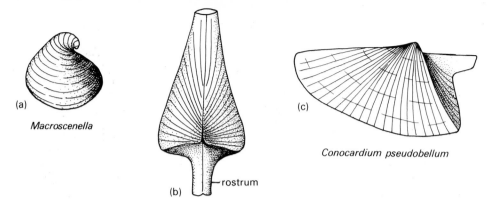

(a)
Macroscenella

(b) ⌐rostrum

(c)

Conocardium pseudobellum

Figure 8.14   Rostroconch morphology shown by: (a) *Macroscenella* (Ord.) protoconch (x 8); (b), (c) *Conocardium* *pseudobellum* (Dev.) in dorsal and lateral view (x 1·8). (Redrawn from Pojeta and Runnegar, 1976)

bilaterally symmetrical protoconch. From this the adult, likewise bilaterally symmetrical shell (**disso-conch**) grew down as a pair of valves. But there is no true hinge for some or all of the shell layers are continuous across the dorsal margin. The valves must have been held rigidly together, the dorsal margin functioning at best as a poorly elastic structure.

The earliest known genus, *Heraultipegma*, is Lower Cambrian. Later rostroconchs reached their maximum development in the early Ordovician, almost rivalling bivalves at that stage, but they declined thereafter and only one order, the Conocardioida, continued until the Permian. In the early rostroconchs (e.g. *Riberoia*) all the shell layers traverse the dorsal margin, whilst in the advanced forms the outer layer does not cross it, suggestive of an independent step towards the condition already achieved by bivalves. *Conocardium* (Fig. 8.14), one of these advanced forms, has a gape at one end and very pronounced **rostrum** at the other.

It has been suggested that rostroconchs occupy a key position in molluscan phylogeny. Pojeta and Runnegar (1976) have proposed that helcionellans (p. 163) are extinct monoplacophorans and that these gave rise to the rostroconchs, losing their segmentation in the process. These in turn produced the bivalves on the one hand (by separation of the valves and development of a proper hinge), and possibly the Scaphopoda on the other. The cephalopods and gastropods, according to these authors, were probably derived independently from the monoplacophorans. When the rostroconchs had given rise to

the more adaptable bivalves competition may well have been an important factor in their demise.

These hingeless 'bivalved' molluscs have helped in an unexpected way to bridge a large morphological gap in the phylogeny of early molluscs. Perhaps more discoveries of similar kind will illuminate the relationships of molluscan classes even further.

Class Gastropoda

Introduction and anatomy of gastropods

Gastropods include all snails and slugs living in the sea, in fresh waters and on land and also the pteropods of the marine plankton. The earliest genera are Lower Cambrian, and though they probably are more abundant now than at any other time they can be found in sedimentary rocks of all ages.

The majority of present gastropods, and the only ones preserved, have coiled shells. Gastropods all have a true head, usually equipped with tentacles, eyes and other sense organs, which is more or less continuous with the elongated body; this typically has a flat sole-like lower surface upon which the animal creeps by small scale waves of muscular contractions, lubricated by slime from **mucous glands.** The visceral part of the body largely resides inside the shell, which in such an example as *Buccinum* (Fig. 8.15a, b), the common whelk, is helically coiled. The **head-foot,** i.e. the protrusible part of the body, can be withdrawn inside the shell by **retractor muscles** (the only attachment of the soft part to the shell) and

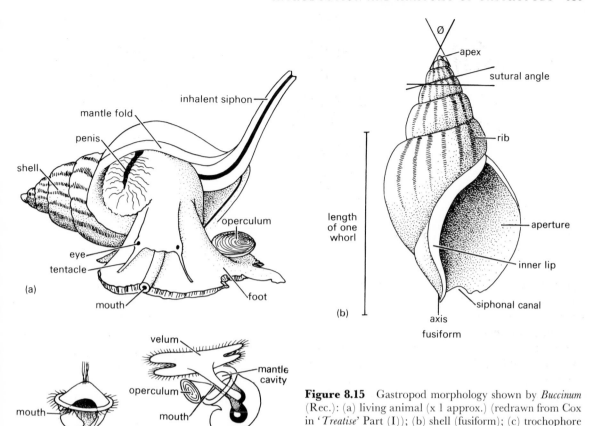

Figure 8.15 Gastropod morphology shown by *Buccinum* (Rec.): (a) living animal (x 1 approx.) (redrawn from Cox in '*Treatise*' Part (I)); (b) shell (fusiform); (c) trochophore larva; (d) veliger larva of gastropod.

closed off from the outside by a plate-like trapdoor (the **operculum**).

The mouth contains a rasping jaw (the radula) in its lower part, composed of a multitude of tiny teeth which scrape against a horny plate in the upper part of the mouth to shred food material. *Buccinum* has a tubular extension of the mouth (the **proboscis**) which is present in the more advanced gastropods. Like most snails, the whelk is hermaphrodite, with male and female organs in the one individual. But individuals copulate with a complex mating pattern, involving the use of a very large male penis and the expulsion of calcareous darts, which are shot out of special apertures and embedded in the body of the other individual; this apparently acts as a stimulating procedure.

The mantle cavity lies anteriorly and in *Buccinum* communicates with the external environment by means of an inhalant siphon: a tubular organ formed from a fold in the mantle and occupying an indentation in the shell margin. Many gastropods do not, however, have such a siphon, and water is drawn in along the edge of the shell. Within the mantle cavity lie the gills, anus, mucous glands and also specialised organs (the **osphradia**) whose function seems to be to sample the water entering the cavity.

It is characteristic of gastropods that the mantle cavity faces anteriorly. Associated with this the internal organs, including the nervous and pallial systems, are peculiarly twisted. This internal asymmetry, known as torsion, is fundamental to gastropod morphology, and though it has been lost in some of the shell-less forms this return to a more normal condition is clearly secondary.

This curious displacement of the internal organs is in no way connected with the asymmetrical coiling of the shell but has a quite different origin. Many suggestions have been proposed to account for it, the most generally accepted being that of the English zoologist Walter Garstang (1951), who has suggested that such torsion gave a singular protective advantage to the gastropod when in the larval state and was retained in the adult.

When gastropod eggs hatch they turn into planktonic larvae known as **trochophores** (Fig. 8.15c) which closely resemble the larvae of certain marine worms. Trochophores are very small and more-or-less globular, with a fringe of cilia round their widest part and another tuft at the apex. The mouth, situated at one side just below the ciliary ring, leads to a simple gut terminating in an anus at the lower pole. The next stage in larval development is the **veliger** stage (Fig. 8.15d), during which a thin shell is secreted over the upper pole and the region under the gut becomes expanded into a large sail-like **velum:** a flat bilobed organ covered with cilia which propel the veliger through the water by their co-ordinated action. It is at this stage that torsion occurs: a rather sudden twisting through 180°, as a result of which the large velum can be withdrawn inside the shell in case of danger. Garstang has suggested that the advantage conferred in being able to tuck away the velum was so great that the resultant displacement and asymmetry of the internal organs was a minor price to pay for the safety rendered. (In addition to his serious papers on the subject, Garstang also wrote about his concepts in the unusual medium of comic verse, and his poem 'How the Veliger got its Twist' crystallises the arguments in a memorable fashion.) The main problem that the gastropods acquired through torsion was that excreta from the anus (located in the mantle cavity) would be expelled just over the mouth. In many gastropods there are devices that cope with this by separating the inhalant and exhalant water. Exhalant siphons are present in some, and in others there is an indentation (**slit-band**) in the shell, forming a channel through which the exhaled water is carried dorsally away from the shell and head region (Fig. 8.16m).

The velum of larval gastropods eventually becomes the foot upon which tle adult glides, by the loss of cilia and the development of an internal system of longitudinal muscles which allow very small rhythmic waves of contraction to pass backwards along the foot; the lubrication of the gastropod's passage is assisted by the supply of copious slime from the mucous glands.

Gastropods may feed in a variety of ways. Some are actively carnivorous and can rasp away the shell of a bivalve to reach the flesh. Others feed on detritus of decayed material. Still others have specialised modes of feeding, such as the coprophilic (faeces-ingesting) gastropods found clustered round the exhalant spires of Carboniferous blastoids.

Classification of gastropods

The classification of gastropods is largely based upon soft parts. Gill and osphradial morphology is most important, as is the structure of the nervous system, heart, kidneys and reproductive system. A condensed version of that in the *Treatise* is given here:

SUBCLASS 1. PROSOBRANCHIATA (L. Cam.–Rec.). Shelled gastropods in which torsion is complete.

ORDER 1. ARCHAEOGASTROPODA (L. Cam.–Rec.): The gills here are **aspidobranch** (i.e. have their filaments arranged in a double comb on either side of the axis and are free at one end). There may be two gills, or one may be lost. The shell structure is variable; some symmetrical forms exist or are found as fossils, but the shells are normally helical spires. Nearly all marine, e.g. *Bellerophon, Euomphalus, Pleurotomaria, Patella, Platyceras, Trochus, Maclurites.*

ORDER 2. MESOGASTROPODA (Ord.–Rec.): Mainly have **pectinibranch** gills: a much more elaborate and efficient system than the aspidobranch condition, which permits free flow of water through the mantle cavity. Living mesogastropods are classified on radular structure, e.g. *Strombus, Cypraea, Natica, Cerithium, Nerinea.*

ORDER 3. NEOGASTROPODA (Cret.–Rec.): The meso- and neogastropods are often combined in the single ORDER CAEONOGASTROPODA. Have pectinibranch gills. An inhalant siphon leads into the mantle cavity, characteristically with a short or long groove in the shell carrying a siphonal tube anteriorly from the shell, e.g. *Murex, Buccinum, Voluta, Conus.*

SUBCLASS 2. OPISTHOBRANCHIATA (?Carb.–Rec.): Marine gastropods which have largely or completely lost the shell. They have undergone detorsion and straightened themselves out. These include the planktonic pteropods and nudibranch sea-slugs, which carry secondary gills on the dorsal surface.

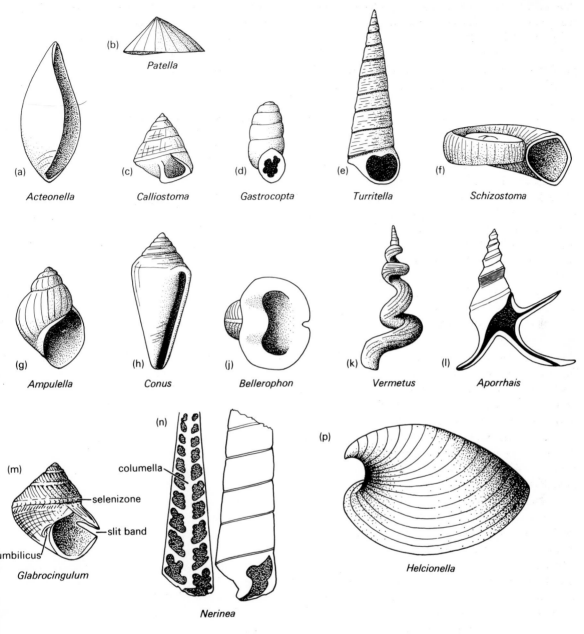

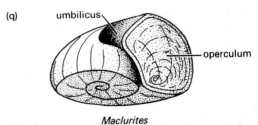

Figure 8.16 Gastropod shell shapes: (a) convolute; (b) patellate; (c) trochiform; (d) pupiform, with constricted aperture; (e) turretted; (f) discoidal (near planispiral); (g) holostomatous turbinate; (h) biconical; (j) isostrophic; (k) irregular; (l) digitate; (m) *Glabrocingulum* (Carb.) showing slit band for exhalent siphon; (n) *Nerinea* (Jur.–Cret.) with columella and internal folded thickenings in the whorls; (p) *Helcionella*, an early Cambrian mollusc, (gastropod or monoplacophoran); (q) *Maclurites*, an Upper Cambrian to early Ordovician operculate gastropod with flattened base (mainly redrawn from '*Treatise*' Part (I); not to scale).

SUBCLASS 3. PULMONATA (Mesozoic–Rec.): In these land-dwelling (and secondarily fresh-water-dwelling) slugs and snails, the gills are lost and the whole surface of the mantle cavity is modified as a lung, liberally supplied with blood vessels and kept permanently moist. The pulmonates are the only molluscs that have made a really successful transition to land. Most of them have retained their shells though the land slugs have lost them altogether.

Shell structure and morphology, which are all that is left to the palaeontologist, are not generally high on the list of criteria for determining correct systematic placement. Furthermore, the common tendency for shells of unrelated stocks to acquire similar forms through homeomorphic evolution causes additional problems.

Shell structure and morphology of gastropods

The shell of a gastropod is basically an elongated cone, rarely septate, which may be coiled in a number of ways. The most characteristic shell shape is a helical spire, coiled about an axis, which is usually illustrated as vertical. There are some gastropods with planispiral and hence symmetrical shells. Usually the whorls all touch one another, but in some cases they do not embrace. Coiling is characteristically dextral but may be sinistral; in very rare cases equal numbers of individuals in a population may be dextral and sinistral.

The basic terminology of gastropod shells is given in Figure 8.15b. It is largely self-explanatory, as are some of the basic shell forms illustrated in Figure 8.16. It is worth noting that the earliest juvenile whorls of certain gastropods (the protoconch) are often of peculiar form and do not necessarily coil in the same axis as the rest of the shell; furthermore, they usually have a different surface ornament.

The external whorls may be ornamented with various characteristic features: ribs, spines or vertical bars (**varices**). The aperture may be entire and unmarked by any feature (**holostomatous**) or, as in neogastropod genera, may be equipped with a **groove** which is the outlet for the exhalant siphon (**siphonostomatous**). The siphon may be set in a deep slit-band which separates it far from the mouth. The calcified strip representing the 'track' of the slit-band as it is calcified is the **selenizone** (Fig. 8.16m). In such genera as the sand-dwelling *Aporrhais* and *Phyllocheilus* the aperture may be enlarged into a flange which helps to stabilise the shell whilst the gastropods are feeding.

· There is a very clear distinction between those shells in which the later whorls do not meet centrally, and thus have an **umbilicus,** and those in which they touch. In the latter case they are welded to a central rod (the **columella**). Likewise, high-spired forms, in which the apical angle is low, are very distinct from low-spired genera, which have a higher apical angle. The shape of the whorls, any unusual coiling patterns, the presence or absence of different kinds of external ornament, apertural shape, and the presence or absence of slit-bands or of siphonal grooves are all used in classification and identification.

Gastropod shell composition

The shells of gastropods have an outer horny layer, the periostracum, below which lies the shell proper, a structure normally composed of layers of aragonite. In fossil shells the aragonite either recrystallises to calcite or may be dissolved, especially where there is leaching in the rock.

Many shells have an inner nacreous layer of very thin aragonite leaves parallel with the internal shell surface, each separated by equally thin organic layers. External to this is the crossed-lamellar layer in which thin lamellae (0·02–0·04 mm thick) are arranged normal to the shell surface. But each lamella itself consists of strips of extremely thin aragonite, arranged obliquely, the strips of adjacent lamellae being arranged at right angles to one another.

There are other kinds of shell structure in gastropods, all of systematic importance, but, as in the case of the layers described above, they almost always disappear in fossils and no trace of them remains.

Evolution of gastropods

General features of evolution

Gastropods must have arisen from a bilaterally symmetrical molluscan ancestor, approximating in general morphology to the hypothetical archimollusc described earlier. The protection given by torsion and the resultant ability to withdraw inside their shells must have given gastropods a high advantage, for from their first appearance in the Lower Cambrian they seem to have been remarkably successful.

The earliest known gastropods (e.g. *Coreospira*, *Helcionella*) are found in the Lower Cambrian. They have coiled shells with a wide expanded aperture and are planispiral, hence bilaterally symmetrical. These very early genera are normally classified with the early archaeogastropod Suborder Bellerophontacea (L. Cam.–Trias.), named after the characteristic genus *Bellerophon* (which was the first Palaeozoic mollusc ever to be described, by the conchologist de Montfort in 1808).

The bellerophontaceans are a most important group of fossil gastropods, and over seventy valid genera have been described. But some of the Cambrian bellerophontiform shells have peculiarities uncharacteristic of the bellerophontaceans as a whole (Yochelson 1967). *Helcionella* (Fig. 8.16p) and its relatives have a simple aperture without any indentations, but all other bellerophontaceans have a pronounced notch or emargination in the plane of symmetry. There are other reasons for believing that *Helcionella* is not a normal bellerophontacean, and some authorities have suggested that it may not be a gastropod at all but rather may be a representative of the monoplacophorans or of an extinct class of molluscs.

Some other Palaeozoic bellerophontiform molluscs, the tightly coiled *Cyclocyrtonella* and the more cap-shaped *Tryblidium*, have paired muscle scars inside the shell and do not appear to have undergone torsion; their muscle scar pattern indicates that rather than being bellerophontaceans they are actually monoplacophorans.

Hence bellerophontiform shells are not necessarily all gastropods, but such simple morphology rather seems to have been common in diverse molluscan groups during the earlier Palaeozoic.

In the Upper Cambrian are found the first asymmetrical, helically coiled shells, belonging to the important archaeogastropod Suborder Pleurotomariina (Cam.–Rec.) There is also an exclusively Palaeozoic suborder, the Macluritina (Fig. 8.16q); all these early forms are low-spired. Recent pleurotomariides are 'living fossils' in which the grade or organisation present in some of the earliest gastropod genera can be seen by direct

homology. Internally the gills of modern pleurotomariids are unmodified (aspidobranch). The reproductive organs are likewise in a relatively primitive condition, since the capacity for internal fertilisation is not present and eggs and sperm are merely shed into the water. It was only the development of internal fertilisation that rendered possible the later invasion of the land and fresh waters.

By Carboniferous times gastropod faunas were very rich and diverse. In one recently described fauna of finely preserved gastropods, the Visean Hotwells Limestone fauna of the Mendip Hills in Somerset, England, no less than forty-five genera and upwards of eighty species occur in association with a typical Lower Carboniferous coral–brachiopod association (Batten 1966). In this fauna there are many archaeogastropods (bellerophontids, pleurotomariids and limpet-like genera), but there are some caenogastropods as well; the ancestry of the latter can be traced back to the Ordovician.

All the species of the Hotwells Limestone were apparently adapted to microniches within their environment.

There is some evidence of caeonogastropods having migrated into a non-marine habitat by the Carboniferous: a first invasion of the habitat so successfully colonised by the pulmonates much later on in the Jurassic and Cretaceous.

In Carboniferous times too there are some reef-like masses of vermetiform gastropods, which have modern though not necessarily related counterparts in the modern vermetids. These are gastropods with peculiarly uncoiled shells, which may live permanently attached to branching corals and feed by straining off food particles from water passed between the edge of the shell and the operculum. In *Vermetus* (Fig. 8.16k) and *Vermicularia*, which are typical examples, two advantages seem to be given by such uncoiling (Gould 1969b). One is rapid upgrowth towards the source of food particles raining down as detritus from the surface. The other is the considerable flexibility rendered possible to individuals when growing round obstacles. Evidently the potential of gastropods to grow in such an unlikely manner for specific purposes was realised in the Carboniferous too.

The gastropods were affected, as were most other organisms, by the great extinction period at the end of the Permian, but they continued to evolve throughout the Mesozoic. Many characteristic groups of

Mesozoic age became very important for a while, such as the Nerineidae (Fig. 8.16n). These are a family of high-spired Mesozoic mesogastropods found in carbonate sediments; in them the inside of the spire is thickened by folded calcium carbonate. Spiral calcite rods running within the spire may be the original duct system within the digestive gland. It has been argued that many nerineids were infaunal and that since they lived in organic-rich carbonate mud, on which they fed, they would not have needed the nutrient storage units that other gastropods possessed in the spine. Hence the space was taken up instead by calcite which followed the contours of the digestive gland and gonad. The calcite taken in with the food was thus conveniently disposed of. Other groups, such as the gigantic neogastropod Family Strombidae and the cowrie-shell Family Cypraeidae, appeared for the first time in the later Mesozoic. But the real acme of gastropod evolution was reached in the Tertiary, continuing until the present, with the great success of the long-siphoned neogastropods which dominate today's gastropod fauna.

Gastropods seem to have been a stable and constant component of the marine fauna since early times, but though they are long-ranged and evolved slowly, new structural developments allowed important advances at different times in geological history.

In the early Tertiary times there appeared the pteropods which are small (c. 2 cm) planktonic opisthobranchs. These may have thin shells which can be coiled or straight; alternatively they may have no shells at all. They spend all their lives afloat and are important components of the plankton. It is often suggested that the pteropods had a neotenous origin from floating veliger larvae, which seems an eminently reasonable proposition. Their primary geological importance is as one of the main components of pteropod ooze in the deep oceans.

The earliest authenticated pteropods came from the Eocene; genera described as 'pteropods' from the Cambrian and other systems are now known to be shells of hyolithids, which are extinct animals of unknown affinities, unrelated to gastropods.

A microevolutionary study of gastropods (Fig. 8.17)

Gould (1969a) elucidated out the evolutionary history of the Pleistocene pulmonate gastropod genus *Poecilozonites* which lived in Bermuda during the last

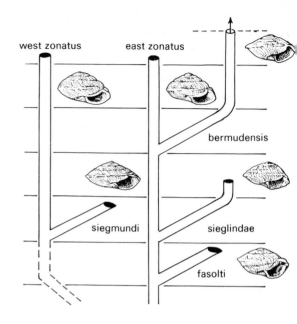

Figure 8.17 Evolution of *Poecilizonites bermudensis* by iterative development of paedomorphic subspecies in two separate areas. (Redrawn from Gould, 1969)

300 000 years of the Pleistocene; one subspecies is living today. Fossil shells of this gastropod occur abundantly in a sequence of alternating reddish soils and windblown sand, which Gould carefully documented. The shorelines oscillated through the later Pleistocene, so that islands on which the gastropods lived were alternately partially submerged and isolated, and exposed and hence linked, though often in different combinations. In his study of the changing populations throughout this time Gould found that two populations of *P. zonatus bermudensis*, distinguished primarily by colour banding, lived in eastern and western regions and were presumably isolated throughout most or all of the time. These eastern and western *P. zonatus* populations formed the parent stocks from which other subspecies were derived. There are four derived subspecies: one originating from the western *P. zonatus* population, the other three from the eastern group. The three eastern forms appeared successively, each becoming extinct before the appearance of the next, and the last-produced subspecies is the one still living today. Gould was able to show that all four subspecies arose from isolated small groups on the periphery of the main population and thence spread to colonise other

regions: a classic case of allopatric speciation in each instance. He also made clear that the origin of each population was the result of a paedomorphic and hence instantaneous change, emphasising its role as an important control of evolution.

Had the stratigraphical, geographical and morphological documentation been less complete, it would have been very easy and tempting to fit together the three eastern *P. zonatus* subspecies as part of a single evolving plexus rather than as three quite distinct iterative populations which started as peripheral isolates.

Therefore, in order to understand the principles whereby population differentiation takes place, it is at this level of detailed analysis that studies have to be undertaken.

Class Cephalopoda

The cephalopods, which are entirely marine, are the most highly evolved of all molluscs, Within this class are included the modern *Nautilus,* the argonauts, squids and octopuses, and the extinct ammonoids and belemnites. All modern cephalopods are distinguished by having a properly developed head with a good brain and elaborate sensory organs; the structural and functional parallels between the eyes of cephalopods and those of vertebrates, which are so well known, serve to illustrate the possibilities of evolutionary attainment inherent in the molluscan archetypal plan.

It may seem remarkable that the highly mobile cephalopods are constructed upon the same basic plan that is found also in the headless and mainly benthic bivalves. Yet as Figure 8.1 shows, all the components of the archetypal mollusc are present in the cephalopods as they are in other molluscs. It is largely because the cephalopods were able, early in their evolutionary history, to develop an effective means of buoyancy using the chambered shell that they were able to colonise the nektic habitat, with its rich food resources of actively moving large-sized prey. Their evolutionary history and functional morphology shows how well they were able to exploit it, for they are nearly all active carnivores and, other than fish, the most accomplished swimmers in the sea.

There have been considerable difficulties in classifying cephalopods, especially in erecting large natural categories. Whilst many modern authorities still use the broad divisions Tetrabranchiata and Dibranchiata, first erected by Owen in 1832, it is recognised that these are only provisional categories, since they are based on gill morphology which may be inferred but not confirmed in fossils.

TETRABRANCHIATA: Forms with four gills present or postulated within the mantle cavity. Includes SUBCLASSES 1. ENDOCERATOIDEA (Ord.–?Sil), 2. ACTINOCERATOIDEA (Ord.–Carb.), 3. NAUTILOIDEA (Cam.–Rec.), 4. BACTRITOIDEA (Ord.–Perm.), and 5. AMMONOIDEA (Dev.–Cret.).
DIBRANCHIATA: Forms with two gills within the mantle cavity. Includes SUBCLASS COLEOIDEA (Carb.–Rec.).

Subclass Nautiloidea and relatives

Nautilus (Figs 8.1, 8.18a–d)

The only living cephalopod genus with a coiled external shell is *Nautilus,* of which there are six living species confined to the Indo-West Pacific faunal province between the Phillipines and Samoa. The *Nautilus* shell, some 20 cm in diameter, is planispiral and ornamented externally with a radial colour banding of irregular and bilaterally symmetrical orange–brown stripes. This shell is divided into internal gas-filled chambers (or **camerae**) by septa, concave towards the aperture; the animal resides in the last chamber **(body chamber)** and moves forwards each time a new septum is secreted. A single tube (the **siphuncle**) passes through the centre of each septum and connects the chambers. Each septum meets the inner wall of the external shell along a slightly curved line **(suture line).** Fossil *Nautilus* and its relatives are usually preserved with the shell dissolved and the chambers filled with spar or matrix. In such cases the position of each septum is marked by a suture line. Growth is very rapid in living *Nautilus,* new septa being emplaced about every two weeks on average. The living animal itself is separated by a fluid cushion from the last septum. The soft parts can be considered (cf. Fig. 8.18b) as two separate units: (a) the body, which is fully enclosed by the mantle and contains the viscera and the mantle cavity with its contents; and (b) the head-foot, a cartilage-supported structure with thirty-eight tentacles surrounding the mouth with its horny parrot-like jaws.

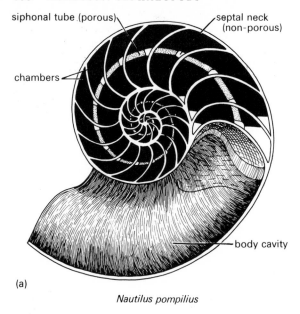

Nautilus pompilius

(a)

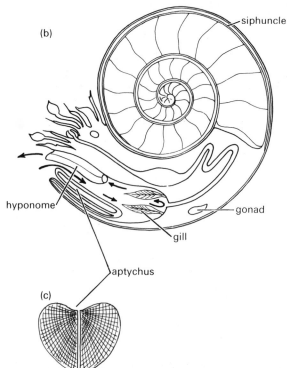

(b)

(c)

The eyes are placed laterally on the head-foot. The dorsal part of the head-foot, above the tentacles, has the form of a **hood** which normally extends some way up the shell. This has a tough warty outer skin, so that when the tentacles are withdrawn for protection the hood closes the aperture, presenting a largely impenetrable and uninviting surface to any predator. The full extent of the unretracted hood is marked by a black film on the shell, only exposed when the hood is closed down over the tentacles. Below the tentacles is the **hyponome** or funnel: a long tubular structure which can be turned in any direction.

Nautilus swims by jet propulsion; water enters the mantle cavity through a slit-like inhalant passage and, passing over the four gills within the mantle cavity so that respiratory exchange is effected, is squirted through the hyponome by the contraction of powerful muscles within the mantle cavity. (Unlike squids the mantle of *Nautilus* does not have muscular walls, and the musculature is confined to a specialised sac: the **branchial chamber**.) As in other cephalopods there is a co-ordinated system of rhythmic flow by means of which water is pumped through the mantle cavity with fairly gentle regular movements. *Nautilus* also has an escape reaction involving a violent contraction of the branchial chamber, so that the animal literally jumps out of the way of any predator; an equivalent mechanism in squids is often accompanied by the emission of ink into the water

through the hyponome from an **ink sac** within the mantle cavity. This biological jet propulsion is the standard means of locomotion in cephalopods.

Nautilus, being an active carnivore, has a highly organised brain coupled with good sense organs and a complex behaviour pattern. When hunting, *Nautilus* spreads its outer tentacles in a 'cone of search', but when it captures food it uses the inner tentacles to handle it. The food, which consists mainly of large crustaceans and fish, is cut up by the beak and stored in an expanded oesophagus prior to being passed to the stomach and digested.

Nautilus has a rather complex reproductive pattern. The sexes are separate; testes and ovaries are to be found at the posterior extremity of the body. During copulation the male transfers a ball of sperm (**spermatophore**) to the mantle cavity of the female using a specially adapted erectile group of tentacles (the **spadix**). Other cephalopods normally have only one modified tentacle (the **hectocotylus**), but their patterns of display, courtship and copulation are often quite complex. All species of *Nautilus* have alternate periods of rest and activity. Individuals are active at night, and during the day they sink to the sea

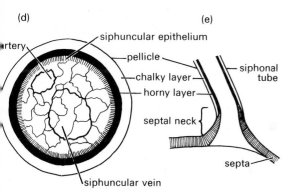

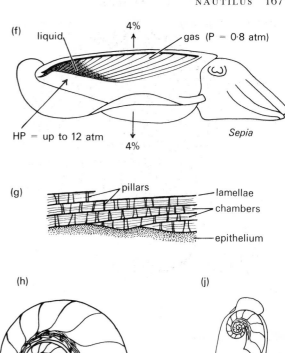

Figure 8.18 (a) *Nautilus pompilius* (Rec.) shell sectioned subcentrally (x 0·5) (redrawn from Denton and Gilpin-Brown, 1966). (b) Ammonoid morphology reconstructed on the basis of the anatomy of *Nautilus*, showing marginal siphuncle and inferred organisation of soft parts. The aptychus is shown in face view (c). The arrows show the direction of water currents into the mantle cavity and out through the hyponome (based on Trauth in '*Treatise*' Part (L)). (d) Anatomy of *Nautilus* siphuncle; transverse section of siphonal tube. (e) Longitudinal section showing junction of septal neck with siphonal tube (based on Denton and Gilpin-Brown, 1966). (f) Buoyancy mechanism in the cuttlefish *Sepia:* cuttlebone with older chambers filled with liquid and younger chambers with gas at about 0·8 atm. pressure. A lift of some 4% is imparted balancing the excess weight of the animal. Hydrostatic pressure of sea up to 12 atm. (g) Structure of lower part of the cuttlebone with chambers supported by pillars and separated by lamellae (based on Denton, 1961). (h) Shell structure with siphuncle (modified from Denton, 1961). (j) *Spirula:* in life position showing the location of the shell.

floor where they rest with the hood pulled down and almost covering the eyes.

The shell of *Nautilus*. The shell of *Nautilus* is made of aragonite in a conchiolin matrix. It consists of two main layers: the outer porcellanous ostracum, and the inner nacreous layer. The outer layer is formed first and grows from the mantle at the edge of the shell, beginning as aragonite seeds in a conchiolin matrix; these become larger and closely packed together as they grow into vertical prisms. The inner nacreous layer is later deposited by the mantle in a series of films; rather like the nacreous layer of bivalves it consists of a brick wall structure of hexagonal aragonite crystals with conchiolin layers between them. The septa have fundamentally the same structure. An empty *Nautilus* shell (Fig. 8.18a)

shows short backward-pointing **septal necks** piercing each septum ventrally. In life these carry the siphuncle: a single calcified strand of living tissue extending from the body to the protoconch and carrying a rich blood supply.

Buoyancy of the shell in *Nautilus* and other cephalopods. Since the whole success of the cephalopods has been so intimately bound up with their possession of buoyant shells, it is appropriate to consider how modern cephalopods of different kinds actually achieve such buoyancy.

The researches of Denton (1961) and of Denton and Gilpin-Brown (1966) and others have shown clearly that the buoyancy of cephalopods works on a quite different principle from that of fishes. In fishes the swim bladder contains gas at an equal pressure to that of the surrounding sea; in deep water fishes the internal pressure of the swim bladder can be enormous.

But the shells of cephalopods such as *Nautilus* and the squids *Sepia* and *Spirula* (which have internal shells; see p. 183) all contain gas at a pressure of less than 1 atm, and neutral buoyancy at different depths is achieved not by pressure equalisation but by density control. In both *Nautilus* and *Sepia* the chambers contain gas at pressures ranging from about 0·3 atm in the more recently formed chambers to about 0·8–0·9 atm in the older chambers of the shell. Some of the chambers also contain liquid (**cameral liquid**), which can be added or extracted through the siphuncle. Addition of more water to the chambers gives the shell a slightly higher density, and it will begin to sink. Removal of liquid from the chambers will lighten the shell, and it will rise. Thus the animal can adjust its density to the depth at which it is living and so remain neutrally buoyant.

The only real disadvantage to the cephalopod is that the hollow non-pressurised shell will implode under external pressure below a certain depth. This depth varies with different cephalopods.

In the living *Nautilus* only the more recently formed chambers contain liquid, and this is present in diminishing amounts from the newest to the older chambers. Thus the chambers beyond the tenth from the body chamber are empty of liquid.

It is the siphuncle that extracts or secretes this liquid. The siphuncle (Fig. 8.18d, e) consists of two parts: the impermeable septal necks, and the permeable strands between them (**siphonal tube**). This tube has an inner core of living material with arteries and veins running the whole length of it, with a cylinder of epithelial cells. Outside this is a horny tube of conchiolin fibres, and surrounding this again is a concentric tube of irregularly arranged aragonite crystals. A very thin external pellicle of conchiolin completes the ensemble. Where the siphuncle joins with the septal necks the horny tube becomes continuous with the nacreous material of the neck. Both the aragonite and horny layers are very porous and permit the passage of liquid through them. There is also a thin layer of scattered aragonite crystals on the concave wall of the septa, which acts like blotting paper and renders the wall wettable and so able to retain the cameral liquid.

The body of *Nautilus* is not in direct contact with

the last septum but is separated from it by a cushion of liquid. When a new septum is formed it encloses a chamber which initially retains all this liquid. But when the septum is fully formed and is strong enough to withstand the pressure of the sea, the siphuncle begins to pump out the liquid, leaving only a small residual amount which is reduced in salt and hypotonic to sea water. Though this pumping process works against osmosis, the water actually in the siphuncle at any one time is 'decoupled' from that in the chamber, so that the actual work done is not very great. Gas very slowly diffuses into the space left, which may explain why the gas pressures in the most recently formed chambers are low. To what extent *Nautilus* can adjust its buoyancy quickly is not known. Observations of animals kept in tanks indicate that they can probably make fairly rapid adjustments.

When the *Nautilus* squirts water through its hyponome the shell does not go into a spin, for the centre of gravity is located a few millimetres directly below the centre of buoyancy (i.e. the centre of gravity of the displaced water). This imparts a remarkable stability to the shell, and it would need a very strong couple to turn the animal on its side or through 90°.

A final point concerns the depth to which the living *Nautilus* can sink before the shell implodes. The shell is very strong and rigid, and experiments have shown that a shell would not implode until a pressure of 5 kg/mm² was applied. This corresponds to the pressures expected at a depth of 600 m. But most living *Nautilus* specimens appear to live at a depth of less than 200 m, though there are reports of living individuals venturing to depths approaching implosion limits.

Evolutionary diversification of early cephalopods

General considerations. Very early cephalopods, such as the Upper Cambrian *Plectronoceras*, have small curving shells with marginal siphuncles. From such ancestors came the very many Ordovician genera so common in the fossil record, for the early Ordovician was marked by an evolutionary explosion of cephalopod genera. These had mainly straight (**orthocone**) or curving (**cyrtocone**) shells and were generally of much larger size than their little Cambrian ancestors. From the one original order, the Ellesmerocerida (Subclass Nautiloidea), there arose by the early Middle Ordovician no less than nine other orders, which persisted through all or part of the Palaeozoic and are grouped in the subclasses Nautiloidea, Endoceratoidea, and Actinoceratoidea. The latest of the orders to appear, in latest Silurian or early Devonian time, was the Nautilida in which the living *Nautilus* is classified. Representatives of all three subclasses are sometimes referred to as 'nautiloids', but this unfortunate usage obscures their taxonomy; here they are called non-ammonoid cephalopods.

Most of these Palaeozoic cephalopods became extinct by the end of the Permian, only the coiled-shell Nautilida surviving through the Mesozoic to the present. The decline of Palaeozoic cephalopods seems to bear some relationship to the success of their progeny: the ammonoids, and especially to the ammonites of the Mesozoic, though as in all such cases of apparent 'takeover' many factors are involved and direct competition is not necessarily the only issue.

Palaeozoic non-ammonoid cephalopods (Fig. 8.19). The post-Cambrian cephalopods are very diverse and abundant and often quite different in shell shape from the living *Nautilus*. As in their sole modern representative, the suture lines, as seen in internal moulds, are nearly always straight or slightly curving and are very rarely more complex. On this and other factors even the planispiral nautiloids can readily be distinguished from the ammonoids.

Only a few of the Palaeozoic cephalopods have planispiral shells. The majority are of either cyrtocone or orthocone shape. These may be very elongate (**longicone**) or short and rather swollen (**brevicone**). It has been generally assumed that all fossil cephalopods with straight or curved shells were **ectocochlear,** i.e. that the shell was wholly external to the body. The X-ray photographs of Stuermer (1970) of certain genera from the Lower Devonian Hünsrückschiefer show very clearly that in some cases there were living tissues, including lateral fins, external to the shell. Some of the orthoconic Palaeozoic cephalopods were apparently ancient squids, precursors of the later belemnites. But at present, with only the Hünsrückschiefer 'window' to look through, it is not known how many of the Palaeozoic orthocones were really **endocochlear,** for in Stuermer's photographs both internal- and external-shelled kinds are evident.

Palaeozoic non-ammonoid cephalopods are classified on various characters, but perhaps the most important are the form and arrangement of the various siphuncular structures. The eleven orders are grouped into three subclasses on the basis of these:

SUBCLASS 1. ENDOCERATOIDEA (Ord.–Sil.)
SUBCLASS 2. ACTINOCERATOIDEA (Ord.–Carb.)
SUBCLASS 3. NAUTILOIDEA (U. Cam.–Rec.)

Modern *Nautilus* (Subclass Nautiloidea) has only septal necks with which to support the siphuncle. In the ancient groups there are a variety of other structures, some of them of very complex form, and in particular there are often structures within the siphuncle itself (**endosiphuncular** structures).

In Subclass Endoceratoidea the shells are either orthocones or cyrtocones and the siphuncles are usually large and normally marginal, and most genera have endosiphuncular conical sheaths or alternatively radial lamellae running the length of the siphuncle.

Representatives of Subclass Actinoceratoidea all have orthoconic shells. The septal necks are short but have inflated balloon-like septal necks between them, within which is a delicate and complex system of radial canals.

The siphuncular structures of Subclass Nautiloidea are extremely diverse. In the Palaeozoic genera there are normally connecting rings between the septal necks, but in the later genera these become very thin and disappear. Members of this subclass have shells ranging from ortho- or cyrtocone to partially or fully coiled shells. Some genera have shells of decidedly odd form, such as the Ordovician Family Lituitidae in which the early chambers are coiled and the rest of the shell is a rapidly expanding orthocone. Even stranger are the shells of Order Ascocerida (Ord.–Sil.) (Fig. 8.19f). In these the shell consists of two quite distinct parts. The first-formed part is a small, rather standard orthocone or slightly curving cyrtocone, with a thin straight siphuncle. This is sometimes found joined to a much thicker swollen brevicone where the internal structure is highly modified. In this the siphuncle is short and confined to the apical end, with inflated connecting rings; from it the highly modified sigmoidal septa are formed to enclose large chambers located in the dorsal part of the shell only, above the body chamber. The two parts are very rarely found together, and it is generally accepted that the longiconic portion was deciduous and thrown away when the animal was mature. Probably a change in mode of life was involved; it is quite likely that the juvenile shell was nektobenthonic animal whilst the mature ascocerid, which had shed its early stage, was an active nektonic hunter.

The great diversity in shell form in the Palaeozoic cephalopods, other than in these bizarre examples, gave rise to a formerly held conception of a gradual increase in coiling from the straight to the fully coiled condition. This, however, is an incorrect view, for each shell form, even if of unusual appearance, is fully adapted to its own particular mode of life, and buoyancy adaptations in particular have been a primary evolutionary control.

It is generally accepted that most orthoconic cephalopods were free-swimming forms carrying their shells in a horizontal position. Some evidence of this comes from the rarely preserved colour markings that occur on the dorsal side of the shell only and appear to have been a kind of camouflage. But more importantly, the internal structures of the shell also indicate a horizontal position. Within the shell, as mentioned, there are endosiphuncular deposits whose function is unknown. There are also regularly shaped masses of calcareous material known as **cameral deposits** (Fig. 8.19c–e). These cameral deposits were secreted progressively from the apical end as the shell grew; they are concentrated near the apex and developed less and less in the chambers nearer to the aperture. They must have given extra weight to the apical end' throughout the growth of the shell, thus allowing a continued equilibrium in a horizontal structure, with the centres of buoyancy and gravity staying close to one another (Flower 1957). Only orthoconic and cyrtonic shells have cameral deposits, for in the coiled shells of the Nautilida and

Ammonoidea the two centres lie in the same vertical plane so that the problem of stability is solved another way. Presumably, as in the modern *Nautilus*, all Palaeozoic cephalopods had some kind of siphuncular buoyancy control, and a slight change in density would have enabled the living animals to rise or sink.

Post-Palaeozoic nautiloids.   A very few orthoconic nautiloids survived into the Triassic but otherwise only the coiled forms of the Nautilida carried on. The elaborate siphuncles of the other orders are never found here; there is only a thin, subcentrally situated siphuncular strand. Most Nautilida, except for some late Palaeozoic cyrtocones, have coiled shells which may be **involute** or **evolute.** Involute shells have the last whorl entirely covering all the former whorls; in evolute shells the former whorls are all visible. Sometimes there is external ribbing or even spines in the post-Palaeozoic genera, and occasional forms such as the Triassic *Clymenonautilus* and the early Tertiary *Hercoglossa* (Fig. 8.19j) and *Aturia* have sutures reminiscent of the goniatites of the Carboniferous.

Subclass Ammonoidea

Cephalopods of Subclass Ammonoidea (Dev.–Cret.) and especially the Mesozoic forms known in the vernacular as 'ammonites' are amongst the most abundant and well known of all fossils. Their beautiful planispiral shells, often strikingly ornamented with external ribbing, have been aesthetic treasures to innumerable collectors. Yet to stratigraphers their usefulness transcends their visual attraction, for by nature of their rapid evolution, abundance and widespread distribution they are the most valuable of all fossils for zoning the rocks in which they occur. They have proved of special effectiveness in the Triassic, Jurassic and Cretaceous

Figure 8.19 Morphology of fossil 'nautiloids': (a) *Proterocameroceras* (Endoceratoidea) (L. Ord. (x 0·4); (b) *Vaginoceras* (Endoceratoidea) (M. Ord.), with endocones within the siphuncle, and long septal necks (x 0·35); (c) Actinoceratid morphology, shell and siphuncle partially dissected (x 1 approx.); (d) *Campyloceras* (Nautiloidea) (L. Carb. (x 1); (e) *Michelinoceras* (Nautiloidea) (Ord.–Trias), with long septal necks and cameral deposits (x 1); (f) *Glossoceras* (Nautloidea–Ascoceridae) (Sil.) (x 1·8): three growth stages; (i) juvenile cyrtocone; (ii) truncated cyrtocone with 'ascocerid' portion growing; (iii) mature ascocerid which has shed the cyrtocone part; (g) *Pentameroceras* (Nautiloidea) (Sil.), an ovoid form with modified aperture (x 0·75); (h) *Lituites* (Nautiloidea) (Ord.), proximal part partially coiled (x 0·35); (j) *Hercoglossa* (Palaeocene.), with 'goniatitic' sutures (x 0·35); (k) *Lobobactrites* (Dev.) Wissenbacher Schiefer: an X-radiograph of an endocochlear nautiloid; total length 65mm (photo by courtesy of W. Stuermer). *(See facing page)*

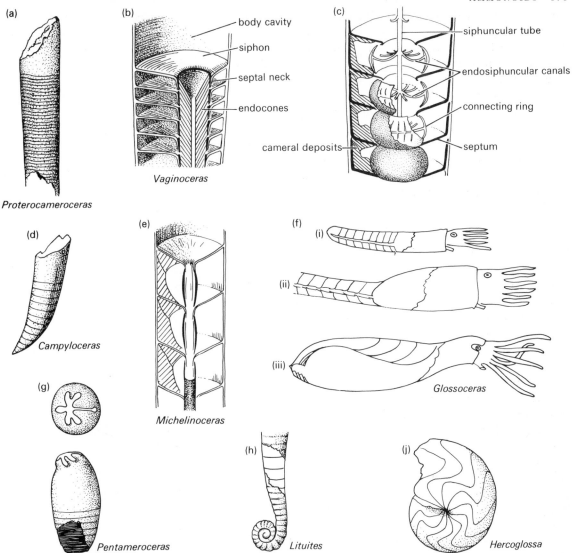

(a) *Proterocameroceras*

(b) *Vaginoceras*
- body cavity
- siphon
- septal neck
- endocones

(c)
- siphuncular tube
- endosiphuncular canals
- connecting ring
- cameral deposits
- septum

(d) *Campyloceras*

(e) *Michelinoceras*

(f) (i) (ii) (iii) *Glossoceras*

(g) *Pentameroceras*

(h) *Lituites*

(j) *Hercoglossa*

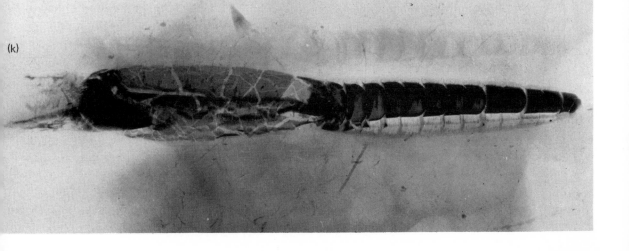

(k)

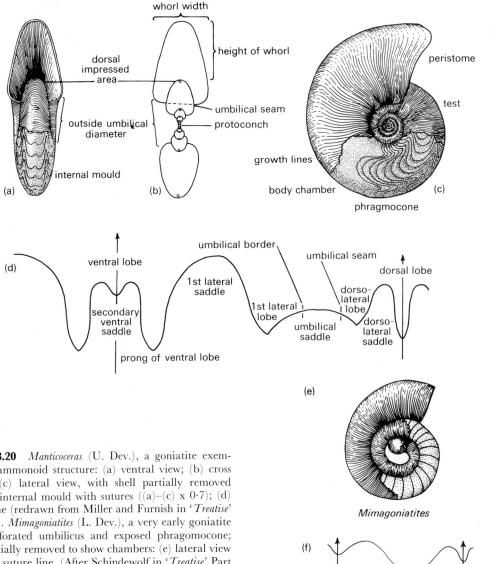

Figure 8.20 *Manticoceras* (U. Dev.), a goniatite exemplifying ammonoid structure: (a) ventral view; (b) cross section; (c) lateral view, with shell partially removed showing internal mould with sutures ((a)–(c) x 0·7); (d) suture line (redrawn from Miller and Furnish in '*Treatise*' Part (L)). *Mimagoniatites* (L. Dev.), a very early goniatite with perforated umbilicus and exposed phragomocone; shell partially removed to show chambers: (e) lateral view (x 5); (f) suture line. (After Schindewolf in '*Treatise*' Part (L))

systems, where their high turnover of species has made it possible to erect zones equivalent to time periods of less than a million years' duration.

The ammonoids are presumed to be tetrabranchiate cephalopods. They may have been derived from the Bactritoidea: a straight-shelled cephalopod subclass which ranged through the Palaeozoic; the shells of these have a bulb-like protoconch and marginal siphuncle, are very similar to those of ammonoids in all respects other than coiling, and like them have no cameral deposits.

How ammonoid shells differ from those of other cephalopods
(Fig. 8.20)

In the ammonoids, except in certain peculiar genera known as **heteromorphs,** the shell is planispirally coiled. Some Palaeozoic nautiloid shells also have this form, but these usually have a central perforation which is absent in all ammonoids except some of the early ones. The suture line normally follows a complex pattern; each suture marks the junction of a septum with the inner surface of the shell wall, and

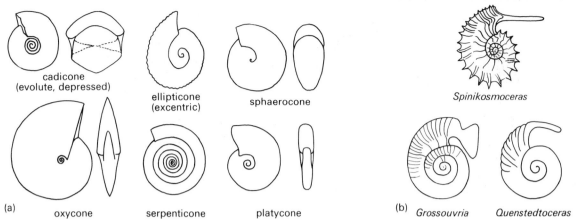

Figure 8.21 Shell shape and morphology in ammonoids: (a) nomenclature for different shell shapes; (b) apertural modifications (lappets and spines) in Jurassic ammonites. (Modified from various authors in '*Treatise*' Part (L))

the array of suture lines is visible in nearly all ammonites since the aragonitic shell readily dissolves in diagenesis. If an individual ammonite septum or a mould of it is examined in face view (i.e. as if looking down the aperture), it appears flat or slightly curving in the centre but becomes increasingly frilled towards its point of attachment to the shell, where it becomes the suture. In the earlier ammonoids, of Devonian and Carboniferous time, the sutures are often simple zigzags, but from the Triassic to the Cretaceous the complexity of the ammonoid suture is considerable.

In most ammonoids the siphuncle is situated near the outer margin (**venter**). One order, the Upper Devonian Clymeniida, is typified by a dorsal suture running along the inner margin of the shell (**dorsum**); in this group too the septal necks are **retrochoanitic** (backwardly pointing), as in nautiloids, and very long. This contrasts with the most advanced ammonoid condition, where the septal necks are short and **prochoanitic** (directed forwards).

Whereas the shells of coiled nautiloids are often unornamented or have only a feeble external sculpture, ammonoid shells are frequently ribbed, and the **ribs** may have knobs, tubercles or spines; in very compressed forms there may be a **keel**. In such genera as the Jurassic *Kosmoceras* such developments are carried to an extreme, and in addition lateral **lappets** are present on either side of a compressed aperture (Fig. 8.21g–j).

Such primary differences in morphology clearly distinguish ammonoid from those of nautiloid and other Palaeozoic cephalopods. Normally the soft parts of ammonoids have been considered basically similar to those of nautiloids, and ammonoid reconstructions have been made very largely on this basis (e.g. Fig. 8.18b). Yet there are some recent discoveries (Lehmann 1971a) that, though allowing for the general correctness of this picture, indicate that the biological affinities of ammonoids are in many ways closer to coleoids (dibranchiate cephalopods such as squids, cuttlefish and octopuses) than to *Nautilus*. Radulae have been found in a few ammonites, and these have seven rows of teeth as in the Recent dibranchiate cephalopods, whereas *Nautilus* has thirteen. And ammonoid jaws, where known, correspond closely with those of recent octopuses; furthermore, there are known ammonoid ink sacs, which the Recent *Nautilus* does not have. Such evidence, though admittedly slender, suggests that ammonoids and coleoids stand rather close together phylogenetically whereas *Nautilus* may be more distantly related. So though the internal anatomy of the extinct ammonoids may be conceived as similar to that of the modern *Nautilus*, such reconstruction remains for the moment no more than a convenient model.

Morphology and growth of the ammonoid shell

Ammonoid shells are characteristically tightly coiled in a planispiral fashion. They form rolled-up cones of which the earliest-formed part is a small bulbous ellipsoidal protoconch. This was probably initially inhabited by a planktonic larva. The protoconch is usually located in the centre of the shell, but in occasional early forms (e.g. Fig. 8.20e) the coiling is looser so that there is an umbilical perforation between the protoconch and the first whorl, like that of the early coiled nautiloids. The septate part of the shell (**phragmocone**) usually grows in a logarithmic spiral from the protoconch, but in some Palaeozoic genera the shell shape may be curiously quadrilateral or triangular. As the shell grows from the apertural end the siphuncle grows with it. The siphuncle starts as a small bulb (the **caecum**) within the protoconch; it is initially thick relative to the first-formed chambers and may wander about in its position. After the first one or two whorls, however, it settles down into its adult, normally ventral position, occupying relatively less space as the shell grows.

The septa were probably secreted in rapid growth episodes, like those of *Nautilus*, followed by resting periods, and the part of the shell where the animal actually resided (body chamber) grew progressively further from the centre during the spiral growth of the phragmocone.

There are several useful terms for describing the final form of the shell (Fig. 8.21). Those shells which are involute have the last whorl covering all the previous whorls; in evolute shells most or all of the previous whorls are exposed. The **umbilicus,** which is the concave surface on each side through which the spiral axis runs, may be almost flat or deeply excavated. The shell may be of normal rather flattened form (**planulate),** very compressed (**oxy-cone),** very fat and inflated (**cadicone)** or almost globular (**sphaerocone).**

Some ammonoids have pronounced ventral keels (Fig. 8.21a), which may be blunt or sharp. The keel may have a single, double or triple parallel ridge system, running along the venter. In cross-section the whorls may take many forms, which are often characteristic of particular ammonoid families and thus very useful in taxonomy.

The outer surface of an ammonoid shell is usually marked by faint growth lines, but in addition there are usually ribs, radially arranged and projecting above the surface. There are so many patterns that only a few can be illustrated here. Some ribs go over the venter; others are interrupted; some are straight whilst others curve or branch; they may be united at **nodes;** and shorter ribs may be intercalated between longer ones.

Tubercles may erupt on points on the ridge or be independent of them; they may be prolonged into spines or united by flat nodes. The external pattern of ribs and associated structures is of great value in classification and identification, but since homeo-morphy is always a possibility their use must be coupled with that of other characters, particularly the nature of the suture.

The curious compression of some ammonoid apertures, and their frequent ornamentation by lappets or spines, have already been mentioned. Some of the more peculiar forms are illustrated here (Fig. 8.21b), but these are unusual and the majority of ammonoid apertures, and especially those of Palaeozoic genera, are simple like those of nautiloids.

In ammonoids of various kinds the body chamber may be long or short as is discussed in the section on buoyancy.

The ammonoid suture (Fig. 8.22)

Most ammonoids are preserved as internal moulds, and in these the junctions between septa and shell walls show up clearly as suture lines, which in ammonoids are always more complex than in nautiloids.

In the early ammonoids (Devonian and Carboniferous) the sutures were normally of a fairly simple form and lack accessory crenulations, but by Permian times some genera were showing more complex sutures of a kind that reached their full flowering in the Mesozoic, when genera with extremely complex sutures were the norm.

To clarify ammonoid suture morphology, the sutures are usually represented graphically. Neglecting surface convexity of the ammonoid shell the sutures are drawn from the venter to the **umbilical seam** (the external suture) and thence to the **dorsum** of that septum. Involute ammonoids have a distinct internal suture, but in the evolute forms the internal suture is small and insignificant and is normally not drawn. Since the full suture is symmetrical about the dorso-ventral axis only half a septum is normally drawn. And because no account is taken of the surface convexity the suture diagram is only a projection, which can never be entirely accurate, but this is unimportant for purposes of comparison of different suture lines. Such diagrams are not only useful for comparing the sutures of

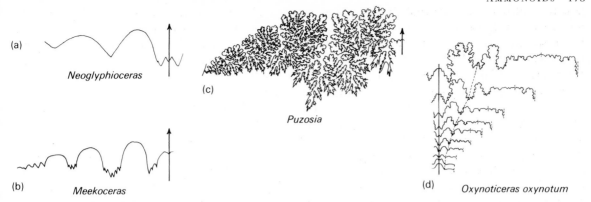

（a）

Neoglyphioceras

（b）

Meekoceras

（c）

Puzosia

（d）

Oxynoticeras oxynotum

Figure 8.22 Suture morphology in ammonoids: (a) goniatitic suture *(Neoglyphioceras)* (L. Carb.); (b) ceratitic suture *(Meekoceras)* (Trias.); (c) ammonitic suture in *Puzosia* (L. Cret.); (d) sutural ontogeny in *Oxynoticeras oxynotum* (Jur.). (Modified from various authors in '*Treatise*' Part (L))

different species and genera, but also essential in tracing the changes in sutural morphology throughout the growth of an individual, which has various uses.

In drawing a suture the venter is shown on the left hand side with an upwardly directed arrow pointing towards the aperture; the umbilical seam is shown as a curving line, and another vertical arrow marks the dorsum on the right hand side. Inflections on the suture line pointing upwards (the apertural direction) are the **saddles** (easily remembered since a horse's saddle faces forwards), whereas the backwardly pointing inflections (facing downwards on the diagram) are the **lobes.**

The suture is most important in taxonomy; particular kinds of sutures characterise distinct ammonoid families and are very useful in classification and identification. There are some broad and general terms defining ammonoid groups, based on sutural morphology, which are often used to characterise different grades of organisation. Thus the **goniatites** are Palaeozoic ammonoids with sharply angular and generally zigzag sutures, without any accessory crenulations. Not all ammonoids with such sutures are goniatites; there are some unrelated Mesozoic genera with similar sutures. In the Triassic **ceratites** the lobes are frilled though the saddles are entire. Some curious Cretaceous genera (**pseudoceratites**) have similar sutures. The Mesozoic **ammonites** all have finely subdivided and complex lobes and saddles, though there are some Permian ammonoids with sutures also of this kind. Though

there is thus a general stratigraphical increase in sutural complexity, the progression from goniatite through ceratite to ammonite is not, however, a direct phylogenetic line.

In many Palaeozoic and Mesozoic ammonoids the ontogeny of the suture has been worked out (Fig. 8.22d), normally by breaking off the chambers one by one as far as the protoconch and drawing the sutures as the inner whorls are exposed. The early sutures in all ammonoids are far less complicated than the later ones, and the whole history of development of a mature suture can be traced using serial diagrams of successive sutures. Such ontogenetic series have several uses. They have been used, for instance, to distinguish homeomorphs in which the mature suture in two distantly related ammonoids looks similar but has been arrived at ontogenetically in quite different ways.

Perhaps the greatest value of sutural ontogeny is in unravelling phylogenies, especially in Palaeozoic forms, for in these ammonoids the earlier stages in development of an 'advanced' suture closely resemble the mature septa of a more primitive ammonoid type. Thus ancestor–descendant relationships can be postulated, and phylogenies have been drawn up on this basis which have later been shown to hold firm in the light of more evidence. At one time ammonites were thought to demonstrate Haeckel's 'biogenetic law' in which their ontogeny was supposed to recapitulate their phylogeny, but this has been shown to be quite wrong. The suture-line does show parallel trends in ontogeny and phylogeny, but in most other characters the ontogenic and phylogenetic sequences occur in the reverse order. Systems of classification built up by such experts as Hyatt and Buckman (who both subscribed to the Haeckel theory) have had to be abandoned.

During ontogeny, in Palaeozoic and Mesozoic genera, primary lobes and saddles appear early, and normally the first and second lateral saddles are very distinct. Though these persist into adult suture, **adventitious** lobes or saddles may appear in between the primaries and eventually grow as large as the primaries themselves. The only way to distinguish which is which is then to trace them back to the earliest ontogenetic stages.

In the ammonites, where the suture reaches its maximum complexity, the lobes and saddles are all crenulate, and the saddles all have accessory lobes. The terminology of all these, as given in Figure 8.20d, becomes rather complex. Some rather peculiar sutures are present in a few Mesozoic genera. In certain examples each mature suture is apparently truncated against the one in front, though in fact the 'missing' parts continue below the preceding septum. Septa may be asymmetrical, and even taxonomically unstable within the one genus.

Siphuncle, suture and buoyancy in ammonoids

Ammonoid shells are always thinner than the shells of *Nautilus;* their siphuncles are ventral (except in Order Clymeniida); the septal necks are normally prochoanitic in ammonoids and retrochoanitic in nautiloids; and the septa in ammonoids are usually convex, not concave, towards the aperture.

It is generally held that ammonoids achieved buoyancy in much the same way as does *Nautilus* and that the fluted septa increased the strength of the shell so that it was able to resist implosion at depth. But to be more specific than this, and most particularly to be able to perform mathematical analysis, various other factors must be known, including the nature and composition of the siphuncle, to see if the shell could have functioned like that of *Nautilus*. There is good evidence from rare unaltered specimens that calcium phosphate is the primary constituent of ammonoid siphuncles, though juveniles have more calcium carbonate and probably the phosphatisation was preceded by a calcitic stage. The calcitic outer tube of *Nautilus* is missing. Now in certain Mesozoic

ammonites the strength of the siphuncular tube (against explosion) has been calculated. Needless to say, the tube strength is an important limitation on depth. The relative strength index $(h/r) \times 100$, where h is wall thickness and r is the tube radius, has been calculated for several Mesozoic ammonoids, and it is most interesting to see that two Orders Phylloceratida and Lytoceratida ($h/r \times 100 = 10$ to 19) have a strength index like *Nautilus*, whilst their derivatives the Ammonitida have a significantly lower value (3 to 6·5). This information accords with the recognition that the Phylloceratida and Lytoceratida were deeper water groups, whose derivatives repeatedly invaded shallow water by a classic process of iterative evolution. The calculations show that the Phylloceratida and Lytoceratida could withstand a water column of some 450 m, about the same as that of *Nautilus*, whereas the Order Ammonitida could probably not withstand depths of more than 100 m. Other data which have been used in attempts to infer bathymetry in ammonoids, such as the thickness of the shell wall and of the septa, have not proved so useful since the relationships involved are very complex and elude simple analysis.

A very ingenious indication of buoyancy control in ammonoids was made recently by Heptonstall (1970). He used a specimen of the ceratite *Buchiceras* (Fig. 8.23) which had overgrowths of oysters. It had previously been shown conclusively by Seilacher (1960) that this specimen had been able to remain afloat in spite of the increasing weight of the oysters; some had grown on the venter, and from their large size they must have grown for several months before the animal died. Heptonstall was able to calculate the weight of the oysters individually and collectively, which was relatively large in relation to the weight of the shell. In order to retain neutral buoyancy whilst the oysters were growing there must initially have been normal large quantities of cameral liquid in the shell, which the ammonoid was able to remove to counteract the weight of oysters. The calculations used in this case may have some avoidable inaccuracies, but even so it is hard to escape the conclusion stated that significant quantities of liquid must have been required in these shells and that they functioned in a manner somewhat similar to the shell of *Nautilus*.

Kennedy and Cobban (1976) have discussed many aspects of ammonite distribution in great detail.

Buchiceras

Figure 8.23 *Buchiceras* covered with epizoic oysters (x 0.7 approx). (After Seilacher, 1960, and Heptonstall, 1970)

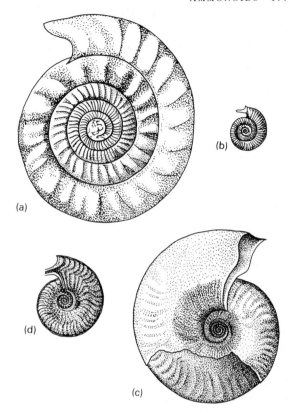

Figure 8.24 Sexual dimorphism in Jurassic ammonites: (a) *Perisphinctes (Arisphinctes) ingens*, (b) *Perisphinctes (Dichotomosphinctes) rotoides*, both from the same horizon in Corallian (x 0.15) – these are probably dimorphs; (c) *Graphoceras cavatum*; (d) *Ludwigina cornu*, a Bajocian dimorphic pair (x 0.65). (Redrawn from Callomon, 1963)

Sexual dimorphism in ammonoids (Fig. 8.24)

The question of whether or not ammonoids were sexually dimorphic has been debated at great length for a very long time. As early as 1869 Waagen, one of the early German palaeontologists, pointed out that in ammonites of Family Oppeliidae (Order Ammonitina) there were apparently two parallel phylogenetic lineages developing throughout the Middle and Upper Jurassic. Furthermore, he found that in some ammonites pairs of 'species' could be distinguished, differing only in the character of their outer whorls; the nuclei could not be told apart. Waagen, however, did not believe that sexual dimorphism could be invoked as an explanation, and it was nearly 100 years later that Callomon (1963) and Makowski (1963) convincingly demonstrated the criteria that could be used in establishing dimorphism unequivocally.

In trying to work out dimorphic pairs it is essential that all the ammonite shells studied are mature specimens, i.e. adult shells in which growth had ceased. Such maturity in ammonite shells is marked by:

(a) crowding (approximating) of the last few septa, due to diminishing growth rates;
(b) a change in the external sculpture near the aperture – the ribbing may be different, or the aperture may be constricted or marked by lappets or horns;
(c) some slight uncoiling of the body chamber from the rest of the shell.

Not all these features are found in any one ammonite, and indeed some of them are confined to one sex only; nevertheless, the presence of only one or two such characters will enable a mature shell to be distinguished. Most ammonoid shells found are in fact mature, and in specimens collected from the same bed mature shells of the same kind are normally very much alike in size.

When mature shells from the same bed and locality are separated into 'species', they often come out into two quite distinct groups, without intermediates. These are distinguished primarily by size, one group being two to four times larger than the other so that the two kinds are generally referred to as **micro-conchs** and **macroconchs.** In microconchs and macroconchs from the same horizon, the inner whorls are normally indistinguishable and only the outer whorls differ in number, size and morphology. If there are any peculiar features of the aperture, these are normally found only in the microconchs whilst the macroconchs have simple apertures, and the more extreme cases of detachment of the body chamber from the preceding whorl are likewise characteristic of the microconchs.

It is theoretically possible that the apparent sexual pairs are merely closely related taxa living in the same place at the same time, but since in such lineages new characters (such as modifications of external sculpturing) appear in both groups at the same time, then sexual dimorphism appears to be the only likely possibility. But all these criteria have to be used with caution in interpreting ammonoid population, for they do not always, by any means, occur in pairs, and in some cases the variation within a population collected from a single locality may be enormous. Hence we have to agree with Lehmann (1971b) who adjures us 'to be extremely careful when it comes to maintaining sexual relationships between hitherto blameless ammonites'.

It is hard to know which morph is male and which is female. Some authors have considered the question so academic as to be unanswerable; others have suggested positively that the microconch is the male, and the macroconch the female. *Nautilus* shows very little dimorphism and is thus no help, but some Mesozoic nautiloids have been described as clearly dimorphic.

Since all known dimorphic pairs were originally described as separate species, what should be done with them taxonomically when they are recognised as dimorphic? Should dimorphic 'species' be grouped together under the one name as true biological species, or should the established names continue in use as 'morphospecies'?

Undoubtedly the former practice is more technically correct, but the latter is perfectly admissible if it is clear that we are talking about 'morphospecies' and not biospecies, and it does have the advantage of greater flexibility. For the moment palaeontologists are divided on this question, though the definition of true biospecies appears to be gaining ground.

Other structures in Ammonoids

In recent years exceptionally well-preserved material has yielded fossilised ink sacs and possibly egg masses inside a few specimens (Lehmann 1971a). Furthermore, both radulae and horny jaws have been found; it is presumed that the ammonoids cut their food up with their jaws but used the radulae to help with swallowing. These various structures are important not only from the biological point of view but also in that they resemble those in coleoids, pointing to a fairly close relationship between ammonoids and coleoids, whereas the modern *Nautilus* stands further apart.

Aptychi and anaptychi (Fig. 8.18c)

Commonly found in association with Mesozoic ammonites are paired calcitic plates known as **aptychi,** which superficially resemble bivalves. Aptychi are usually scattered on bedding planes where ammonites are abundant, and where the aragonitic shells of the latter are crushed as in so many Jurassic shales; the calcite aptychi retain their original slightly curving relief. Sometimes no ammonites are found, only aptychi, where the aragonite shells have been dissolved completely. Rarely aptychi have been found inside body chambers of ammonoids, meeting like double doors with their straight edges vertical and their external edges fitting the shape of the aperture. They have a convex ornamented outer surface and presumably functioned as a pair of opercula, protecting the body of the ammonoid when it was withdrawn inside the body chamber. In some Cretaceous genera the aptychi are fused into a single calcite plate.

There are also **anaptychi,** which consist of a single plate of chitin or other organic material, normally flattened and in a butterfly shape. Anaptychi do not fit the aperture properly and thus presumably were not opercula. Some rare Liassic anaptychi have been discovered inside the body chamber of ammonites; here they have been shown to consist of two opposing components very like the jaws of coleoids, and as such they are interpreted. The little that is known about ammonoid feeding habits suggests that they fed on slow-moving live animals (gastropods, for instance), carrion and vegetable matter.

Phylogeny and evolution in the Ammonoidea (Fig. 8.25)

The earliest ammonoids are of Lower Devonian age. They belong to an ancestral stock, Order Anarcestida, which may have been derived from the straight-shelled Subclass Bactritoidea. In both groups there are similar, almost straight sutures, and there is a large bulbous protoconch which is accommodated in the Anarcestida by a perforate umbilicus. Nevertheless, other early genera do not have the perforate umbilicus, and very rapidly there arose more 'advanced' features, such as involution of the shell and increased sinuosity of the sutures leading to the goniatitic condition. All the early forms have ventral retrochoanitic siphuncles, but a second Palaeozoic Order is distinguished from all other ammonoids by a dorsal siphuncle. In this Order, the Upper Devonian Clymeniida, the shells may be involute or evolute, and the sutures are goniatitic. The third of the Palaeozoic Orders is the Goniatitida, which includes by far the majority of all Palaeozoic forms. All these have a prochoanitic siphuncle, apart from a few primitive genera, and typical goniatitic suture lines.

These Palaeozoic groups continued throughout the Devonian and Carboniferous and into the Permian; only in the Upper Permian were there any signs of a real decrease in numbers, for many of the dominant stocks of the Upper Palaeozoic had by then become extinct. At the end of the Permian, as happened with so many groups, widespread extinction was the rule, and only a very few genera belonging to fourth Order Prolecanitida crossed the Permian/Triassic boundary.

Shortly after the beginning of the Triassic the few Permian survivors gave rise to a vastly successful and 'explosive' adaptive radiation. The Triassic ammonoids nearly all belong to Order Ceratitida, which has several superfamilies. They are broadly known as 'ceratites', and indeed in most members the suture is some variant on the established ceratitic suture theme, though the external ornament became quite complex in some groups (e.g. Superfamily Tropitacea, whose members are generally ornamented with strong ribs or nodes). But a ceratitic suture is not characteristic of all Ceratitina. Indeed, as has often been pointed out, the suture in *Pinacoceras* (Superfamily Pinacocerataceae) reaches a degree of 'ammonitic' complexity barely rivalled by that of Jurassic or Cretaceous genera. A few peculiarly coiled 'heteromorphs' are known.

In the early Triassic there also arose the ammonite stock that was to give rise to all post-Triassic ammonoids. This is Order Phylloceratida, an almost smooth-shelled group with a characteristic 'phylloid' suture (Fig. 8.26a). The history of this Order is interesting, on account of both its stratigraphical persistence (it continued until the Cretaceous) and its remarkable conservatism. Within the group there was extraordinarily little evolution, but its evolutionary offshoots became the very diverse ammonites of the Mesozoic.

From the Phylloceratida there arose another persistent Order, Lytoceratida (Fig. 8.25b), which originated around the Triassic/Jurassic boundary. Though the main root-stock genera (such as *Lytoceras*) in parallel with the Phylloceratida became little differentiated, they produced a greater number of radiations than did the parent Phylloceratida. Both the Phylloceratida and Lytoceratida produced radiating lineages, but the phylogeny is so complex and the difficulties in reconstructing phylogeny are so great that in many cases it is not known for certain which of the parent stocks gave rise to which descendants. Hence the taxonomy is unclear, and ammonite students have normally retained a polyphyletic (or 'ragbag') Order Ammonitida to accommodate them, more as a matter of convenience than for scientific accuracy.

The patterns of evolution illustrated in the chart display **iterative evolution,** in which an ancestral stock from time to time gives rise to short-lived groups which replace each other successively. These new groups expand and diversify for a while, but they are geologically speaking ephemeral and in due course become extinct. The niche they vacate is then occupied by descendants of the same long-lived ancestral stock.

The two ancestral Orders, Phylloceratida and Lytoceratida, continued throughout the Mesozoic and gave rise to superfamilial side branches (the Eoderocerataceae, Hildocerataceae, etc.). Each of these dominated the scene for a while and during its 'little hour of grace' constituted a miniature adaptive radiation, rapidly producing short-lived families so that the phylogenetic tree resembles the prongs of a toasting fork. Iterative evolution as shown by the Mesozoic ammonites has also been described under the well-chosen term 'the palaeontological relay'. Because of the rapidity of turnover, wide distribution,

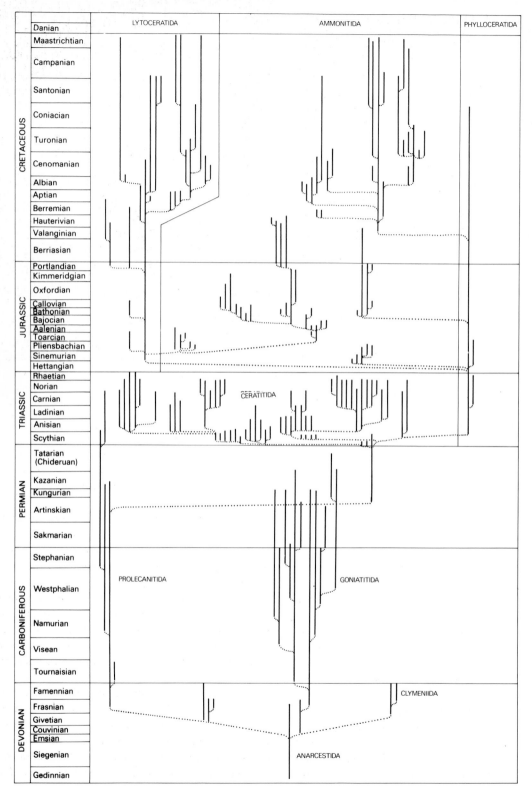

Figure 8.25 Evolutionary pattern of ammonoid families and superfamilies (after Moore in '*Treatise*' Part (L)).

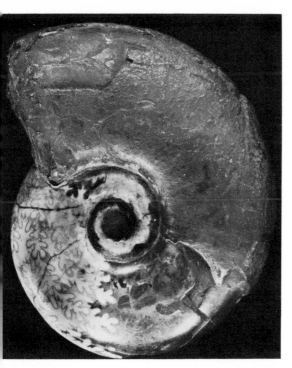

Figure 8.26 (a) *Phylloceras*, an Upper Triassic phylloceratid from the red limestones of Hallstatt, Austria (x 0.7). Shell removed and surface polished to show body chamber (filled with brecciated debris), and sutural patterns, picked out by crystallisation within chambers. This group represents the rootstock of all post-Triassic ammonites, but itself remained very conservative. (Royal Scottish Museum) (b) *Lytoceras fimbriatus* (L. Lias.) Lyme Regis, England (x 0·35): a representative of one of the two deepwater stocks from which many Ammonitina arose.

abundance and ease of recognition of the ammonites, they are of outstanding zonal value.

Heteromorphs and extinction in Ammonites (Fig. 8.27)

It is a well-known fact that at certain periods in geological history some groups of ammonites evolved shells of highly aberrant form. Such shells are known as heteromorphs. Some appeared during the late Triassic, some in the Jurassic, and there was a more extensive development of heteromorphs during the later Cretaceous. Heteromorphs may be loosely coiled with the whorls wholly or partially separated, e.g. *Choristoceras* (Trias.), *Spiroceras* (Jur.) and *Lytocrioceras* (Cret.). Other genera are almost straight, e.g. *Bochianites* (Cret.), resembling baculitids. Others have the body chamber hooked over the top like a walking stick, e.g. *Hamulina* (Cret.). In such bizarre genera such as the Cretaceous *Macroscaphites* and *Scaphites*, though the early whorls are of normal form, the shell is then straight and finally sharply recurved near its termination, so that the aperture faces the first-formed whorls. The Triassic *Cochloceras*, and the Cretaceous *Turrilites*, *Ostlingoceras* and other genera, are helically coiled like gastropods, whilst the Cretaceous *Heteroceras* has its first few whorls of helical shape and the rest of the shell like *Scaphites*. *Nipponites*, from the Cretaceous of Japan, is perhaps the most extreme of all heteromorphs, having a very long tubular shell coiled in a series of U-bends into an unlikely tangle. There are many other such irregularly coiled genera in the Cretaceous, and their existence has given rise to very much evolutionary speculation.

For a long time, the palaeontological literature on heteromorphs was dominated by the idea that such shell forms were degenerate, retrogressive and biologically inadaptive. Furthermore, because the Triassic and Cretaceous episodes of heteromorphy took place shortly before major extinction periods for

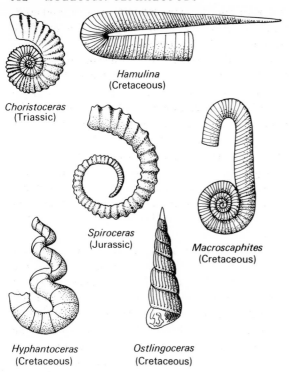

Choristoceras
(Triassic)

Hamulina
(Cretaceous)

Spiroceras
(Jurassic)

Macroscaphites
(Cretaceous)

Hyphantoceras
(Cretaceous)

Ostlingoceras
(Cretaceous)

Figure 8.27 Heteromorphic ammonoids showing variety in form (redrawn from '*Treatise*' Part (L)).

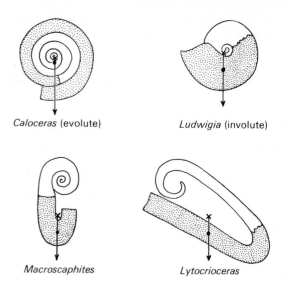

Caloceras (evolute)

Ludwigia (involute)

Macroscaphites

Lytocrioceras

Figure 8.28 Possible life attitudes of normal and heteromorphic ammonoids, as calculated by Trueman, 1941. The body chamber is stippled.

the ammonoids, it seemed to many palaeontologists that there was a definite relationship between heteromorphy and extinction. During the 1920s and 1930s the view was quite widespread that internal rhythms in evolution eventually culminated in a kind of 'racial senescence' during which bizarre and overspecialised forms were produced, as a last and final extravaganza before inevitable extinction overtook the degenerating stock.

But the concept of heteromorphs as degenerate and inadaptive phylogenetic end-forms, plausible though it seemed to Schindewolf and its other proponents, is now no longer generally held. To begin with, it has been shown (Wiedmann 1969) that heteromorphs appeared in certain lineages only and were not characteristic of the ammonoids as a whole at any one time. In the Triassic there were only four closely related heteromorphic genera, quite long-ranged and probably specialised for bottom living, which became extinct at about the same time as eight normally coiled superfamilies; clearly there is no likelihood of a

causal connection between heteromorphy and extinction. Likewise the seven known genera of Jurassic heteromorphs were probably monophyletic and were specialised bottom dwellers, whereas Cretaceous heteromorphy, which reached its maximum development in the early Cretaceous, seems to have been a polyphyletic phenomenon. Curious shell forms appeared in several lineages quite suddenly, often associated with reduction of the primary suture. Many of the uncoiled or partially coiled heteromorphs that had been present in the Lower Cretaceous produced descendants with normal or near-normal coiling; this again falsifies the view that heteromorphs were overspecialised end-forms from which no further evolution was possible. Presumably their ability to return to normal coiling when appropriate ecological niches were available was a factor in the subsequent success of the re-coiled genera.

It has been shown, furthermore, by density/buoyancy calculations (Trueman 1941) that heteromorphs such as *Macroscaphites* and *Lytocrioceras* were well adapted for floating in the water in particular attitudes (Fig. 8.28). Whilst the apertures do not necessarily face in the same directions as those of normally coiled involute or evolute genera, there is no reason to believe that the forms were in any way unfunctional.

Though there were some heteromorphs at the end of ammonite history the majority of the ammonites at that time were normally coiled. In the early Upper Cretaceous the ammonites went into a slow decline over a long period of time, and towards their final end they had become restricted to certain parts of the world only. The number of genera became fewer, and finally no new characters appeared. There is no question of an internally weakened stock; the decline was rather the adverse effect of long-continued environmental conditions, of a kind detrimental to the ammonites and probably associated with a series of marine regressions. The last major Cretaceous regression, at the Maastrichtian/Danian boundary, coincided with the final demise of the last ammonites; there are cephalopods in the succeeding Danian stage, but they are coiled nautiloids of the genera *Aturia* and *Hercoglossa* (Fig. 8.19j) which have zigzag suture lines. These presumably were occupants of one of the ecological niches vacated by the ammonites, but Danian nautiloids apparently did not invade the multiplicity of microniches that their ammonite precursors had done.

Subclass Coleoidea: dibranchiate cephalopods

Modern coleoids

The dibranchiate cephalopods or coleoids include squids, cuttlefish, octopuses, the paper-nautilus or *Argonauta,* and various extinct groups, notably the belemnites. They all have only a single pair of gills within the mantle cavity, and it is this feature that separates them from all the other cephalopods.

The modern squids and cuttlefishes range from tiny animals only 4 cm long to the gigantic *Megateuthis* which is (including the tentacles) 18 m long. Yet their geological importance is limited; only the belemnites are normally preserved as fossils, and in any case coleoids are more diverse and abundant now than they were formerly.

One kind of dibranchiate structure is shown by the cuttlefish *Sepia* (Order Sepiida) which is illustrated in Figure 8.18f, g. It differs from *Nautilus* in having only eight arms, provided with suckers and hooks. These hooks, like the jaws, are horny and are not usually preserved fossil, though some rare fossilised examples are known. In the mantle cavity with its two gills there is an ink sac, used for clouding the water as an escape reaction from predators. Otherwise, and apart from shell structure, the internal organisation of the body is broadly comparable.

The shell of *Sepia* is wholly internal. It is a large oval body, known in the vernacular as the **cuttlebone,** located dorsally and composed of closely-spaced oblique calcareous partitions supported by pillars. The cuttlefish controls its buoyancy by secretion and extraction of liquid from within the spaces between partitions. Such a shell as this seems at first sight very dissimilar to that of a nautiloid, but it is morphologically the dorsal half of the nautiloid phragmocone, which has become flattened and expanded. In the squid *Loligo* the shell (sea-pen) is horny.

Since the chambers within the partitions are very narrow, the change in buoyancy through the pumping out of water can be a very slow process, at least as regards water deep within the chamber. Nevertheless, buoyancy regulation by the alteration of water balance within the chambered shell is not the only system in modern squids. Quite a large number of squids e.g. *Cranchia* have instead adopted the use of ammonium chloride within the tissues, which is isotonic and isomotic with sea water and thus gives neutral buoyancy. In these cranchids the coelomic cavity is vastly distended and filled with ammoniacal liquid of s.g. $1 \cdot 010–1 \cdot 012$, which is less than that of sea water (s.g. $1 \cdot 026$). Though such ammoniacal squids are very abundant and the use of ammonia is not confined to the cranchids, the system cannot apparently be used for rapid buoyancy changes.

Packard (1972), in discussing the evolutionary convergences and interactions between cephalopods and fish, has put the ammoniacal and other squids into an interesting evolutionary perspective. He has shown how by early Palaeozoic times cephalopods were the most advanced and mobile of all marine animals. With their chambered shells and buoyancy mechanism along the lines of *Nautilus* they were highly successful, though limited in how deep they could go due to danger of implosion. But when fish began to diversify and invade the upper waters of the sea, in the later Palaeozoic, one effect of such competition was to put pressure on living space. The origin of the 'ammoniacal' buoyancy system was one answer to the problem, for it frees those cephalopods which possess it from the limitations of the shell, and it is perhaps not surprising that the deeper parts of the sea have been extensively colonised by ammoniacal squid. Another solution to competition from fish was that many cephalopods became more fish-like, structurally and functionally, so that the two were then competing on more equal terms. Packard's statement that 'cephalopods, functionally are fish' is deeply significant to the issue.

Spirula (Fig. 8.18h, j) is a little squid some 10 cm long which has an open spiral chambered shell enclosed entirely within the body. The shell has a ventral siphuncle which connects with the chambers only through a very small porous region in each. The lower chambers are largely filled with liquid which is normally inert within them, but this can be added to or subtracted from through the porous region so that the shell's buoyancy can be controlled in a similar manner to that of *Nautilus*.

The shell of the female *Argonauta* (paper nautilus) (Fig. 8.29n) is a very thin spiral shell, but it is formed by secretion of the tentacles and carried external to the body. It is used as a brood pouch, more rarely for carrying about a captured male! The male, incidentally, is shell-less and only about one-twentieth the size of the female. It has a very long sexual organ (the **hectocotylus**) which carries the sperm; this can detach itself from the male and swim actively towards the female – a kind of 'guided missile copulation' which seems to be unique in nature.

Extinct coleoids and their evolution (Fig. 8.29)

In very many Jurassic and Cretaceous sediments there may be an abundance of fossil belemnites. These are the internal shells of fossil squid-like cephalopods, but they do not closely resemble the shells of any modern squid or cuttlefish at all.

In the typical genus *Belemnites* the shell has three parts. The largest and most posterior section is the **guard:** a massive bullet-shaped cylinder of solid calcite. It is parallel-sided, tapers posteriorly to a point and is indented at its anterior end by a conical cavity (the **alveolus**). If the guard is cut transversely the structure is seen as radially oriented needles of calcite, with concentric growth rings, which are also apparent in horizontal section. The axis of growth is not central, however, but placed towards the ventral margin. Usually the surface of the guard is smooth, but it may be granular or pitted. Some genera, such as *Hibolites,* have a long ventral groove extending two-thirds of the way from alveolus to point, and the posterior part of the guard itself is somewhat swollen. *Actinocamax* is more parallel-sided but likewise has a ventral slit below the alveolus. Some species of *Duvalia* have a curious laterally flattened guard.

Within the alveolus and fitting it exactly is the **phragmocone.** This is a conical thin-walled aragonitic structure which projects outside the alveolus. It is septate with its septa concave anteriorly and separating fair-sized camerae. A slender siphuncle threads through the septa at the ventral margin. These structures, together with a tiny bulbous protoconch, leave no doubt that the coleoid phragmocone is the direct homologue of the shell of a nautiloid, ammonoid or *Spirula*.

The guard, which is such a substantial part of the belemnite, has no direct homologue in other coleoids. It seems to have acted as a necessary counterweight to maintain the belemnite body level when swimming, i.e. fulfilling in a different way the same function as the cameral deposits of nautiloids. Since the guard consists of thick and unrecrystallised calcite, it has proved most useful in palaeotemperature analysis of the Jurassic and Cretaceous by O^{16}/O^{18} isotope ratios (Fig. 8.29j).

The third component of the belemnite shell is the **pro-ostracum:** a long flat expanded tongue projecting forwards and presumably covering the anterior part of the body. It seems to be homologous to the 'pen' of the squid *Loligo*. This is rarely preserved, however, and its function is poorly understood. Sometimes specimens with eight radiating sets of hooks have been found forwards of the guard, testifying to the former presence of arms, and even fossilised ink sacs have been located in place.

Belemnites seem therefore to have been a kind of fossil squid, with a different hard-part construction to others and therefore with a unique system of buoyancy control.

Geological history of coleoids

Belemnites have been reported from erratic boulders in North America believed to be of Mississippian age; these are currently the oldest belemnites known. It is generally agreed that the forerunners of the Mesozoic belemnites were derived from orthocone cephalopods about this time through the expansion of the thin covering of the tip of the phragmocone into the

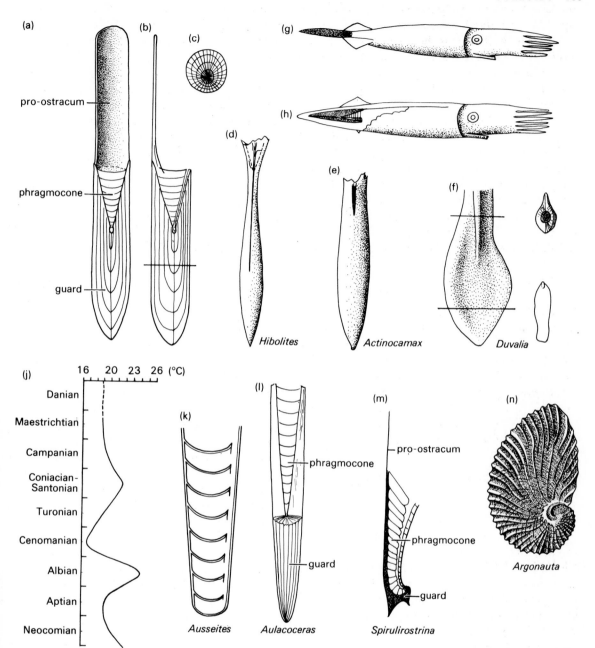

Figure 8.29 Coleoid morphology: (a) longitudinally sectioned belemnite in ventral view; (b) in lateral view; (c) guard cut transversely (x 0·75); (d) *Hibolites* (Cret.) (x 1 approx.); (e) *Actinocamax* (Cret.) (x 0·75); (f) *Duvalia* (Cret.) (x 0·75); (g) reconstruction of swimming belemnite, with the guard projecting from the body and exposed (based on Seilacher, 1968, *Palaeogeog., Palaeoclimatol. Palaeoecol.* **4**, 279–85); (h) the more conventional interpretation of Naef and others, with the guard wholly internal; (j) climatic fluctuations in the Cretaceous, plotted from oxygen isotope ratios in belemnite guards (after Bowen, 1961, *J. Palaeont.* **35**, 714–18); (k) *Ausseites* phragmocone (Trias.) (x 0·75); (l) *Aulacoceras* (Trias.) (x 0·75); (m) *Spirulirostrina* (Mio.) in section (x 0·75); (n) 'shell' of Recent paper-nautilus *Argonauta* (x 0·35). (Mainly based on illustrations of Naef and others in Roger, J. in Piveteau (ed.) *Traité de paleontologie*, II. Paris: Masson.)

massive guard by a simple process of relative growth. Little is known about the earlier stages in belemnite evolution, and only in the Triassic are there encountered well-known forms which in some ways have characters intermediate between those of orthocones and belemnites proper. The Triassic *Ausseites,* a genus from Austria, has a small guard and a long phragmocone with a marginal siphuncle. *Aulacoceras* has a larger, externally ribbed guard, though the phragmocone otherwise resembles that of *Ausseites.* It is possible here to assume an evolutionary progression from an orthocone with external thickening, through the Triassic genera to the true belemnites. The latter expanded and diversified throughout the Jurassic and Cretaceous and then, like the ammonites, declined exceedingly, though a very few belemnites continued into the early Tertiary before becoming extinct.

A possible side branch of Mesozoic 'belemnites' is represented by the Triassic *Phragmoteuthis* and Jurassic *Belemnoteuthis.* In these the guard is present but is very thin and delicate, whilst the phragmocone is short and broad with a very long pro-ostracum extending from it. The first octopuses are known from the Cretaceous of Lebanon, and there are some fossil argonauts known from the Tertiary, but otherwise the fossil record of these modern types is so poor that it is not worth mentioning further. But there are some

extinct coleoids of Tertiary age whose fossil history is interesting if a little confusing. Some genera have a somewhat reduced guard and a spiral coil to the early part of the phragmocone. The Eocene *Belosepia* and Miocene *Spirulirostina* are of this kind; in both the siphuncle is much expanded and the ventral part of the phragmocone reduced. Though this kind of structure is in some ways intermediate between that of the belemnites and *Sepia,* which has a very reduced guard, the fossils are very few, and to relate the end members directly in a sequence *Belemnites–Belosepia–Sepia* is at present unjustified.

The Order Teuthida, to which *Loligo* belongs, seem to have developed along another line, and the pro-ostracum has become an important internal structure at the expense of the phragmocone and guard. Teuthoids are first found fossil in early Tertiary time.

The ancestry of the little squid *Spirula* with its coiled internal shell is unknown. It has no guard or pro-ostracum, but only a loosely coiled phragmocone with a large protoconch and marginal siphuncle.

Since the buoyancy mechanisms of modern coleoids are now beginning to be well understood, it may be hoped that this will give a correspondingly broader conception of the means whereby fossil cephalopods, including coleoids, were able to control their buoyancy. It is unquestioned that this has been the most important single factor in their evolution.

BIBLIOGRAPHY

Books, treatises, symposia

Moore, R. C. (ed.) 1969. *'Treatise'* Part (I). Mollusca 1. (Principles, gastropods)

Moore, R. C. (ed.) 1964. *'Treatise'* Part (K). Mollusca 3. (Nautiloids)

Moore, R. C. (ed.) 1957. *'Treatise'* Part (L). Mollusca 4. (Ammonoids)

Moore, R. C. (ed.) 1971. *'Treatise'* Part (N). Vols. 1–3 (1969–71) (Bivalves)

Morton, J. E. 1967. *Molluscs.* London: Hutchinson. (Valuable summary of morphology and relationships)

Piveteau, J. (ed.) 1952. *Traité de paleontologie. II Mollusques.* Paris: Masson & Cie. (Shorter than the *'Treatise'*, but an essential compilation)

Individual references

Bathurst, R. G. C. 1975. *Carbonate sediments and their diagenesis.* Amsterdam: Elsevier. (Fine structure of molluscan shells)

Batten, R. L. 1966. *The Lower Carboniferous gastropod fauna from the Hotwells Limestone of Compton Martin, Somerset.* Paleontogr. Soc. monogr., no. 119, 1–52; no. 120, 53–109. (Monographic study of a diverse fauna)

Callomon, J. H. 1963. Sexual dimorphism in Jurassic ammonites. *Trans Leics. Lit. Phil. Soc.* **57,** 21–56. (Many examples cited, with methodology)

Denton, E. J. 1961. The buoyancy of fish and cephalopods. In *Prog. biophys. biophys. Chem.* **11,** 178–234. (Essential reference on attainment of negative buoyancy)

Denton, E. J. and J. B. Gilpin-Brown 1966. On the buoyancy of the pearly nautilus. *J. Mar. Biol. Ass. UK* **46,** 723–59. (Siphuncular structure and function)

Fischer, A. G. and C. Teichert 1969. Cameral deposits in cephalopod shells. *Univ. Kansas Paleont. Contrib.* **37,** 1–30. (Cameral morphology and use as counterweights)

Flower, R. H. 1957. Nautiloids of the Palaeozoic. In *Treatise on Marine Ecology and Palaeoecology* 2, H. S. Ladd (ed.). Geol. Soc. Amer. Mem. no. 67, 829–52. (Ecology and annotated bibliography)

Garstang, W. 1951. *Larval forms and other zoological verses.* Oxford: Blackwell. (Gastropod torsion)

Gould, S. J. 1969a. An evolutionary microcosm: Pleistocene and Recent history of the land snail *P. (Poecilozonites)* in Bermuda. *Bull. Mus. Comp. Zool.* **138,** 407–532. (Allopatric speciation)

Gould, S. J. 1969b. Ecology and functional significance of uncoiling in *Vermetularia spirata:* an essay on gastropod form. *Bull. Mar. Sci.* **19,** 432–45. (Uncoiling useful for rapid upward growth)

Heptonstall, W. B. 1970. Buoyancy control in ammonoids. *Lethaia* **3,** 317–28. (*Buchiceras* remained afloat in spite of increasing weight of oysters)

Hudson, J. D. 1963. The recognition of salinity-controlled mollusc assemblages in the Great Estuarine Series (Middle Jurassic) of the Inner Hebrides. *Palaeontology* **11,** 163–82. (See text)

Jefferies, R. P. S. 1960. Photonegative young in the Triassic lamellibranch *Lima lineata* (Schlotheim). *Palaeontology* **3,** 362–9. (Use of lunule as a brood chamber)

Jefferies, R. and R. P. Minton 1965. The mode of life of two Jurassic species of *Posidonia* (Bivalvia). *Palaeontology* **8,** 156–85. (Planktonic mode of life postulated)

Kennedy, W. J., J. D. Taylor and A. Hall 1968. Environmental and biological controls on bivalve shell mineralogy. *Biol. Rev.* **44,** 499–530. (Shell mineralogy subject to many controls)

Kennedy, W. J. and W. A. Cobban 1976. *Aspects of ammonite biology, biogeography and biostratigraphy,* 1–94, Spec. pap. in paleont., no. 17. (Invaluable treatment)

Lehmann, U. 1971a. Jaws, radula and crop of *Arnioceras* (Ammonoidea). *Palaeontology* **14,** 338–41. (Unusual preservation)

Lehmann, U. 1971b. New aspects in ammonite biology. *Proc. North Amer. Paleont. Conv. I,* 1251–2269. (As above, with interpretation)

McAlester, A. L. 1962. Systematics, affinities and life habits of *Babinka,* a transitional Ordovician lucinoid bivalve. *Palaeontology* **8,** 231–41. (Multiple muscle scars suggest affinities with segmented ancestors)

Makowski, H. 1963. Problems of sexual dimorphism in ammonites. *Acta Palaeont. Polonica.* **12,** 1–92. (Comes independently to the same conclusions as Callomon (1963))

Mutvei, H. 1971. The siphonal tube in Jurassic Belemnitida and Aulacocerida. *Bull. Geol. Inst. Uppsala N.S.* **3,** 27–36. (Detailed electron micrography)

Mutvei, H. 1972. Ultrastructural studies on cephalopod shells. Part I: The septa and siphonal tube in *Nautilus.*

Part II: Orthoconic cephalopods from the Pennsylvanian Buckhorn Asphalt. *Bull. Geol. Inst. Uppsala. N.S.* **3,** 237–73. (Detailed micrography)

Packard, A. 1972. Cephalopods and fish: the limits of convergence. *Biol. Rev.* **47,** 241–307. (Evolutionary interactions between fish and cephalopods; a valuable work)

Pojeta, J. and B. Runnegar 1976. *The palaeontology of rostroconch mollusks and the early history of the Phylum Mollusca,* US Geol. Surv. Prof. Pap., no. 968. 1–88. (Rostroconchs as bivalve ancestors)

Raup, D. M. 1966. Geometrical analysis of shell coiling: general problems. *J. Paleont.* **40,** 1178–90. (See text)

Raup, D. M. 1967. Geometrical analysis of shell coiling: the cephalopod shell. *J. Paleont.* **41,** 43–65. (See text)

Raup, D. M. and S. M. Stanley 1971. The cephalopod suture problem. In their *Principles of paleontology,* 172–81. San Francisco: Freeman. (The best treatment of this theme)

Runnegar, B. and D. Gobbett 1975. *Tanchintongia,* gen. nov.: a bizarre Permian myalinid bivalve from West Malaysia and Japan. *Palaeontology* **18,** 315–22. (An extreme case of functional adaptation)

Seilacher, A. 1960. Epizoans as a key to ammonoid ecology. *J. Paleont.* **34,** 189–93. (See text)

Seilacher, A. 1968. Swimming habits of belemnites – recorded by boring barnacles. *Palaeogeography, Palaeoclimatol. Palaeoecol.* **4,** 279–85. (Suggestion that the guard was in life exposed)

Skelton, P. W. 1976. Functional morphology of the Hippuritidae. *Lethaia* **9,** 83–100. (Perforated 'lid' allowed ingress of water currents)

Stanley, S. M. 1968. Post-Palaeozoic adaptive radiation of infaunal bivalve molluscs: a consequence of mantle fusion and siphon formation. *J. Paleont.* **42,** 214–29. (Radiation dependent on one key character)

Stanley, S. M. 1970. *Relation of shell form to life habits in the Bivalvia (Mollusca).* Geol. Soc. Amer. Mem., no. 125. (Extended treatment of Recent bivalve functional morphology)

Stenzel, H. B. 1971. Oysters (Mollusca 6.). *Treatise* Part (N) Vol. 3. (A full account of morphology, evolution and ecology. Deals succintly with evolution in *Gryphaea*)

Stuermer, W. 1970. Soft parts of trilobites and cephalopods. *Science* **1170,** 1300–2. (Soft tissues external to shell of Devonian orthocones)

Taylor, J. D., W. J. Kennedy and A. Hall 1969. The shell structure and mineralogy of the Bivalvia. Introduction, Nuculacea–Trigoniacea. *Bull. Brit. Mus. Zool. Suppl.* **3,** 1–125. (Standard work)

Trueman, A. E. 1941. The ammonite body chamber, with special reference to the buoyancy and mode of life of the living ammonite. *Quart. J. Geol. Soc. Lond.* **96,** 339–83. (Classic study of functional morphology of unusual shell shapes in ammonoids)

Wiedmann, J. 1969. The heteromorphs and ammonoid extinction. *Biol. Rev.* **44,** 563–602. (Full documentation of heteromorphy as a functional phenomenon)

Wilbur, K. M. 1961. Shell formation and regeneration. In

Physiology of Mollusca Vol. 1, K. M. Wilbur and C. M. Yonge (eds), 243–81. New York and London: Academic Press. (Secretion and development of bivalve shell)

Yochelson, E. L. 1967. *Quo vadis Bellerophon?* In *Essays in palaeontology and stratigraphy*, 141–61. Univ. Kansas Spec. Pub., no. 2. (Bellerophontids may not have been gastropods)

Yochelson, E. L., R. H. Flower and G. F. Webers 1973. The bearing of the new later Cambrian monoplacophoran genus *Knightoconus* upon the origin of the Cephalopoda. *Lethaia* **6,** 275–310. (Early diversification of cephalopods)

9 Echinoderms

The familiar starfish and sea-urchins which are so common in shallow waters are representative of an entirely marine phylum, the Echinodermata (Cam.–Rec.), which stands apart zoologically from nearly all other invertebrate groups. Echinoderms of all kinds have internal **mesodermal skeletons** of porous calcite plates, which are normally spiny and covered outside and in by a thin protoplasmic skin. Normally the skeletons have a five-rayed or **pentameral symmetry,** though in some fossil groups this is not so and in some modern and fossil sea-urchins a bilateral symmetry is superimposed upon the radial plan. Another important feature of echinoderms is the **water-vascular system:** a complex internal apparatus of tubes and bladders containing fluid. This has extensions which emerge through the skeleton to the outside as the **tube-feet** or **podia.** Tube-feet have various functions, especially locomotion, respiration and feeding. They may be considered as all-purpose organs used by animals living inside a calcite box for manipulating the environment.

Because of their calcitic skeleton echinoderms are very abundant in the fossil record, and often their remains have greatly contributed to carbonate sediments. Thus crinoidal limestones, composed largely of the stem fragments of sea-lilies, are very common in some rocks, notably the Carboniferous. In such rocks the porous plates have often been impregnated with diagenetic calcium carbonate but are otherwise unchanged. Echinoderms are, however, stenohaline, and their remains are only found in sediments of fully marine origin.

CLASSIFICATION

A modern system of echinoderm classification as defined in the '*Treatise*' and modified by Sprinkle (1976) is:

SUBPHYLUM 1. ECHINOZOA: Radiate echinoderms, usually globose or discoidal.
  CLASS 1. ECHINOIDEA (Ord.–Rec.): Sea-urchins.
  CLASS 2. HOLOTHUROIDEA (Ord.–Rec.): Sea-cucumbers.
  CLASS 3. EDRIOASTEROIDEA (L. Cam. – L. Carb.) There are several other extinct classes.
SUBPHYLUM 2. ASTEROZOA
  CLASS STELLEROIDEA (Ord.–Rec.)
  SUBCLASS 1. ASTEROIDEA (Ord.–Rec.): Starfish.
  SUBCLASS 2. OPHIUROIDEA (Ord.–Rec.): Brittle-stars.
  SUBCLASS 3. SOMASTEROIDEA (Ord.–Rec.)
SUBPHYLUM 3. CRINOZOA: 'Pelmatozoans', i.e. echinoderms having a small plated body (**calyx**) fixed by a **stem,** and with pinnulate arms adapted for food gathering.
  CLASS CRINOIDEA (L. Cam.–Rec.): Sea-lilies.
SUBPHYLUM 4. BLASTOZOA: 'Pelmatozoans', often stalked, lacking free arms, but with biserial **brachioles** for food gathering, and often various respiratory structures in the cup.
  CLASSES 1 and 2. DIPLOPORITA and RHOMBIFERA (?Cam.–Dev.): Cystoids. Extinct groups with perforated plates in the calyx.
  CLASS 3. BLASTOIDEA (Sil.–Perm.): Extinct 'pelmatozoans' with complex respiratory structures.
  CLASS 4. EOCRINOIDEA (L. Cam.–Sil.): Primitive echinoderms with pores along the sutures.
SUBPHYLUM 5. HOMALOZOA (Ord.): Rare peculiar organisms, calcite-plated but with no planes of symmetry. These have been the subject of much controversy and may be a separate chordate subphylum on their own: the CALCICHORDATA.

Only the important groups are given here, but there are many others. In particular there was a marked proliferation of short-lived echinoderm classes in the Lower Palaeozoic. Some of these seem to combine characters typical of many groups and are hard to classify. They are often known only from a single locality and a small number of specimens. But though these are clearly echinoderms, their characters are so different from those of other known echinoderms that separate classes (e.g. the echinozoan Classes Helicoplacoidea, Cyclocystoidea and Lepidocystoidea) have had to be established to accommodate them. Though there are some Cambrian echinoderms, the epoch of their maximum proliferation (at class level) was the Ordovician as can be seen from their times of origin in the above list. At generic level, however, echinoderms were most abundant in the Carboniferous. A whole range of remarkable forms arose at that time in a great burst of adaptive radiation, but only some of these were successful; the others, to which the Cambrian classes belong, produced no new lines of descent and became extinct.

SUBPHYLUM ECHINOZOA

Class Echinoidea

Morphology and life habits of three genera

ECHINUS

The common sea-urchin, *Echinus esculentus* (Order Echinoida), which lives in shallow waters round the North Atlantic, shows the fundamentals of echinoderm structure as organised in an animal well adapted for a benthonic free-living mode of life (Fig. 9.1a, b).

Echinus has a globular **test** some 10 cm in diameter which is slightly flattened at the poles. In life it is covered with short (1–2 cm) spines; if these are removed the plating structure is visible. On the upper (**aboral** or **adapical**) surface there is a central **apical disc:** a double ring of plates surrounding a central hole or **periproct,** which contains the anus. The apical disc is formed of two types of plates: the larger **genital plates,** and the smaller **ocular plates** which are usually outside the ring of genitals. Each is perforated by a pore. The **genital pores** are the outlets of the **gonads,** and the **ocular pores** are part of the water-vascular system. One genital plate (the **madreporite**) is larger than the others. It has numerous tiny perforations which lead into the water-vascular system below. The anus resides in the

centre of a number of small plates attached to a flexible, rarely fossilised membrane extending across the periproct.

The test is divided into ten radial segments extending from the apical disc to the **peristome** (q.v.) which surrounds the mouth on the lower (**adoral**) surface. The five narrower segments are the **ambulacra** (ambs) which connect with the ocular plates, whereas the broader **interambulacra** (interambs) terminate against the genital plates. Both ambulacra and interambulacra consist of double columns of elongated plates which meet along a central suture in a zigzag pattern. In the ambulacrum this is the **per-radial suture.** The interambulacral plates are large and tubercular, without perforations, but the ambulacral plates each have three sets of paired pores near the outer edge of the plate. These **pore-pairs** are the sites where the tube-feet emerge through the test from the internal part of the water-vascular system.

The ambulacra and interambulacra are widest at the **ambitus,** which is the edge of the specimen when seen from above or below. The peristome is a large adoral area, covered in life by a flexible plated membrane, which contains the mouth centrally. In fossil specimens, however, the membrane has normally gone, leaving a large circular or pentagonal cavity. Five pairs of **gill-notches** are found where the interambulacra abut the edge of the peristome, and from these project feathery bunches of gills which provide surfaces for respiratory exchange additional to those of the tube-feet. Inside the periproct the test is turned back into a perforated flange which is the **perignathic girdle** (Fig. 9.2). This girdle forms a support for the masticatory apparatus of the echinoid: the **Aristotle's lantern** (Figs 9.2, 9.26d). This lantern has five strong jaws, each with a single calcitic tooth. The whole assembly is suspended by ligaments and muscles attached to the perignathic girdle. It operated as a kind of five-jawed grab, and though each jaw has only limited play the teeth can rasp away at organic detritus or algal material on the sea floor and pass it inwards to the gut.

Within the test (Fig. 9.2) most of the soft parts are related to the structures already described. Inside the test is a thin layer of protoplasm, and since the gut is only a simple tube running spirally round the inner wall from mouth to anus the body of the test is largely empty. But at breeding time, which is normally in the

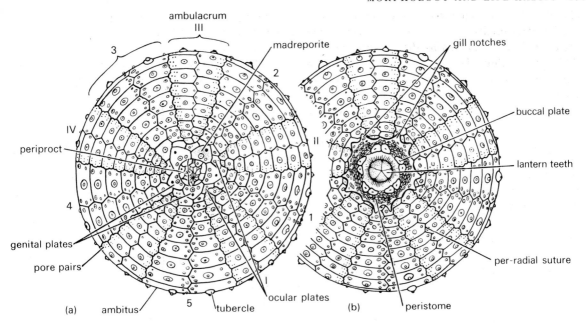

Figure 9.1 *Echinus esculentus* (Rec.), test deprived of spines: (a) aboral (adapical) surface; (b) adoral surface (x 0·8). (Redrawn from Durham in '*Treatise*' Part (U))

summer, the gonads swell enormously before releasing their products through the genital pores. Since echinoids often live in clumps or congregate to spawn, the chances of cross-fertilisation of eggs and sperm from male and female individuals are fair. The ciliated **echinopluteus larvae** which grow from the zygotes swim actively in the plankton and undergo many transformations before finally settling down.

The **coelom** has various tubular elements. Of these the **haemal** and **perihaemal** systems seem to be involved in material transfer, and the **axial organ** seems to be associated with the repair of injury, but their specific functions are unclear.

The water-vascular system of the echinoderms, which is also coelomic, resembles nothing else in the animal kingdom. Its primary function is to operate the tube-feet. In echinoids its only exit from the test is via the madreporite. From this a calcified tube (the **stone-canal**) descends to near the top of the lantern. Here it joins the **circum-oral ring,** from which five **radial water vessels** extend, one running up the centre of each ambulacrum. Each of these passes finally through an ocular pore, but it only forms a tiny closed tube (apparently light-sensitive in some

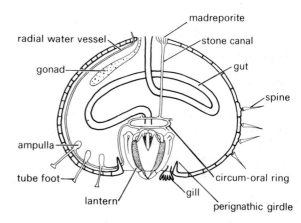

Figure 9.2 Internal morphology of *Echinus* (simplified) passing through an ambulacrum (left side) and interambulacrum (right side).

echinoids). From the radial water vessels there arise, at intervals, paired lateral tubes, each leading to a tube-foot and associated apparatus (Fig. 9.3). At the base of each tube-foot is an inflatable sac (the **ampulla**), and the tube-foot leads outwards from this, dividing as it passes through the pore-pair and reforming on the other side. This device prevents the tube-foot from being withdrawn right inside the test when retracted, and since one of the functions of the

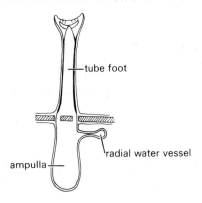

Figure 9.3   Structure of tube foot in section, showing radial water vessel, ampulla, and distal part of tube foot with longitudinal muscles and suction cup.

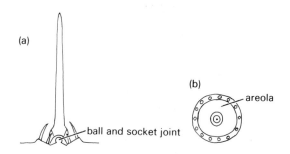

Figure 9.4   Spine attached to tubercle, showing ball and socket joint and areola with protective spines surrounding it: (a) section; (b) tubercle with spine removed.

tube-foot is respiration it also separates incoming oxygen-rich water from the outgoing fluid depleted in oxygen. The tube-foot possesses longitudinal muscles and has a suction cup at the end, rendering it prehensile. Within the water-vascular, haemal and perihaemal systems are many amoeboid cells (**coelomocytes**) which perform numerous functions.

Echinus moves by using its tube-feet, especially those on the lower part of the body. It can extend the elastic tube-feet for a considerable distance, approximately half of those in any one ambulacrum being extended at a given time, the other half retracted. A tube-foot will extend when water pressure within it increases due to contraction of the ampulla, the radial water vessel itself or a neighbouring tube-foot. As water comes into it the longitudinal muscles relax; when these contract water is forced back into the ampulla which correspondingly relaxes. At the time of maximum extension the suction cups on the end of the extended tube-feet adhere to an adjacent part of the sea floor. When the tube-feet contract the echinoid moves along the sea floor, supported by spines. *Echinus* moves slowly over the sea floor in this way, feeding voraciously with its jaws and defended against predators by its armament of spines. Each spine, like the individual plates of echinoids, is a single crystal of calcite. The spine base forms a socket which articulates with the ball-joint of the tubercle below it; the tubercle is crystallographically continuous with the plate (Figs 9.4, 9.24a). Round the tubercle is a ring of muscles, attached to the spine base so that the spine can be moved in any direction.

Amongst the spines are small organs of balance (**spheridia**), on the adoral part of the per-radial suture, and **ophicephalous pedicellariae,** which are tiny spines with their heads modified as pincers (some with poison glands); these clean the surface, discourage predators and prevent larvae from settling (Fig. 9.5). Normally pedicellariae lie recumbent on the surface, but they can be erected and will snap shut on any extraneous object. New pedicellariae are formed when any are dislodged in defence.

Echinus is a **regular** echinoid: one in which the periproct opens in the centre of the apical disc (**endocyclic**). Such regular echinoids are common today and in the fossil record, and they live either on the sea floor or, like *Strongylocentrotus*, in cavities in rocks which they may have excavated themselves. Regular echinoids are normally illustrated according to a conventional orientation (Figs. 9.1a, 9.20a, b). The madreporite is always shown at the right anterior and with its dependent interambulacrum is numbered 2. The numbering proceeds anticlockwise (as seen from the adapical pole) so that genital 5 is always posterior. Roman numerals designate the oculars and ambulacra, likewise numbered anticlockwise, but starting to the right of the genitals. The same system is used in numbering the plates of **irregular** echinoids: those with a dominant bilateral symmetry, marked particularly by the position of the periproct which is no longer within the apical system (**exocyclic**). *Echinocardium* and *Mellita*, described below, are two very dissimilar irregular echinoids, with different life habits.

Figure 9.5 Opened ophicephalous pedicellaria with sensory hairs and poison spines.

ECHINOCARDIUM (Fig. 9.6)

Echinocardium (Order Spatangoida) is abundant in shallow water but unlike *Echinus* lives in a burrow within the sediment. Its morphology is highly modified in accordance with this burrowing mode of life. Living below the surface gives it protection against predators, but it has to maintain an adequate connection with the surface for food supply and respiratory exchange, and it also has to be able to cope with sanitation. It has in fact essentially the same problems of life as a subsurface-living bivalve. These problems have been solved mainly by reshaping of the test and by extreme modification of the tube-feet and spines in different ways for doing different jobs.

In *Echinocardium* a bilateral symmetry is superimposed upon the radial symmetry. The test is covered with a mat of short spines, but when these are removed it is seen as heart-shaped in plan, a flattened ellipse in profile. The aboral surface possesses an elongate apical disc from which the periproct is absent; the latter is located on the nearly vertical posterior wall of the test. A single ambulacrum (III), dissimilar to the others, is located in a deep anterior groove and goes straight towards the peristome. The other ambulacra are paired (II + IV; I + V). Each of these is in two parts. The aboral parts are expanded into four recessed leaf-like **'petals'**, which terminate above the ambitus. The ambulacra continue below this level but are of more normal form, flush with the surface and less pronounced. In the petals the outer pore of each pore-pair is elongated, slit-like and widely separated from the round inner pore. From these emerge flattened respiratory tube-feet, leaf-like and rectangular. Elsewhere, including within the anterior ambulacrum, the pore-pairs are more normal, though they may carry tube-feet specialised for other functions. Adorally (Fig. 9.6b) the peristome

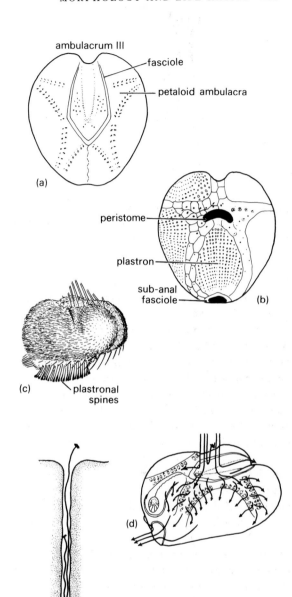

Figure 9.6 *Echinocardium cordatum* (Rec.): (a) aboral view (x 0·8; (b) adoral view; (c) lateral view with all spines present; (d) oblique lateral view showing current directions; (e) in life position in burrow (after Nichols, 1959).

is located far forwards and has a projecting lip (or **labrum**) below. There is no lantern. Adorally the plates are enlarged, and from their large pore-pairs emerge sticky food-gathering tube-feet. Behind the mouth is a flattened area (the **plastron**) formed from the modified posterior interambulacrum and densely covered with flat paddle-shaped spines.

In life *Echinocardium* lives in a burrow up to 18 cm deep, with a single funnel connecting it with the surface (Fig. 9.6e). This funnel is both created and maintained by enormously long tube-feet with star-shaped ends resembling flue brushes. These emerge from the adapical part of the anterior ambulacrum, their bases protected by a pyramid of spines which help in building the lower part of the funnel. A set of similar tube-feet, emerging from non-petaloid regions of the two posterior ambulacra below the vertical rear wall, build a single 'sanitary tube' to receive excreta.

Two regions of the test known as **fascioles** generate currents. The larger fasciole surrounds the anterior ambulacrum as a ribbon-like strip; the smaller (**subanal**) fasciole is an elliptical ribbon located below the anus. In both the surface is covered with small vertical spines (the **clavulae**), each of which is covered by innumerable cilia, as in the intervening epithelium; it is the co-ordinated beat of these cilia that produces the currents whose direction is shown in Figure 9.6d. A strong current goes down the anterior ambulacrum, and food particles are caught by the sticky tube-feet and passed to the mouth. Other currents bathe each of the paired ambulacra, facilitating respiratory exchange, and the current from the subanal fasciole propels waste matter into the sanitary tube. When the sanitary tube is filled up *Echinocardium* moves forwards. The funnel-building spines are withdrawn, the anterior spines are erected and scrape away at the front wall of the burrow, whilst the paddle-like spines attached to the plastron move the echinoid forwards. A new funnel is created whilst the old funnel, burrow and sanitary tube collapse behind it.

The whole organism has the same basic elements as does *Echinus*, apart from the lantern and girdle which appear only in embryo and are soon lost through being resorbed. Other than the modified shape of the test, it is mainly a division of labour between the spines and tube-feet that enables *Echinocardium* to live far below the surface whilst feeding, respiring and excreting effectively.

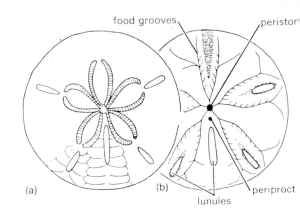

Figure 9.7 *Mellita quinquiesperforata* (Rec.): (a) aboral (b) adoral view. (x 1·5)

Specific bioturbation structures in the sediment often of complex form, result from the forward movement of such heart-urchins as *Echinocardium* These are clearly recognisable and testify to the former presence of sea-urchins moving within the sediment even if the animals themselves have not been fossilised.

MELLITA (Fig. 9.7a, b)

Mellita quinquiesperforata (Order Clypeasteroida) is a flattened sand-dollar common in littoral and sub littoral sands in south-eastern North America. It live either on the surface or buried horizontally within it It is very flat with its ambulacra petaloid aborally and with five perforations (**lunules**) in the test: two pair in the paired ambulacrum, the other being unpaired Adorally the peristome is central, with the anus very close behind it and just in front of the posterior lunule On the adoral surface are five dichotomous channel (the **food grooves**) which are lined with tube-feet They run between the lunules and converge on the mouth. The whole test is covered with a dense mat o fine spines, of which the adoral ones are used for walking. Feeding is carried out by the tube-feet in the food-grooves which are sticky and extensible. They can catch particles of organic material from the substrate and pass it along to the mouth. Long ora spines form a protective mesh over the mouth; they screen out sand but help in the ingestion of organic detritus. *Mellita* normally travels forwards, but it can also rotate and right itself if overturned. As it moves i is usually covered by a thin layer of sediment making characteristic trail, and it sometimes passes sand upwards through the posterior lunule as it progresses forwards.

The lunules are used in filtering sand and selecting food and also serve as short-cuts for food from the aboral surface to the mouth. Excreta pass upwards through the posterior lunule onto the aboral surface and are left behind as *Mellita* moves forwards.

M. sexiesperforata has six lunules but lives in a very similar manner. *Dendraster*, another sand-dollar which is non-lunulate, lives inclined at a high angle to the substrate with most of the posterior part showing above the surface.

CLASSIFICATION OF ECHINOIDS (Fig. 9.8)

The taxonomic division of Class Echinoidea into Subclasses Regularia and Irregularia, originally proposed in 1825, was used until the 1950s. Nevertheless authorities such as Mortensen (1928–52), who continued to employ it, recognised that it was not a phyletic classification and that irregularity had probably arisen more than once. The classification used by Durham and Melville in the *Treatise* and later amended (Durham 1966) is based upon a wider variety of stable characters, defining major groupings. These include the overall structure and rigidity of the test, the number of ambulacral and interambulacral columns, and the structure of the ambulacral plates, lantern and girdle.

At lower taxonomic levels, i.e. the ordinal, the structure of the apical disc and whether or not the echinoid is regular have been used. Thus the classification used in the *Treatise* and adopted here, though not agreed upon by all authorities, is as follows:

SUBCLASS 1. PERISCHOECHINOIDEA (Ord.–Rec.): Regular endocyclic echinoids with interambulacra in one to many columns, ambulacra in two to twenty columns, no compound plates, perignathic girdle simple or absent, and lantern with simple grooved teeth. Includes all Palaeozoic echinoids and the cidaroids which have an aulodont lantern.
SUPERORDER 1. PALAECHINACEA (U. Ord.–Rec.): Three perischoechinoid orders of 'true' echinoids.

ORDER 1. ECHINOCYSTITOIDA (Ord.–Perm.): Have flexible tests.
ORDER 2. PALAECHINOIDA (Sil.–Perm.): Have rigid tests.
ORDER 3. CIDAROIDA (Dev.–Rec.): Have only two columns of plates in the ambulacra, and in the interambulacra of later genera. The interambulacral plates are ornamented with a single large central tubercle.
SUPERORDER 2. MEGALOPODACEA (Ord.): Contains three genera alone: *Bothriocidaris*, *Neobothriocidaris* and *Eothuria*, whose characters place them apart.
SUBCLASS 2. EUECHINOIDEA (U. Trias.–Rec.): Regular or irregular echinoids, with bicolumnar ambulacra and interambulacra. Lantern present or absent. There are eighteen orders arranged in four superorders.
SUPERORDER 1. DIADEMATACEA (U. Jur.–Rec.): Euechinoids with aulodont lantern (like cidaroids) and usually with simple plates, though there may be some compounding. There are many deep-water forms (including ORDER ECHINOTHUROIDA, which has flexible tests), but shallow water species include coral reef inhabitants with poison spines. Of the five orders all are regular except for ORDER PYGASTEROIDA.
SUPERORDER 2. ECHINACEA (U. Trias.–Rec.): Regular echinoids with gill slits, compound plates, a complex perignathic girdle, and stirodont or camarodont lantern. There are seven orders including Order Echinoida, (the taxonomic position of the peculiar ORDER ORTHOPSIDA is unknown; it may belong here).
SUPERORDER 3. GNATHOSTOMATA (Jur.–Rec.): Irregular echinoids devoid of compound plates. Lantern with keeled teeth and girdle present. Apical system and peristome more-or-less opposite one another. There are two important orders, the ORDER HOLECTYPOIDA and ORDER CLYPEASTEROIDA, the latter of which includes the sand-dollars.
SUPERORDER 4. ATELOSTOMATA (Jur.–Rec.): Irregular echinoids with no lantern and girdle in the adult. Apical system and peristome rarely opposite. Ambulacra petaloid or subpetaloid. Three important orders are:

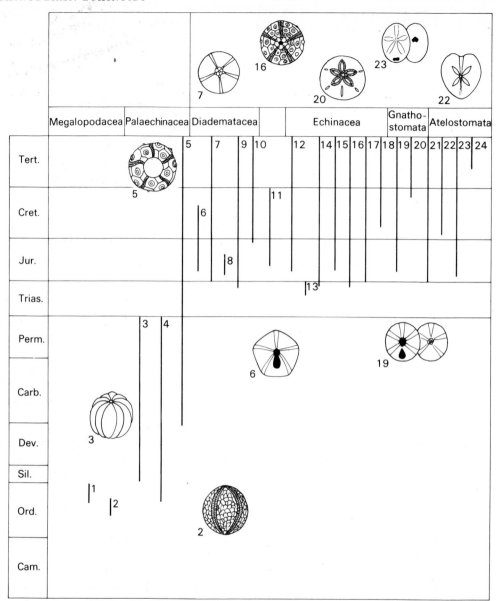

Figure 9.8 Time ranges and classification of Echinoidea: (1) Megalopoda; (2) Bothriocidaroida; (3) Palaechinoida *(Melonites);* (4) Echinocystitoida *(Aulechinus);* (5) Cidaroida *(Cidaris);* (6) Pygasteroida *(Pygaster);* (7) Pedinoida *(Pedina);* (8) Heterocidaridae; (9) Diadematoida; (10) Echinothuroida; (11) Orthopsida (Order uncertain); (12) Arbacoida; (13) Pleisiocidaroida; (14) Salenoida; (15) Temnopleuroida; (16) Hemicidaroida *(Hemicidaris);* (17) Phymosomatoida; (18) Echinoida *(Echinus);* (19) Holectypoida *(Holectypus);* (20) Clypeasteroidea *(Mellita);* (21) Holasteroida; (22) Spatangoida *(Micraster);* (23) Cassiduloida *(Catopygus);* (24) Neolampadoida.

Order 1. Cassiduloida (Jur.–Rec.): Sub-globular with phyllodes and bourrelets well developed.

Order 2. Holasteroida (L. Jur.–Rec.): With many 'sub-spatangoid' features.

Order 3. Spatangoida (L. Cret.–Rec.): e.g. *Micraster, Echinocardium.*

In the following discussion of morphology, function and habit the two primary echinozoan subclasses, the Perischoechinoidea (primitive) and Euechinoidea (advanced), are taken separately.

SUBCLASS PERISCHOECHINOIDEA

Perischoechinoids (Ord.–Rec.) are the primitive echinoid stock which includes all extinct Palaeozoic echinoids and the living and fossil Order Cidaroida. Only thirty-seven genera and about 125 species of Palaeozoic echinoids are currently known. Yet since their preservation may be good, phylogeny and evolution have been studied in detail, especially in the marathon works of Jackson (1912) and Kier (1965).

Palaeozoic echinoids differ from their later counterparts in several respects:

(a) Many of them have flexible tests, often of large size and with the plates not rigidly united so that they are often found in a collapsed state. In some cases the plates are thick and have bevelled edges so that they can slide over one another at the margins.

(b) Either the ambulacral or interambulacral columns or both may consist of many columns of plates, but compounding of the ambulacral plates is unknown; that is, there is never more than one pore-pair per column.

(c) The perignathic girdle is normally absent or of simple construction, and the lantern is flattish and less elaborate than that of later echinoids.

(d) The test is usually globular and invariably regular, with the periproct in the centre of the apical disc **(endocyclic).** Only a few flattened genera are known.

(e) Some of the earlier genera have the radial water vessel enclosed internally by inwardly projecting flanges from the ambulacral plates.

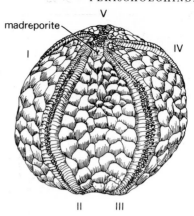

Figure 9.9 *Aulechinus* (Ord.), lateral view (x 2). (Redrawn from MacBride and Spencer in '*Treatise*' Part (U))

Palaeozoic echinoids seem to have been a relatively small and unimportant component of the Palaeozoic biota and were probably environmentally restricted.

SUPERORDER PALAECHINACEA

Orders Echinocystitoida and Palaechinoida

The flexible echinocystitoids include the very earliest true echinoids: the Upper Ordovician *Aulechinus* and *Ectenechinus. Aulechinus* (Fig. 9.9) has only two columns of plates in the ambulacra. This is characteristic of the primitive Family Lepidocentridae to which it belongs. The ambulacral plates are curious in that the per-radial suture is situated in a deep groove with the single unpaired pore on the aboral margin of each plate close to it. Furthermore, the radial water vessel was enclosed by a tubular covering arising from the lower surface of the ambulacral plates; evidently the ampulla lay below this. The related *Ectenechinus* (Fig. 9.10b) has paired pores, though one pore is smaller than the other, and in later lepidocentrid genera, e.g. the Silurian *Aptilechinus* (Fig. 9.10a), the pore-pairs are identical though still close to the per-radial suture. The apical system of *Aulechinus* and other lepidocentrids is peculiar in that there is only one genital plate with a single pore, though there are five oculars. Jaws and other parts of the lantern have been found, but the lantern is much less complex than in the euechinoids. In these early echinoids the interambulacra are smooth or have

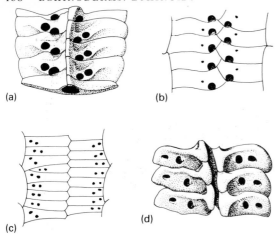

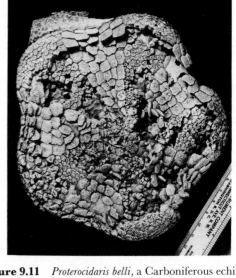

Figure 9.10 Portions of the ambulacra of perischoechinoids: (a) *Aptilechinus* (Sil.), internal view; (b) *Ectenechinus* (Ord.), with pores near the per-radial suture; (c) *Pholidechinus* (L. Carb.), with pores remote from the per-radial suture; (d) *Hyattechinus* (L. Carb.), internal view, with partial enclosure for the radial water vessel. (Not to scale); (redrawn from Kier, 1973, 1974)

Figure 9.11 *Proterocidaris belli*, a Carboniferous echinoid, aboral view, (x 0·5 approx.). (See also Fig. 9.12; photograph by courtesy of P. M. Kier)

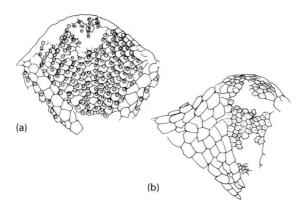

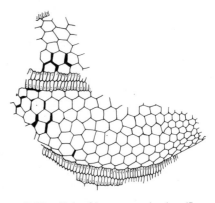

Figure 9.12 *Proterocidaris belli* (L. Carb.) (Echinocystoida: (a) showing expansion of ambulacra adorally; (b) and then contraction adapically (x 0·5 approx.). (Drawing of top of specimen shown in Fig. 9.10, from Kier, 1965)

Figure 9.13 *Palaechinus merriami* (L. Carb.) (Palaeochinoida), preserved fragment of test in lateral view (x 2 approx.). (Redrawn from Kier, 1965)

only small tubercles, but *Aptilechinus* has large spines on the ambulacra (Kier 1973). The lepidocentrids lasted until the Carboniferous but gave rise to another family, the Echinocystitidae, in which the number of ambulacral plates increased markedly and the plates on the adoral surface became expanded so that opposite surfaces of the same echinoid appear very dissimilar. *Proterocidaris* (Figs 9.11, 9.12b) illustrates this clearly, but there the apical system is more

'normal' with five ocular and five genital plates. Some of the later echinocystitoids, e.g. the Permian *Pronechinus*, had flattened tests. A third family, the Lepidesthidae, split off from the Echinocystitidae in the Devonian.

In the Palaechinoida (Fig. 9.13), which were probably derived from the Echinocystitoida, the test is always globular and the plates are thick, rigidly united and arranged in more-or-less regular vertical columns.

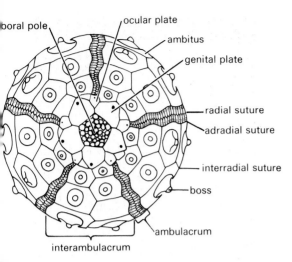

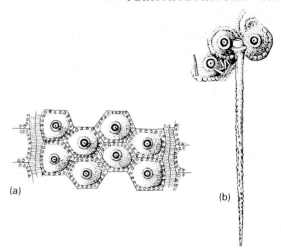

Figure 9.14 Cidaroid morphology, aboral, with terminology (x 1). The anus is located within the periproctal membrane at the aboral pole. (Redrawn from Fell in '*Treatise*' Part (U))

Figure 9.15 Ambulacra and interambulacra of *Archaeocidaris* (L. Carb.): (a) reconstructed; (b) with spine preserved. (Based on Jackson, 1912)

Various evolutionary trends have been noted within the echinoids of the Palaeozoic. Some, such as the flattening of the test and adoral expansion of the ambulacra, are found in the one Order Echinocystitoida alone. Other trends are found in more advanced Palaeozoic echinoids, such as a general increase in the size and number of ambulacra; an increase in the complexity of the lantern; the development of regularity in the interambulacral plates; and the loss of the enclosure of the radial water vessel. Most changes are allometric, but some aristogenetic change is seen, i.e. in the increased complexity of the lantern and in the development of tubercles and spines for the first time.

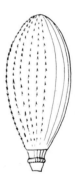

Figure 9.16 Club-like spine of Jurassic to Recent *Cidaris* (x 2).

Order Cidaroida

The cidaroids are the only echinoid group to survive the Palaeozoic and they still persist today. They formed the root stock of all post-Palaeozoic echinoids whilst themselves showing relatively little important change. The modern cidaroids, which have changed very little at least since the Cretaceous, have often been regarded as 'living fossils'. All cidaroids have relatively narrow and frequently sinuous ambulacra composed of small plates each with a single pore-pair (Fig. 9.14, 9.15). The interambulacral plates are very large with a single large tubercle in the centre of each

plate, to which a strong spine is attached (Figs 9.15a, b; 9.16). The **mamelon,** or central boss, is surrounded by a wide smooth **areola** around which is a ring of tiny **scrobicular tubercles.** Outside these a series of small secondary tubercles is irregularly dispersed.

The apical system is usually large and has five ocular and five genital plates, but in most cidaroids these are not rigidly united to the test and normally drop out on fossilisation. The lantern is relatively simple; the perignathic girdle has **apophyses** only (Fig. 9.17), i.e. flanges reflected from the interambulacra.

The oldest known cidaroids, other than possibly some Silurian forms represented only by spines which

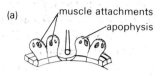

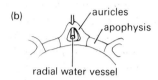

Figure 9.17 (a) Perignathic girdle of cidaroid with apophyses only, showing muscle attachments (viewed from within the peristome); (b) perignathic girdle of Recent *Paracentrotus* (Echinoida) with apophyses and auricles. (Simplified from Cuénot in '*Treatise*' Part (U))

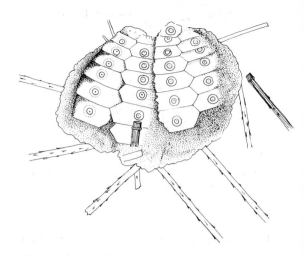

Figure 9.18 *Miocidaris* (Perm.), collapsed so as to conceal the ambulacra (x 1·75). (Based on Kier, 1965)

have been referred to this group, belong to Family Archaeocidaridae (U. Dev.–Perm.), which like other Palaeozoic echinoids have flexible tests and multiserial interambulacra. *Nortonechinus* (Dev.) and *Lepidocidaris* (Carb.) have six to eight columns of plates; the common Carboniferous *Archaeocidaris* (Fig. 9.15a) has four. But cidaroids of Family Miocidaridae (including *Miocidaris* (Fig. 9.18), the only known genus to survive the Permian extinction) have only two columns of plates in ambulacra and interambulacra, settling the pattern for all later echinoids, for which *Miocidaris* alone was the ancestor. In Palaeozoic cidaroids adaptive trends have been documented which all relate to improved locomotion on a harder substrate.

There are a few small-sized cidaroid genera in the Triassic, including the first cidaroids of modern type, but the cidaroids reached their acme in the Mesozoic, declining later though they are still abundant today in the Indo-West Pacific province, in the North Atlantic and – as an endemic family, the Ctenocidaridae – in the Antarctic.

Cidaroids are noteworthy for the remarkable development of their spines. Fusiform, club-shaped and peculiarly shaped spines are found in the shallow water genera. Most of these are sluggish and the heavy spines seem to be used for stabilising them in rough water. Slender elongate spines are more usual in the deeper water or mud-living species.

SUPERORDER MEGALOPODACEA

The Megalopodacea, the only other perischoechinoids, are of Ordovician age and are known from a few puzzling specimens alone. The notorious *Bothriocidaris* (Fig. 9.19), which has been one of the most controversial of all fossil echinoderms, comes from the Ordovician of Estonia, but recently specimens of a related genera have been found in Scotland (Paul 1967) as well as a single true *Bothriocidaris*. It has five single-columned interambulacra and five double-columned ambulacra which do not extend as far as the peristome, each ambulacral plate having two large spines and single pore-pair. The peristome has a primitive lantern but no girdle, and is surrounded by a single or double ring of plates with pore-pairs. On one interpretation, there are no genital plates, and the madreporite is on one of the oculars; alternatively, the apical plates traditionally regarded as ocular plates could actually be genitals (Durham 1966).

Bothriocidaris has been variously regarded in the past as an echinoid, a diploporite cystoid, or an echinoderm of doubtful affinities, but since the discovery of a specimen with a true lantern its echinoid nature is no longer in doubt. Its taxonomic position, on the other hand, is far from clear, but it is not believed that any later echinoids were directly descended from it. *Eothuria,* another Ordovician

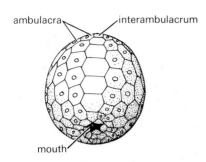

ambulacra interambulacrum

mouth

Figure 9.19 *Bothriocidaris* (Ord.), oblique adoral view (x 2·5). (Redrawn from Tasch, 1973)

oddity, was originally described as a plated holothurian (sea-cucumber), but other than the presence of many tiny openings in each ambulacral plate it seems to have echinoid features, albeit deviating from the norm. This again was an early offshoot of the great Ordovician radiation of echinoderms and does not seem to have produced any known descendants.

SUBCLASS EUECHINOIDEA

The euechinoids (Trias.–Rec.) include all Mesozoic to Recent echinoids other than cidaroids. They all have five bicolumnar ambulacra and five bicolumnar interambulacra; they may be regular or irregular; the lantern may be present but in irregular forms can be secondarily lost. All euechinoids seem to have descended from the early Mesozoic cidaroids. They began to diversify in the later Triassic and early Jurassic, undergoing a great adaptive burst and evolving into many families. At this time there were all manner of functional innovations (Kier 1974), which allowed ecological differentiation denied to the conservative perischoechinoids. The echinoids of the Palaeozoic were all regular and endocyclic. Though the regular endocyclic system has been retained by many of the Mesozoic and later taxa, numerous other groups quite independently became irregular, and all of these adapted for a wholly or partially infaunal existence. The difference between two kinds of echinoid is so pronounced that for a very long time Class Echinoidea was divided simply into two subclasses: the Regularia and Irregularia. There are, however, different 'grades' of irregularity in the echinoids. Some are rather simple with the periproct still in the adapical region, though outside the apical

disc, whilst in others it has migrated to the lower surface. The extremes of functional differentiation, however, were not reached until the Cretaceous with the origin of Order Spatangoida.

The early Mesozoic adaptive radiation of euechinoids gave rise to representatives of all four superorders. In some of these the archetypal plan tended to remain rather conservative after its foundation in the early Mesozoic, and there was very little functional evolution thereafter. In others each stage, though in itself highly functional, was capable of yet further functional modification. The adaptive differentiation of the more extreme forms of burrowing echinoid, as represented by *Echinocardium*, is a far cry from the morphology of its remote cidaroid ancestor. Yet once the origin of irregularity had provided the initial stimulus, there were then plenty of new morphological characters for evolutionary processes to work on.

Morphological characters of euechinoids

Apical disc (Fig. 9.20)

In regular euechinoids the apical disc is entire and contains the periproct. The latter bears the anus, located centrally within a series of small plates which cover the periproctal membrane. The disc may be **exsert,** with the circlet of ocular plates outside the apical disc, or **insert,** with ocular and genital plates alternating in a single ring around the periproct. But even within the same species there may be individuals with a fully insert, exsert or intermediate condition. Genera such as *Salenia* (Fig. 9.20n) and *Peltastes* have greatly expanded apical discs covering up to a third of the total surface area of the echinoid. These have an extra **(suranal)** plate by the periproct, and the ocular and genital plates are often highly ornamented.

In irregular echinoids the periproct may have wholly or partially escaped from the apical disc and be located outside it. One can even see this actually happening in the development of modern irregular sea-urchin larvae. The periproct in these is initially apical, but after a short period of growth it moves externally, disrupting the embryonic plates as it does so; these then re-form, though in an asymmetrical fashion, and genital V is usually missing. The irregular patterns are, of course, retained in the adult.

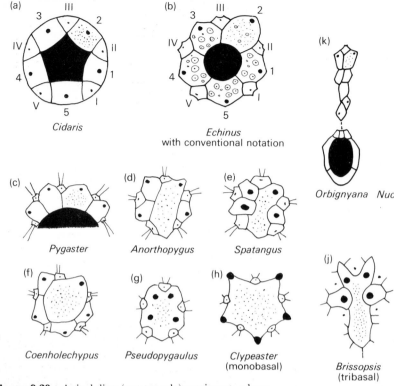

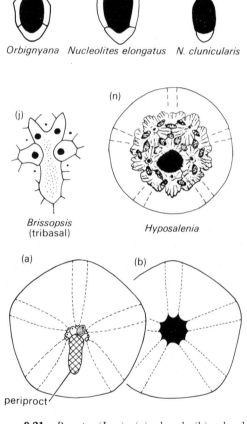

Figure 9.20 Apical discs (not to scale): periproct, where present, shown in black. (Redrawn mainly from Hawkins, 1912, *Zool. Soc. Lond. Proc.*, 440–97; and Kier, 1974)

Various kinds of apical discs in fossil and recent echinoids are here illustrated in a morphological series showing all stages in the migration of the periproct, though this is not intended to display the actual course of evolution: merely grades of organisation. The Pygasteroida show some interesting evolutionary trends in this respect. The earliest pygasteroid, *Plesiechinus hawkinsi* from the Lias, has an apical disc like that of a regular echinoid and is purely endocyclic. In slightly later pygasteroids the periproct is displaced outside though still enclosed by elongated plates and genital V is still present (an equivalent condition is seen in some early *Nucleolites* species). In more 'advanced pygasteroids genital V is absent, and the periproct lies at least partially outside the apical disc, forming a keyhole-shaped depression on the adapical surface (Fig. 9.21). And in the majority of irregular echinoids only four genital plates remain, though occasionally a fifth is found.

Other than the breakout of the periproct, a

Figure 9.21 *Pygaster* (Jur.): (a) aboral; (b) adoral (x 0·75).

reduction in the number of genital plates has been common in various irregular echinoid lineages. In cassiduloids and holectypoids there are defined trends from a system with four distinct genital plates (**tetrabasal**) to a system with a single large plate with four pores (**monobasal).** A monobasal system has been adopted by all clypeasteroids. Most spatangoids are tetrabasal, some **tribasal,** a few monobasal; it seems that these too are undergoing a general reduction to a monobasal system.

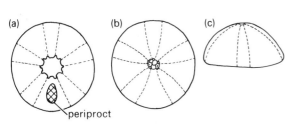

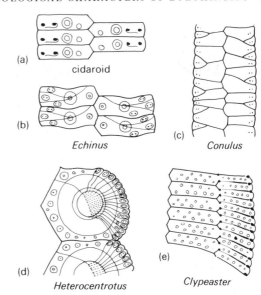

Figure 9.22 *Holectypus* (Jur.): (a) adoral; (b) aboral; (c) lateral view (x 1 approx.). (Redrawn from Black, R., 1970, *Elements of palaeontology.* (Cambridge: C.U.P.)

Figure 9.23 Ambulacral plating showing simple (cidaroid) and compound plates. (Not to scale) (Redrawn from Durham *et al.*, in '*Treatise*' Part (U))

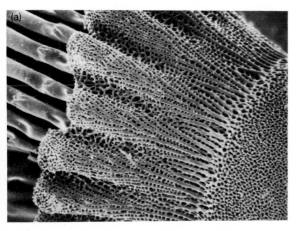

Figure 9.24 Structure of recent echinoids: (a) *Centrostephanus nitidus* (Rec.), Indian Ocean. Ambital spine base. The spine muscles attach to the raised wedges on the upper part (x 100 approx.). (b) *Psammechinus miliaris* (Rec.), Torquay, England. Aboral pore pair with neural groove (x 50 approx.). (Scanning electron micrographs by courtesy of Andrew B. Smith)

Ambulacra

Compound plates (Fig. 9.23). In euechinoid ambulacra there are never more than two columns of plates, but there often are a very large number of pore-pairs (Fig. 9.24a) and hence tube-feet. This has been achieved in different ways by various groups of echinoids. Thus in the cidaroids the plates have single pore-pairs but are very small, hence numerous. Alternatively compounding of plates is common in many euechinoids. Compound plates have two, three or more **demiplates** within the confines of a single ambulacral plate. Each demiplate has its own pore-pair. Such morphology is most pronounced in genera like *Heterocentrotus* (Fig. 9.23d), in which the large central tubercle is traversed by numerous demiplates. Compounding results from the subjugation and incorporation of embryonic plates into a smaller number of 'master plates' and is important in

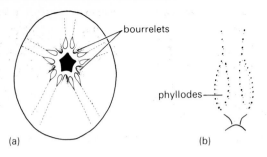

Figure 9.25 *Catopygus* (Cassiduloida) (Cret.) in adoral view showing: (a) phyllodes and bourrelets (x 2); (b) phyllodes enlarged.

allowing a larger number of tube-feet per unit area. It is also commonly associated with the development of a large central tubercle, and another advantage of the system must be that a larger tubercle could be supported by larger plates formed by the fusion of smaller ones. Compounding also separates the rows of tube-feet within one ambulacrum so that the test bears ten widely spaced columns of tube-feet rather than five closely spaced pairs of columns as in the cidaroids.

Petaloid and subpetaloid ambulacra. Irregular echinoids such as *Echinocardium* have ambulacra forming 'petals', i.e. with the adapical parts expanded to form a flower-like rosette. Within the petals the pore-pairs are always elongated and the individual pores widely separated, one being slit-like; they are always associated with flattened respiratory tube-feet. Some of the earlier and 'primitive' irregular echinoids (e.g. *Holectypus* (Fig. 9.22), *Pygaster* (Fig. 9.21)) have no petals but they do, incidentally, have gills. The next grade in organisation is a slightly expanded subpetaloid condition, and a more advanced grade still is the system of some clypeasteroids and cassiduloids in which the petals are large but flush with the surface. In the early flattened echinoids so many tube-feet would have been located on the lower surface, and used for food-gathering, that respiratory exchange might have been impaired had it not been for the development of petals on the upper surface. These petaloid ambulacra were then pre-adapted for respiratory use in deeper burrowing echinoids. Some holasteroids and spatangoids have depressed petals, especially the geologically later ones (*Echinocardium* being a prime example), and the deepening of the

petals is such that they form an effective channelling structure for the respiratory currents as well as providing a basis for a protective cover of flat-lying spines. The differentiated anterior ambulacrum of holasteroids and spatangoids also arose through relatively gradual stages from an originally flush and undifferentiated type.

Phyllodes and bourrelets (Fig. 9.25a, b). Certain irregular echinoids, especially the cassiduloids, and a few regular genera have specialised ambulacral structures in the vicinity of the mouth. These are the **phyllodes**: areas in which the ambulacrum expands into a leaf-like shape close to the periproct. The phyllodes are separated by interambulacral swellings known as **bourrelets.** In the phyllodes the plates are very crowded, a much larger number of tube-feet occur than elsewhere in the ambulacra and, furthermore, the pores tend to be larger. In the irregular echinoids these extra tube-feet are used primarily for feeding. Some, such as the spatangoid *Meoma*, feed on organic-rich sand, and the tube-feet are used as 'sticky shovels' to deposit sand in the mouth. Regular echinoids may also have phyllodes, particularly those living in high energy environments where the tube-feet are used for clinging to the substrate. Presumably the extra adoral tube-feet in *Proterocidaris* and other flexible echinoids may have served a similar purpose. A tendency to develop phyllodes has been noted in many lineages. Later members often have broad phyllodes with larger but fewer pore-pairs. In many species the tube-feet emerge from single pores, the respiratory function which necessitates separate channels for oxygen-rich and oxygen-poor coelomic fluid having been lost.

Interambulacra and spines

Cidaroids are characterised by interambulacral plates having a single large central tubercle. This system was retained in such genera as *Hemicidaris* and *Salenia*, though the tubercles are generally smaller. But in the majority of other euechinoids the size of the primary tubercles decreased markedly, and most particularly in the irregulars. On the interambulacral plates of sand-dollars there may be up to ten primary tubercles per square millimetre and hundreds per plate.

This tendency, which so greatly altered the whole

appearance of the interambulacral plates in the evolution of euechinoids, is functionally sound, for the modern cidaroids can neither bury themselves in detritus for camouflage nor burrow. Short spines (Fig. 9.24b) are a prerequisite for burrowing echinoids; they also provide cover and shade. Perhaps the most striking modification of all echinoids is in the modern reef-dwelling regular echinoid *Podophora* in which the adapical spines are mushroom-shaped, their tops being flat polygonal plates forming a continuous basaltiform mosaic, which gives protection in heavy surf; these echinoids can also cling to the rocks on which they live by their phyllodal tube-feet.

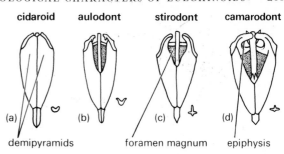

Figure 9.26 Lantern pyramids of increasing complexity from left to right: grooved teeth (cidaroid, aulodont) and keeled teeth (stirodont, camarodont) are shown in section. (Redrawn from Jackson in '*Treatise*' Part (U))

Lantern and perignathic girdle (Figs 9.2, 9.26)

The lantern is most important in echinoid taxonomy at the superordinal level, though it is rare to find it in fossils. The details of the complex structure and mode of operation of the lantern are beyond the scope of this work, but it is important to appreciate the four main stages in its evolutionary development. The lantern consists of five identical **pyramids,** each composed of two **demipyramids.** Between these is a depressed **foramen magnum** and above are short rod-like **epiphyses.** Each pyramid contains a single calcitic tooth which grows down continually as it is abraded. Isolated pyramids here illustrated show the four main lantern types: **cidaroid, aulodont, stirodont** and **camarodont.** In the primitive cidaroid stage the teeth are grooved and the foramen magnum shallow; the epiphyses are small and short. Later stages show the deepening of the foramen magnum, the growth of the ephiphyses until they join across the top, and the change from grooved to much stronger keeled teeth. These changes are associated with improvements in function. The lantern of cidaroids can only move up and down and open and shut like a grab. The camarodont lantern, however, is capable of sideways scraping which is much more effective in bottom feeding, and its whole organisation seems to be related to this end. The most modified lanterns are those of the flattend clypeasterioids which are highly expanded with wedge-like teeth. Though these cannot move sideways they are very powerful and capable of pulverising the bottom detritus on which the echinoid feeds. The lantern has been dispensed with in the Atelostomata, such as *Echinocardium*, which are microphagous feeders. It is unfortunate that such a taxonomically important structure as the lantern is only rarely found fossilised, but where it has been discovered it supports the conclusions drawn on other grounds as to relationships between groups.

Palaeozoic echinoids other than cidaroids do not have a true perignathic girdle, and the muscles and ligaments were attached directly to the inside of the test. The presence of a girdle raises the attachment of the muscles and mechanically improves their line of action. Cidaroids have a primitive girdle with projections (**apophyses**) in the interambulacrum. Euechinoids have both apophyses and **auricles:** supports at the base of the ambulacrum which, in more advanced echinoids, may arch over the radial water vessel. The development of auricles in addition to apophyses probably increased the spread of muscle attachment and thus allowed the possibility of lateral movement of the lantern.

Gills

Gill notches are found in echinoids from the early Jurassic onwards. They are present in nearly all regulars and in some early or primitive irregulars (e.g. *Holectypus, Pygaster*) but not in the later irregulars, in which petaloid ambulacra having efficient respiratory tube-feet are developed.

Marsupiae (Fig. 9.27)

Some twenty-eight species of modern echinoids have been described in which the females have developed special brood-pouches (**marsupiae**) in which to

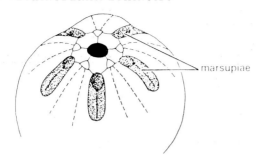

Figure 9.27 *Pentechinus* from the Tertiary of south-eastern Australia showing marsupiae (x 1). (Based on Philip and Foster, 1971)

incubate the fertilised eggs. These then develop directly without the free-swimming larval stage. Nearly all of them live in cold Antarctic waters where development and growth are slow. Such brood-pouches occur in recent cidaroids and spatangoids, but they are also found in some Cretaceous and Tertiary echinoids of Orders Temnopleuroidea, Spatangoidea and Clypeasteroidea from south-eastern Australia (Philip & Foster 1971), which from palaeo-climatic evidence seems to have been very cold at that time. The presence of such genera in the Australian Tertiary and their absence in modern offshore Australian fauna accord well with geological evidence on the northward drift of the Australian continent away from Antarctica in the later Tertiary.

The temnopleuroid *Pentechinus* has (in the female) each of the five genital pores opening into an elongated depression extending some way down the interambulacrum, whilst in *Paradoxechinus* a deep annular depression surrounds the apical disc, and the genital pores open directly into this. *Paraspatangus* has a shallow depression in which the apical disc lies. Normally in marsupiate echinoids the brood-pouches are adapical, but an adoral marsupium is present in the clypeasteroid *Fossulaster*. The independent origin of marsupiae in so many cold water forms testifies to an evident advantage of viviparity in low temperature conditions.

EVOLUTION IN ECHINOIDS

The first recognisable echinoids are Ordovician, appearing at the same time as a whole plethora of other echinoderm types, some at least of which, in certain features of their organisation, are reminiscent of echinoids. Amongst these could be numbered holothuroids and the echinozoan Class Ophiocystoidea which have enormous plated tube-feet and a lantern structure. (Apparent resemblances between echinoids and edrioasteroids or diploporites are superficial.) The features of true echinoid organisation became more clearly defined a little later, and by the Carboniferous the large regular flexible-tested perischoechinoids with many columns of plates were becoming diverse though never really abundant. With the great extinctions of the Permian only *Miocidaris* survived into the Mesozoic; the fortunes of the whole Class Echinoidea rested on a single genus (Hawkins 1943). From this ancestor came first of all the small Triassic cidaroids; then, with a great burst of adaptive radiation, came eighteen families of which many were 'improved regulars' and the rest irregulars, adapted for many modes of life. Today echinoids are a vital part of the invertebrate marine realm and are probably as abundant now as at any time in the past. In the Recent fauna, 25 per cent are spatangoids, 16 per cent cidaroids, 14 per cent clypeasteroids and 14 per cent temnopleuroids. The fortunes of various families have varied greatly during the Tertiary; for instance, cassiduloids were very important until the Pliocene when they declined greatly, as did the clypeasteroids, and many old groups such as the holectypoids are very nearly extinct.

Modern echinoids are very successful feeders. Some even devour other echinoids; the process of such feeding has been shown recently by time lapse photography, which showed three large regulars devouring a sand-dollar, eating it from the edges like a biscuit.

The precise relationship between structural and functional differentiation is perhaps more clearly seen in the echinoids than in any other invertebrate group. By contrast with the evolutionary pattern of most fossil groups, evolutionary changes and repeated trends in the Mesozoic and later echinoids have been rather slow and spread out. So often in other taxa the vitally important intermediate stages are compressed in time, and there are no clear links between one grade of organisation and another. Abrupt changes of this kind are seen in trilobites, molluscs and brachiopods, and the whole process of understanding phylogeny and its functional significance is bedevilled by morphological discontinuities. Even though there

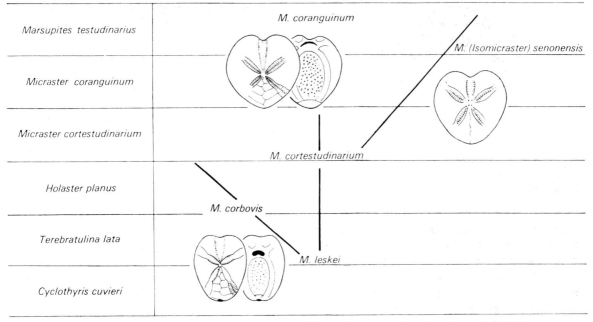

Figure 9.28 Major features of evolution in *Micraster* in the Upper Cretaceous (based on Kermack, 1954).

are gaps in echinoid phylogeny, such as in the establishment of euechinoid orders, they are not all that abrupt, and such an important stage as the breakout of the periproct can be seen as a relatively gradual process. The evolutionary trends within both Palaeozoic and Mesozoic to Recent echinoids, which Kier (1965, 1974) has carefully documented, occur in several groups testifying to continued selection leading to 'the gradual improvement of the animal as a living mechanism'. The remarkable functional differences between the early Mesozoic cidaroids and, say, the spatangoids or clypeasteroids has been achieved only by modification of existing organs or of elements already present in the ancestor. Even the fascioles of the spatangoids are merely tracts of modified spines, the clavulae, in which the surface is covered with a ciliated epithelium of a kind native to echinoderm organisation. In fact the only new kinds of organ in the euechinoids not possessed by their cidaroid forebears are the auricles of the girdle, some small components in the lantern and the ophicephalous (pincer-like) pedicellariae.

Furthermore, the functional significance of any structural changes in extinct genera may be interpreted because of comparable functional adaptations in many modern representatives, using similar

modifications of organs generally present in echinoids, e.g. spines, fascioles, etc. (for the same general purposes). In this context the record of evolution in *Micraster*, a Cretaceous spatangoid, is especially good. It has been elucidated stratigraphically, tested statistically and interpreted functionally with reference to its modern relative *Echinocardium* and others.

Evolution in *Micraster* (Figs 9.28, 9.29)

One of the best known of all evolutionary series, showing an apparently continued and directional change through time, has been documented in *Micraster*, a spatangoid from the Upper Cretaceous of north-western Europe.

In 1899 Rowe, working in southern England, carefully collected some 2000 specimens of *Micraster* in rigid stratigraphical sequence through six successive zones of the Upper Chalk. He observed some clear changes in going from the lower zones to the upper, confirming previous suggestions; this was all the more striking in view of the monotonous lithology throughout. The main changes Rowe noticed on going up the sequence were:

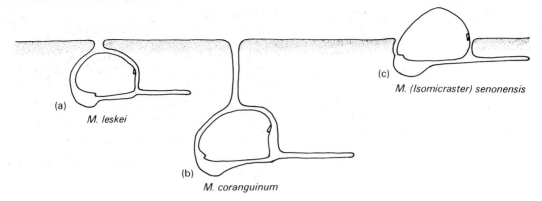

Figure 9.29   Inferred life positions and burrowing depths of Cretaceous micrasters. (Based on Nichols, 1959)

(a) the test becomes broader, and both the tallest and broadest parts move posteriorly;

(b) the anterior groove deepens and increases in tuberculation;

(c) the mouth moves anteriorly, and the labrum or lip below it becomes very pronounced eventually covering the mouth;

(d) the paired ambulacra lengthen and become straight, whilst the interporiferous areas change from being smooth through a variety of intermediates until the final pattern is one with a deep groove per-radially and inflated lateral areas.

Though Rowe imagined that the original lineage had subdivided more than once, a full understanding of phylogenetic pathways had to wait until Kermack's (1954) statistical work on equivalent populations. Whilst some of Rowe's concepts proved to be invalid, the main thesis was upheld. Kermack showed that in any one population individuals might show different grades of advancement of some characters relative to others; some would be 'average' for the stratigraphical level, others 'retarded' or 'advanced'. According to Kermack (Fig. 9.28) the original stock, represented by the *Micraster leskei* population, consisted of relatively small, rather globular echinoids, with a shallow anterior groove and the mouth rather far back, without a labrum. The main line of descent led to *M. cortestudinarium* and finally *M. coranguinum,* in which the main shift of the characters was as Rowe had described (a–d above). An early side branch led to the conservative *M. corbovis* which retained many of the primitive features

of the parent stock. Another side branch originated after the extinction of *M. corbovis* and led to *M. (Isomicraster) senonensis*, in which the subanal fasciole is lost, the test is tall and pyramidal and the anterior ambulacrum is subpetaloid rather than differentiated.

The evolutionary changes in the main line seem to be independent of environment as the latter was more or less unchanging. They could therefore be associated with either a change in a niche within that environment or an improvement in adaptation within an existing niche. Nichols (1959) shed light upon the biological meaning of these changes by studying the precise relations between structure and function of the different organs in *Echinocardium* and other living spatangoids: the modern equivalents of *Micraster*. He suggested that:

(a) *M. leskei* and its descendant *M. corbovis* were adapted to shallow burrowing just below the surface (Fig. 9.29a);

(b) the changes in the main line *M. leskei* to *M. coranguinum* give evidence of adaptation to progressively deeper burrowing conditions (Fig. 9.29b).

These adaptations are fairly clear. The deepening of the anterior food groove, forward movement of the peristome and development of the labrum are all advantageous for the direction of feeding currents to the mouth in a deep-burrowing echinoid. The increase in number of tube-feet suggests an equivalent adaptation. Other than the subanal fasciole there is no other clearly developed fasciole on the adapical

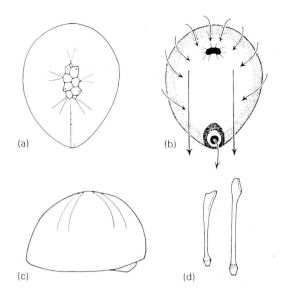

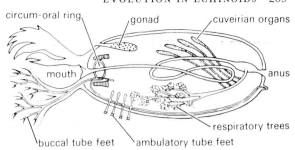

Figure 9.31 *Holothuria*, sectioned longitudinally. (Based on Nichols, 1974)

Figure 9.30 *Echinocorys scutata* (Cret.): (a)–(c) in adapical, adoral and lateral views, showing marginal diffuse and perianal fascioles (stippled) and current directions (x 0·6); (d) detached paddle-like spines from the plastron (x 15) approx.). (Redrawn from Stephenson, 1963)

region, but the whole surface of the animal, which was highly tuberculate, probably acted as a 'diffuse fasciole', being covered with the bases of what were probably clavulae.

Whilst the high zonal micrasters are interpreted as deep burrowers, *M. Isomicraster (senonensis)* was probably a shallow burrower (Fig. 9.29c). The subanal fasciole has gone, and in addition the pyramidal test would help to prevent particles from settling. There are very many respiratory feet, all of small size; this device probably ensured against predation. *M. senonensis* seems to have been derived from the mid-zonal micrasters, possibly as a peripheral isolate, and to have migrated back into the southern England area towards the end of the Upper Cretaceous, when it interbred with *M. coranguinum*, producing a number of hybrids.

M. (Isomicraster), though defined as a subgenus on morphological grounds, cannot truly be so distinct if such interbreeding was possible, but on account of long-continued taxonomic practice it is retained as such.

In the proliferation of spatangoids in the Cretaceous numerous types arose, adapted for deep or shallow burrowing, and it has been possible to interpret some of these functionally. *Echinocorys* (Fig.

9.30a–d) is a highly modified spatangoid with a high steep-sided test, an almost flat base and a sharp ambitus. Short thorny spines were evidently for the protection of the adapical surface; spatulate spines attached to the tubercles of the adoral surface allowed forward progression, and ciliary currents from the marginal diffuse fasciole and the more differentiated subanal fasciole produced feeding currents. The ambulacra are only subpetaloid but are very long. *Echinocorys* was probably a shallow burrower but presumably inhabited a different niche from the isomicrasters (Stephenson 1963).

There is still much to be learned about the ecology and evolution of fossil echinoids; but their potential for palaeobiological study is singularly high.

Class Holothuroidea

'There were nasty green warty things, like pickled gherkins lying on the beach' (H. G. Wells, *Aepyornis Island*, 1927). Thus has the external appearance of holothuroids (holothurians or sea-cucumbers) been most graphically described. Holothuroids (Fig. 9.31) are fusiform or cylindrical echinozoans which generally lie on the sea floor (or burrow into it) with their long axis horizontal. The mouth and anus are at opposite ends. In the type example *Holothuria* the mouth end has a feathery inflorescence of sticky tentacles (the **buccal tube-feet**) which are used in feeding. The skin is warty and leathery, and within it are embedded the calcitic elements that are the only parts to be fossilised. These calcitic plates are normally reduced to little **sclerites,** having the form of anchors, gratings or spoked wheels (Fig. 9.32). Very rarely they form a contiguous cover; more usually the integument is very flexible. It is from such

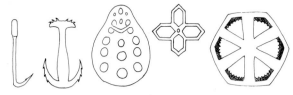

Figure 9.32 Holothurian sclerites (anchors, gratings and wheels). (Based on Frizzell *et al.*, in '*Treatise*' Part (U))

sclerites that fossil genera can be identified and allocated to families. Only in the mouth region is there a rigid series of plates, forming a calcareous **perioral ring,** perhaps homologous with the echinoid lantern.

The five ambulacra are arranged in two sets parallel with the long axis: three ventrally and two dorsally. The tube-feet of the dorsal set form small **sensory papillae,** while those of the ventral set are ambulatory and have suckers. The internal structures are much modified. There are a pair of **respiratory trees** arising from the swollen **cloaca** at the hind end of the intestine. Water coming in through the anus is pumped by rhythmic pulsations of the cloaca into these trees for respiratory exchange with the coelomic fluid. In Subclass Aspidochirota, to which *Holothuria* belongs, there are unusual defensive structures located within the body. These are the **cuvierian organs:** thread-like masses which can be shot out of the anus to entangle stickily any predator that disturbs the holothurian.

Habits and evolution of holothurians

The earliest holothurian remains may be of Ordovician age; the earliest undoubted holothurian spicules are, however, from the Devonian. *Eothuria* (Class Echinoidea, Superorder Megalopodacea) might have been somewhere near the ancestral line and indeed was first described as a plated holothurian. Reduction of the plates to sclerites, giving the body its present flexibility, seems to have been almost universal. Only a few plated genera exist today, all belonging to the Dendrochirota, which with the Aspidochirota and the Apoda form the three defined subclasses of Class Holothuroidea. The first two are mainly benthic holothurians, which creep along on their ventral tube-feet or on a muscular slug-like sole in which the tube-feet are reduced. But one of

the aspidochirote suborders, the Elasipoda, has swimming representatives which keep themselves afloat by pulsation, like jellyfish, and are plank-tivorous. Some of the benthic genera sweep the sea floor ahead of them as they move with their sticky buccal tube-feet and so pick up organic detritus; others spread them out in a **horizontal collecting bowl,** as do the sluggish burrowing Apoda.

Holothurians today are abundant in warm shallow waters, but they are also successful in the deep sea, where undersea photography has shown dendro-chirotes 'congregating in herds like grazing cows'.

Class Edrioasteroidea

The Edrioasteroidea (L. Cam. – L. Carb.) are a small extinct group of echinoderms with five distinct ambulacral and five interambulacral areas, confined to the upper (adoral) surface. These run to a central mouth. Most edrioasteroids are flattened and dis-coidal though some are globular or elongated.

Edrioaster (M. Ord.) (Fig. 9.33a, b) has a flexible **theca** in which five biserial ambulacra are regularly plated and alternate with irregularly plated in-terambulacral areas. The ambulacra are sinuous, four having an anticlockwise twist whilst the fifth turns clockwise and curls round the excentric periproct. A small **hydropore** near the mouth presumably led to the water-vascular system. The ambulacra have two sets of plates: a lower set of **flooring plates,** arranged as in an echinoid, the margins of each pierced by a single pore; and an upper set of **cover plates,** hinged on top of the flooring plates, which could open to let out the tube-feet or close down to protect them. The **peristomial plates** covering the mouth were probably fixed (as in the crinoid **tegmen**) and could not open. The whole ambulacral plating system is strikingly reminiscent of that of the Crinozoa (q.v.). As in that subphylum, the radial water vessel may have run in a median groove along the per-radial suture, though Bell (1976) has suggested an alternative position. There is some evidence for this, for traces of both the **circum-oral ring** and the radial water vessel have been preserved in *E. buchianus*. The aboral surface consists of plates like those of the ambulacra, but the marginal plates are larger forming a distinct and rigid ring.

In spite of the similarity of edrioasteroid ambulacra to crinoid food grooves there were no brachioles

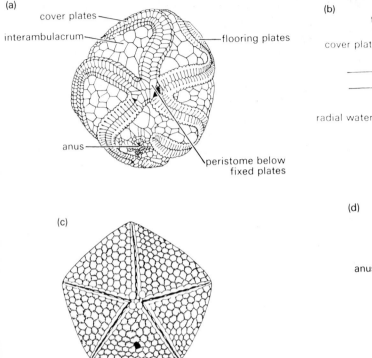

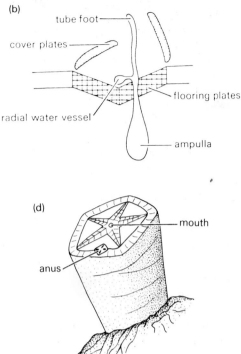

Figure 9.33   (a) *Edrioaster bigsbyi* (Ord.), with cover plates removed from three ambulacra (x 1 approx.) (based on Kesling in '*Treatise*' Part (U)); (b) section through *Edrioaster* ambulacrum with inferred position of soft parts; (c) *Stromatocystites* (L. Cam.), the earliest known edrioasteroid (x 1) (redrawn from Pompeckj in '*Treatise*' Part (U)); (d) *Cyathocystis* (Ord.) a columnar edrioasteroid (x 2) (based on Nichols, 1974).

(short fingers extending from the test). The nature of the tube-feet is unknown. *Edrioaster* may have been sessile but was not fixed to the substrate and might have moved along by pulsations of the flexible theca.

Habits and evolution of edrioasteroids

The superficial resemblance of edrioasteroids to the Precambrian *Tribrachidium* has often been noted, though this may be no more than coincidental. And though the origin of edrioasteroids is really unknown, the aboral features that they share with early crinozoans seem to suggest a common origin. The earliest known edrioasteroid is *Stromatocystites* (L.–M. Cam.) (Fig. 9.33c) which is amongst the earliest of all echinoderms. It is of pentagonal shape with five straight ambulacra with the flooring plates in four columns and tesselate interambulacral plates. This was apparently a free-living form.

Edrioasteroids reached their acme in the Middle Ordovician. They are not uncommon in the late Ordovician, but thereafter are found only in small numbers until the later Carboniferous with a slight expansion towards the end. Virtually all known species come from northern Europe and North America. Some genera were free-living, but most were permanently fixed by the marginal ring to the substrate. Sometimes living shells were apparently used as a base, and the high selectivity of particular species for one kind of shell suggests that these may have been commensal.

The majority of genera were discoidal and not unlike *Edrioaster*. In *Pyrgocystis* and *Rhenocystis*, however, there is a long tower-like stem of imbricating plates below the flat adoral surface; *Cyathocystis* (Fig. 9.33d) is also columnar. These forms may have been permanently fixed in soft mud. At one locality in the Devonian of Iowa (Koch & Strimple 1968) many

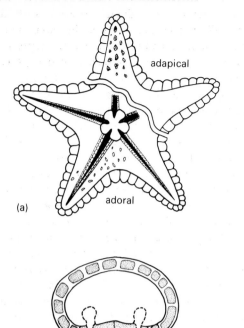

(a) adapical

(b) adoral

radial water vessel

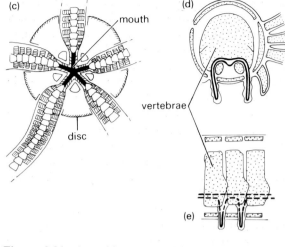

(c) mouth

disc

(d) vertebrae

(e)

Figure 9.34 Asteroid anatomy: (a) external with food grooves and florescent gills; (b) cross section of arm showing plating structure and external position of radial water vessel. (Modified from Sladen in '*Treatise*' Part (U).) Ophiuroid anatomy: (c) adoral surface; (d) cross section of arm with vertebrate with radial water vessel behind; (e) lateral view. (Redrawn from Woods in '*Treatise*' Part (U), and Nichols, 1971)

edrioasteroids *(Agelacrinites hanoveri)* along with the coral *Aulopora* and the cystoid *Adocetocystis* are present in what must have been their life habit, attached to a discontinuity surface. The sea floor at that time consisted of wave-fretted limestone, eroded into knobs. Edrioasteroids are found fixed to the sloping surfaces of these knobs; all stages in growth are found. Another edrioasteroid species lived on the sides of the cystoid thecae. The whole assemblage was overwhelmed by silt, and all specimens are preserved in place.

SUBPHYLUM ASTEROZOA (Figs 9.34a–e)

Asterozoans (Class Stelleroidea) are grouped in three subclasses and include the primitive Subclass Somasteroidea, starfish (Subclass Asteroidea), and brittle-stars (Subclass Ophiuroidea). They are relatively rare in the fossil record since they break up very easily after death, and the paucity of their fossil remains is no guide to their former abundance. Nevertheless they are amongst the most abundant of

animals living today on rocky shores of continental shelves and in the deep sea.

In all asterozoans the central part of the body (**disc**) extends laterally into five or more **arms,** the mouth faces downwards and the anus, where present, is aboral. In asteroids the arms are not sharply marked off from the central disc, whilst in ophiuroids the central disc is clearly delimited and bears five flexible snake-like arms (Greek: οφιυρος = snake). Most of the visceral elements in asterozoans, such as the intestine and gonads, are broadly homologous with those in other echinoderms. The water-vascular system resembles that of crinoids in that the radial water vessels lie in deep grooves in the ventral (adoral) surface. This is one of the many features that suggest a relationship with crinoids rather than echinoids.

SUBCLASS ASTEROIDEA (Fig. 9.34a, b)

Asteroids usually have five arms though these may be further subdivided. These arms normally extend like digits from the disc, but some starfish have a very pentagonal outline with short arms. On the upper surface there are a series of plates, which from the top down are the **carinals, dorso-laterals** and **marginals.** These may interlock so that the test is rigid, but more often they are flexible, allowing the arms to curl, which is very useful during climbing. On the lower surface the plates are larger and organised in each arm into a double column of ambulacral plates, through which the tube-feet connect by pores to the ampullae even though the radial water vessel is on the outside of the test. Outside these are the **adambulacral plates** which adjoin the marginals. All the adoral plates unite around the mouth in a rigid **peristomal ring.**

Between the plates (**ossicles**) of the upper surface project extensions of the coelom, known as **papulae,** which are respiratory; since asteroids are very active animals such 'extra' respiratory structures are essential.

Evolution and habits of asteroids

It is generally recognised that asterozoans derived from some kind of crinozoan ancestor, the early members moving freely across the sea floor but probably feeding on detrital particles. Most of the subsequent diversification of asteroids can be related to a change in feeding habits. Most modern starfish are voracious predators and feed on molluscs, in particular preying actively on bivalves. When a starfish such as *Asterias* locates a bivalve it climbs on top of it, engages the two valves with its suckered tube-feet, and for several minutes or longer pulls the two valves in opposite directions. Eventually the stress set up becomes too great for the bivalve's adductor muscles, and the valves open slightly. The starfish then everts its stomach out of its mouth and squeezes it bit by bit inside the opened valves. Digestion of the soft parts takes place within the bivalve shell, and when feeding is complete the stomach returns inside the starfish again. After feeding, the starfish can go without food for months without ill effects. There is good evidence from characteristic associations of disarticulated bivalve shells with fossil starfish that this extraoral feeding habit is ancient, extending as far back as the Devonian and probably further. In a well-known example (Clarke 1912) a large bedding surface of Devonian rock revealed some 400 specimens of *Devonaster eucharis* in intimate associate with the bivalves *Grammysia* and *Pterinea,* the environment having been 'invaded by starfish which congregated in vast numbers in order to feed on the clams'. Not all asteroids feed in this way; some feed intraorally, as do many ophiuroids, by taking the prey, whether gastropod or bivalve, into the mouth. Even here, however, intrusion of the stomach lobes into the bivalve appears to accompany digestion. Other starfish are deposit feeders, and basket-stars are suspension feeders. Since some starfish have preyed actively on bivalves for at least 400 million years, it is highly probable (Carter 1967) that some of the characteristic physiological and structural characteristics of bivalve shells (e.g. the interlocking commissures) can be considered as defensive structures against asteroids, the ever present enemy. Indeed the morphological innovation of interlocking commissures in bivalves arose around the same time as the great expansion of extraorally feeding asteroids, in the later Ordovician and early Silurian.

SUBCLASS SOMASTEROIDEA

The earliest known 'starfish' (e.g. *Villebrunaster, Chinianaster*) are Tremadocian and come from southern France and Bohemia. These belong to the Somasteroidea, a primitive group probably ancestral to both the Asteroidea and the Ophiuroidea, though primitive ophiuroids occur in the same fauna. The Somasteroidea, now elevated to equal rank with the Ophiuroidea and Asteroidea, were thought to be extinct until 1963, when a Recent deep-water genus *Platasterias* was first described as a somasteroid (Fell 1963). The organisation of any somasteroid is closely reminiscent of a stemless biserial inverted crinoid. Thus the arms of *Villebrunaster* have a biserial ambulacral structure from which arise at 45° to the axis lateral rod-like **virgalia,** which resemble and are probably homologous with the pinnular plates of crinoids. Furthermore, *Platasterias,* and indeed a few other primitive Asteroidea, do not have suctorial tube-feet. They feed microphagously, or selectively on small amphipods. In the later evolution of crinoids from somasteroids the dominantly transverse growth gradients were gradually transferred to longitudinal ones, so that asteroids lost their crinoid-like appearance, becoming typically asterozoan with the addition of marginal plates.

The first true asteroid is of early Arenig age, and by the Middle Ordovician the somasteroids had largely disappeared. Most Recent species of the Asteroidea, like those of the Ophiuroidea, belong to long-ranged genera which have persisted for a long period of geological time.

SUBCLASS OPHIUROIDEA (Fig. 9.43c–e)

The most distinctive feature of ophiuroids, namely their thin snake-like arms, has been an important factor in their great success, since they use sinuous movements of the arms for locomotion.

All the viscera are contained in the central disc, and all that lie within the arms are the greatly enlarged **vertebrae,** homologous with the fused ambulacral plates of asteroids. These have special articulating hinges so that the arm is very flexible. A sheath of plates surrounds the vertebrae, and the strong muscles that move the arms are found in the space between. Though the earliest ophiuroids are found in the same beds as the first somasteroids, the latter were probably their ancestors. The ambulacra enlarged to form vertebrae, and the radial water vessel became enclosed by the growth of a protective **ventral shield.** There are two orders: the Ophiurida (Ord.–Rec.) in which the arms can bend only in the horizontal plane, and the Euryalae (Carb.–Rec.) in which the arms can move in all planes; the latter are often climbing forms.

Ophiuroids are immensely successful in modern oceans, especially at bathyal and abyssal depths where they are often found crowded together in great numbers. They are mainly suspension feeders, but there are carnivorous species which feed on small bivalves, though these do not have the extraoral feeding habit of asteroids. Ophiuroids seem to have retained much the same organisation since the Ordovician, and their genera, like those of asteroids, are very long-ranged. They do not seem to have been badly affected by any major extinction periods.

Starfish beds

The remains of fossil asterozoans are all too scanty, since the plates normally disarticulate after death, not being bound together as are those of echinoids. Several well-known starfish beds are known, however, in North America, in Great Britain and in the Devonian of Germany, where starfish are almost always associated with the strange arthropods *Mimetaster* and *Vasconisia* (Stuermer & Bergstrom 1973) (see Bibliography following Ch. 12).

There are three or four British starfish beds which have long merited attention. Of these, one in the Lias of Dorset, is full of ophiuroids (two species), but there is no other fauna. Sedimentological criteria (Goldring & Stevenson 1972) make it plain that these ophiuroids were smothered and rapidly buried by a thick cloud of silty sediment. Modern ophiuroids cannot escape from sediment more than 5 cm thick, and these Liassic specimens probably all died in their life position.

The famous Ashgillian starfish bed at Girvan has an extremely rich fauna of starfish, trilobites, brachiopods, molluscs and the early echinoids *Aulechinus*, *Ectenechinus* and *Eothuria*. From the sediment inside the echinoid tests and other criteria it has been concluded that the fauna was shifted some distance and rapidly buried following turbulence in a shallow marine environment. Other cases which have been considered confirm that only as a result of 'catastrophic' conditions are starfish likely to be preserved whole.

SUBPHYLUM CRINOZOA

Crinozoans are primitively stalked echinoderms ('pelmatozoans') with long arms and normally lack complex respiratory structures. The comatulid crinoids have, however, lost their stalks and become secondarily free.

Class Crinoidea

Crinoids are very diverse and important in Palaeozoic faunas, and their remains have contributed substantially to Palaeozoic limestones. Complete crinoids (e.g. Fig. 9.35), however, are rarely preserved. Crinoids are less abundant than they were, but at the present they are represented by twenty-five stalked genera and by some ninety genera of unstalked comatulids, which are the dominant crinoids of modern oceans. One of these, the modern free-living *Antedon* (Fig. 9.36), is taken as a type example showing the basics of crinoid morphology. *Antedon* has a stalk in its early life, formed of

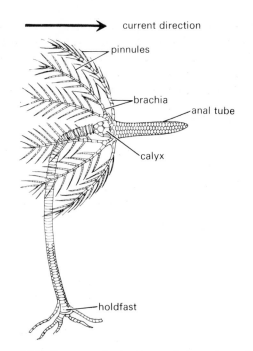

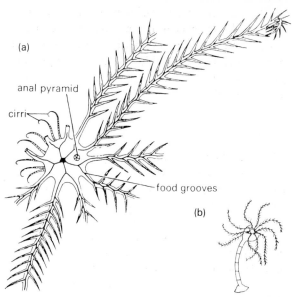

Figure 9.35 Features of crinoid morphology shown by Ordovician *Dictenocrinus*, reconstructed as a rheophile. (Substantially modified from Bather in Piveteau (eds), *Traité de Palaeontologie*, Part III)

Figure 9.36 (a) Calyx of Recent *Antedon* seen from above with brachia and pinnules; (b) larval *Antedon* still attached to its stalk.

columnar plates but it soon breaks free of it before it is fully grown and is then free to swim or crawl over the sea floor.

The body consists of a globular plated cup (the theca) from which long plated arms (**brachia**) arise. The theca has two parts: a lower region of thick rigid plates, pentamerally symmetrical, and a domed flexible roof (tegmen) with a central mouth and lateral anus. Inside the theca is a spirally twisted gut and a circum-oral ring as in echinoids, but the madreporite is replaced by ciliated funnels. The gonads are borne not within the theca but on the arms.

Since there is no stem the base of the cup (or calyx) is made of a single large plate (the **centro-dorsal**) which is morphologically the top plate of the stem. Above this is a ring of five **basal plates** with another ring of five **radial plates** above it. Attached to the radial plates are the **brachial plates** of the lower part of the arms. These arms subdivide almost immediately so that there are ten arms in all, each armed with many **pinnules** constructed of **pinnular**

plates. In *Antedon* and other comatulids the rigid plates of the calyx are very small and contain only a chambered part of the coelom. A number of flexible **cirri** articulate on the centro-dorsal plate and can be used for temporary fixation of the living crinoid.

The arms are flexible, having articulated brachial ossicles and two segmented strands of muscle with which the arm can move in any direction (Fig. 9.37c). From the circum-oral ring a radial water vessel runs down each arm, which as in starfish, but by contrast with echinoids, lies outside the plates. A double row of tube-feet arises from this vessel, with a median food groove in between. When the crinoid is feeding, the pinnules with these tube-feet are extended and, being sticky, catch organic detritus and small organisms which are then carried to the mouth by a **ciliary mucus tract** along the food groove. The pinnular food grooves all join with those of the brachia, which in turn unite so that five primary grooves cross the tegmen and enter the mouth. When disturbed the tube-feet can retract into the food groove, and as they do so small calcitic cover plates (**lappets**), normally held erect, come down like trapdoors to protect them.

Antedon has only ten arms, but other comatulids may have as many as 200. The deeper water species seem on the whole to have fewer arms.

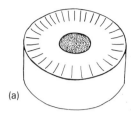

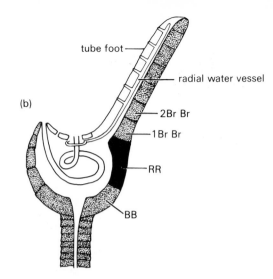

(a)

(c)

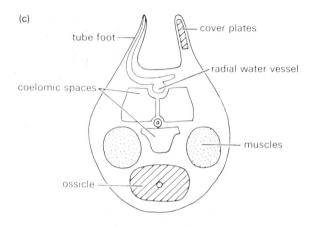

Figure 9.37   (a) A solitary crinoid ossicle showing central canal; (b) stylised cross section of crinoid with stem ossicles – basals (BB), radials (RR), primibrachs (1Br Br) and secundibrachs (2Br Br) and water-vascular system (redrawn from Moore, Lalicker, and Fischer, 1953, *Invertebrate paleontology*, New York: McGraw Hill)): (c) cross-section of crinoid arm, with muscles, r.w.v. ossicles and cover plates with coelomic spaces (redrawn from Nichols, 1974)

Main groups of crinoids

Crinoids seem to have originated during the great burst of Lower Ordovician adaptive radiation which gave rise to so many echinoderm types. A number of related forms, including blastozoan Class Eocrinoidea, go back to the Cambrian, and these were probably the ancestors of crinoids. The earliest true crinoids (e.g. the Arenig *Dendrocrinus*) belong to Order Inadunata (Fig. 9.38), which have rigid calycal plates. There are two other Palaeozoic orders: the Flexibilia (Fig. 9.45c), in which calycal plates are only loosely united, and the Camerata (Fig. 9.39), where the proximal arm ossicles are incorporated into the theca. Inadunates just crossed into the Triassic; the others became extinct in the Permian. A possible Recent survivor of the inadunates might be *Hyocrinus*, an Antarctic form, though this could merely be an aberrant articulate.

All Mesozoic to Recent crinoids belong to the fourth order, the Articulata, in which the arms are very flexible. Most modern crinoids are unstalked comatulids; the acquisition of brachial flexibility and liberation from the stem were undoubtedly of great importance in determining the evolutionary potential of this crinoid group, allowing an active search for good feeding grounds. The living stalked species are unusually small, inhabiting depths greater than 100 m and having stems rarely exceeding 0·75 m.

Palaeozoic crinoids

All Palaeozoic crinoids have stems, which in the case of some Lower Carboniferous forms are up to 30 m in length. They generally have larger calycal plates than *Antedon*, and a third circlet of plates (the **infrabasals**) may be present under the basals. In describing crinoids, the morphology of the calyx is usually represented by an 'exploded diagram' (Fig. 9.39c), and the plates are distinguished symbolically as brachials (**BrBr**), radials (**RR**), basals (**BB**), and infrabasals (**IBB**). Where there are only RR and BB present the crinoid is **monocyclic; dicyclic** crinoids have IBB as well.

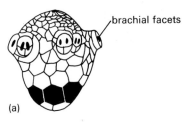

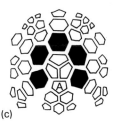

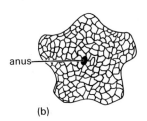

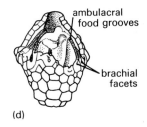

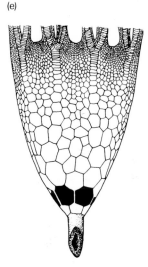

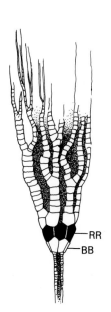

Figure 9.38 *Cupulocrinus* (Ord.), as preserved in the rock (x 2). (Drawn from a photograph by Ramsbottom, 1961, *Palaeont. Soc. Monogr.* 1–37)

Figure 9.39 *Actinocrinites* (L. Carb.), a monocyclic camerate crinoid with brachial plates incorporated into the large calyx: (a) lateral view with brachial facets (x 1); (b) tegminal (dorsal) view; (c) exploded plate diagrams; (d) broken specimen of the related *Cactocrinus*, showing subtegminal enclosed food grooves leading from the brachial facets to the mouth ((a)–(d) largely based on Woods, 1946, *Palaeontology* (Cambridge: C.U.P.)); (e) *Scyphocrinites* (Dev.), a large camerate, with pinnulars incorporated in the calyx (x 0·3) (redrawn from Moore, Lalicker and Fischer, 1953, as in Fig. 9.37).

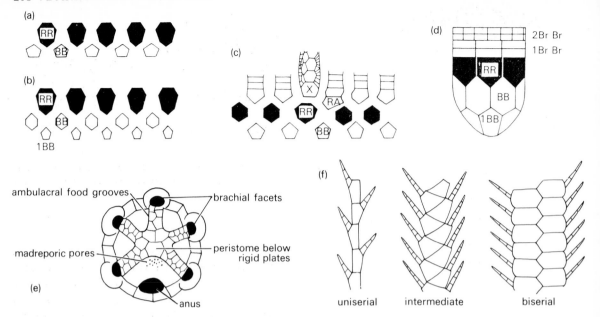

Figure 9.40 Plate diagrams: (a) Monocyclic calyx; (b) dicyclic calyx; (c) *Cupulocrinus* (Ord.), dicyclic calyx with an anal tube (redrawn from Ramsbottom, 1961 (as in Fig. 9.38). (d) Lateral view of dicyclic calyx: 1BB – infrabasals; BB – basals; RR – radials (in black); RA – radianal; X – x-plate; 1BrBr – primibrachs; 2BrBr – secundibrachs; (e) tegmen of crinoid *Cyathocrinites* (redrawn from Bather in '*Treatise*' Part (U)); (f) evolution from a uniserial through an intermediate to a biserial condition in the brachia. (Modified from Moore, Lalicker and Fischer, 1953, as in Fig. 9.37)

Inadunate crinoids

The Inadunata (L. Ord.–U. Perm.) is a very large order with more than 200 species. Inadunates are probably the most primitive of all crinoid groups (Figs 9.38, 9.40), having a rigid calyx with the brachials free or loosely connected above the radials. The mouth is below the tegmen, the food grooves above it. The arms may or may not have pinnules. Commonly the pentameral symmetry of the calyx is disrupted by an extra plate, the **radianal (RA),** within the radial circlet, and there is usually an **anal (X)** plate in the brachial circlet (Fig. 9.40c). One inadunate group is dicyclic; all others are monocyclic. Radial plates may be 'compounded', i.e. divided transversely into **infraradials (IRR)** and **supraradials (SSR),** and there are radianal equivalents **(IRA and SRA).** In some an **anal tube** emerges from above the radianals; this is greatly elongated in such genera as the Ordovician *Dictenocrinus* (Fig. 9.35), and *Dendrocrinus* where it is made of thin folded plates. The earliest inadunates had an elongated straight cup, but in many later genera the calyx became flattened and expanded laterally. Furthermore, the almost vertical direction of the dichotomous arms in the early inadunates was generally replaced by a more flower-like crown.

Numerous other evolutionary trends have been distinguished, and many specialised forms are known. *Petalocrinus* (Fig. 9.41) is a tiny Silurian inadunate resembling a minute palm tree. Here the stem is short, the calyx small, and each of the five brachia fused into a leaf-like divided plate, with a series of dichotomous food grooves impressed upon its surface. *Hybocystis* has only three free arms, the other two being represented only by food grooves extending down the sides of the cup. The Ordovician Porocrinidae (e.g. *Porocrinus* and *Triboloporus*) (Fig. 9.42a, b) are a strange family which have been placed in the inadunates (Kesling & Paul 1968). In these the calyx is strengthened by thickened ridges linking the plate centres. Within this framework, where the plates join there are thin circular areas of the calyx bearing folded calcitic membranes **(goniospires).** Each goniospire consists of three sets of folds, one for each of the adjacent

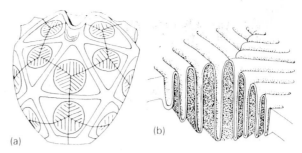

Figure 9.41 *Petalocrinus* (Sil.), a small Silurian inadunate with its brachia welded into flat plates with food grooves. Section across shows the cover plates (x 2 approx.). (Redrawn from Tasch, 1973 (see Bibliography following Ch. 1))

Figure 9.42 *Porocrinus* (Ord.), with folded membrane structures at the plate junctions: (a) lateral view of calyx (x 4 approx.); (b) section through folded membrane. (Redrawn from Kesling and Paul, 1968)

plates. The goniospires probably had a respiratory function, but of greatest interest is their similarity to other types of folded membrane structure in both cystoids and blastoids, which arose quite independently. Some functional and other considerations on such structures are given later.

Flexible crinoids

The Flexibilia (M. Ord.–U. Perm.) (e.g. *Protaxocrinus*) (Fig. 9.45c) are less numerous and diverse than inadunates. They are all dicyclic with three infrabasals, one smaller than the other two. The lower brachials are incorporated into the cup but are not rigidly joined so that the whole cup is flexible. The mouth and food grooves are on the upper surface of the tegmen. There is radianal and an anal plate. Some evolutionary trends, such as the relative widening of the calyx and the tendency for the lower part of the arms to droop over the cup, are parallel with those in the inadunates.

Camerate crinoids (Fig. 9.39)

Order Camerata (Ord.–M. Perm.) is the largest of all the Palaeozoic groups. In this group the cup is rigid, the radianal plate is absent and some of the lower brachial plates are incorporated into the calyx. The mouth and food grooves lie below the tegmen, and in some cases (e.g. *Cactocrinus*) the thin subtegminal tubes linking the brachia with the mouth have been preserved and can be dissected out. The arms always have pinnules. There are both monocyclic and dicyclic genera classified in the suborders Monobathra and Diplobathra respectively. The earliest camerates, such as the Lower Ordovician *Reteocrinus*, have many features reminiscent of inadunates, implying a relationship. Such a crinoid as *Actinocrinites* exhibits the basics of camerate structure clearly, in the exploded plate diagram and in the lateral and oral views. The tegmen is large and dome-shaped and has a subcentral anus. Arms arise from the circular **brachial facets,** which are lateral projections from the calyx. The many **interbrachial plates (IBrBr)** and brachial plates which are incorporated into the calyx make a large rigid theca.

In both monobathral and diplobathral crinoids there are evolutionary trends parallel with those of the Flexibilia. Thus in the early genera such as the diplobathral *Reteocrinus* the interbrachial plates are small and irregular, lying in depressed zones between the arms. In the descendants of *Reteocrinus* they have become large, regular, polygonal and less depressed. Furthermore, the anal and interbrachial plates tend in the later genera to move upwards and out of the cup, permitting a perfect radial symmetry with the radials all in contact.

There are many unusual camerates. The Devonian *Scyphocrinus* is a rather extreme form in which not only the lower brachials but also the lower pinnular plates have become incorporated into a very large theca. Many genera, such as *Barrandeocrinus*, developed curiously downturned arms, drooping over the cup.

The arms here are very broad, with the pinnules bent over to meet medially so that a trough was formed, presumably acting as some kind of food groove.

Mesozoic to Recent crinoids: articulates

The earliest articulate crinoids are Triassic. A well-known species, *Encrinus lilliformis* (Fig. 9.43a), occurs in great numbers in the German Muschelkalk. This and other early genera bear a considerable resemblance to inadunates, from which they probably evolved, though some articulates could possibly have come from the Flexibilia; the order would thus be polyphyletic.

In articulates the calyx is relatively reduced: five basals, five infrabasals and five radials usually being present. The mouth and food grooves are exposed on the surface of the flexible tegmen, but brachials are never incorporated in the cup. Articulates are further and most importantly distinguished by the flexibility of their long arms, which can move up and down, coil up or twist about. It is this character that has given them a great evolutionary advantage, for those many genera of the Order Comatulida which break free from their stalks in the early stages are able actively to move around on the sea floor rather than being fixed. Of modern crinoids only twenty-five genera have stalks, and these are rarely more than 0·75 m long, contrasting with some of the giant 30 m stalked forms of earlier times. The stems frequently have elongate cirri erupting from nodes.

Genera such as *Pentacrinus* are quite abundant in the Jurassic, and there are many mid-Cretaceous forms. Mesozoic and Tertiary genera are generally long-ranged and include a number of micromorphic forms, e.g. *Eugeniacrinus*. The unstalked comatulids are very abundant today, there being ninety genera. In certain regions of the world, Antarctica in particular, there are endemic comatulid faunas which have evolved in isolation since the beginning of the Pleistocene. The Antarctic genera now seem to be spreading northwards along the many submarine ridges radiating from the Antarctic continent. Comatulids first became common in the Jurassic; evidently the potential for movement inherent in the unstalked crinoids' arms was realised early.

Successive comatulid genera continued through the Mesozoic and Tertiary. The Jurassic *Saccocoma* (Fig. 9.43b) from the Solnhöfen Limestone has very long feathery arms and curious flattened lappets, being extensions of the proximal arm plates. This might have been a free-swimming form, the lappets increasing the surface area and thus retarding sinking. Large plated stemless genera (e.g. *Marsupites* and *Uintacrinus*) are well-known Cretaceous fossils. Most living genera are unknown as fossils.

Ecology of crinoids

Modern unstalked crinoids live at all depths from sub-littoral to abyssal; stalked forms are normally found only below 100 m. Crinoids are uncommon in shallow water and tend to be most abundant in rather inaccessible regions, so that until recently little was known about their life and feeding habits. First-hand diving observations have illuminated many unsuspected adaptations and stimulated functional considerations on ancient forms. But knowledge is still very limited, and most observations so far have been made largely on comatulids.

Present day comatulids divide into two quite distinct feeding types: **rheophilic** (current-seeking) and **rheophobic** (current-avoiding). Rheophilic crinoids (Figs 9.44, 9.45a) live in relatively shallow current-swept waters and make use of the currents in feeding. During the day individuals may hide in cavities but come out at night to feed, climbing with their cirri on to a high position on a rock or sea grass and unfurling their arms. The arms spread out in a **vertical filtration fan,** with the polar axis horizontal, the aboral side facing the current and the pinnules spread out to form a grating. The tube-feet are extended from these to form a fine filtering net. Even with a current speed of only 2 cm/sec a surface of 500 cm^2 can filter 1 l./sec. Such a system is remarkably efficient, making use of existing currents.

Rheophobic crinoids live in current-free waters though most deep water crinoids are rheophilic. They rely entirely, as far as is known, on gravity feeding; only through a continuous rain of detrital particles can they obtain enough to eat. In the laboratory *Antedon* behaves as a rheophobe though it is naturally a rheophile. Rheophobes lie on the bottom, the arms outspread to form a **horizontal collecting bowl,** capturing organic particles falling on it. Some rheophobes live in deeper waters, so that a fair amount of detrital material has accumulated by the time it reaches them, but they do not consume

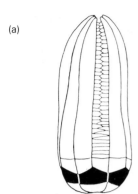 (a)

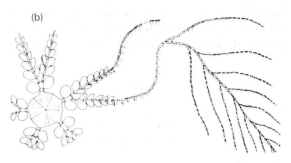 (b)

Figure 9.43 (a) *Encrinus* (Trias.), a possible early articulate with some features reminiscent of inadunate morphology (x 0·75); (b) *Saccocoma* (Jur.) possibly planktonic articulate (x 5). (Redrawn from Moore, Lalicker and Fischer, 1953, *Invertebrate fossils*, New York: McGraw Hill)

anything like the quantity that rheophiles do. They are simply exploiting a different source of food, and in compensation their growth and metabolic rates are probably less. Abyssal crinoids (and others) may filter horizontal currents.

Most fossil crinoids, stalked or unstalked, were probably rheophiles or rheophobes; hence their mode of life may be interpreted from their morphology (Breimer 1969, Breimer & Webster 1975). Camerate crinoids have long and very flexible stems, as do some inadunates. Sometimes crinoid faunas consisting largely of camerates occur in high-energy carbonate sequences. In the Lower Carboniferous Burlington Limestone of Missouri, the many camerates (Families Actinocrinidae, Batocrinidae, etc.) are interpreted largely as rheophiles. The stems are strong but flexible, and their flexibility increases towards the cup. The crown could thus assume the attitude of a vertical filtration fan, with the aboral surface facing the current and the long anal tube facing away from it (Figs 9.35, 9.45a). Since the arms of most camerates are biserial with very many pinnules, these crinoids would seem to have been highly efficient filterers, straining off food particles borne by the currents which, as is clear from the sediments, are known to have swept the Burlington Limestone sea.

Some camerate stalks have cirri on one side only and were probably fixed to the sea floor in a recumbent attitude, with only the crown and upper stalk rising above the sea floor, again being held as a vertical filtering fan.

Most flexible crinoids have short rigid stalks, and the crown could not have bent into a vertical position. Furthermore, the arms of the Flexibilia do not have pinnules, so all the evidence is against a rheophilic

Figure 9.44 *Cenocrinus asterius* Linne, a Recent rheophilic crinoid living at depth of 200–300m off north coast of Jamaica. The arms are of radius 20–30cm and form a vertical filtration fan. (Photograph by courtesy of Dr D. B. Macurda)

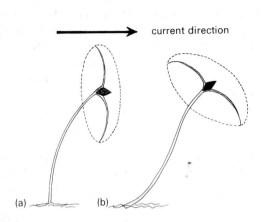

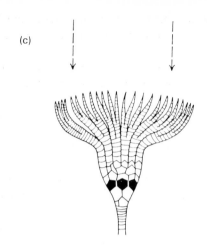

Figure 9.45 Stalked rheophilic crinoids with calyces arranged to act as vertical filtration fans: (a) with no lift from the current; (b) with lift (modified from Breimer and Webster, 1975); (c) rheophobic crinoid (based on *Protaxocrinus*) with head acting as a horizontal bowl.

mode of life. They are best interpreted as rheophobes with the crown acting as a horizontal collecting bowl (Fig. 9.45c), though some may have been macrophagous like modern basket-stars. The Crawfordsville Limestone of Indiana is full of the stalks and crowns of inadunate and flexible crinoids and was deposited in quiet and poorly aerated waters. Most of the crinoids on morphological grounds were probably rheophobes (though they might possibly have been able to create their own currents by undulatory movements of the arms). The very different aspects of the Burlington and Crawfordsville crinoid fauna definitely point to a different ecology and feeding type.

Palaeozoic crinoid gardens may have been stratified like a tropical jungle, partitioning resources by feeding on different food at different levels.

Fossil articulates can likewise be interpreted as rheophiles or rheophobes. But in the case of the enormous Jurassic *Seirocrinus* rather different interpretive criteria have been used (Seilacher *et al.* 1968). Here on a single bedding plane, now in the Tübingen Museum, there lie some fifty crinoid specimens, each with a crown about 80 cm across and a stem up to 15 m in length. They are preserved with the crowns face down in black euxinic shales. One interpretation of these is that they lived a pseudo-planktonic life, hanging downwards from a floating driftwood log which slowly became waterlogged and sank, so that the crowns touched down first followed by the stems which settled in loops as the log drifted slightly before coming to rest. In *Seirocrinus* the basal part of the stem is the more flexible, by contrast with the norm. This too is in accordance with a pseudo-planktonic mode of life, for the stem needs to be flexible at the base to withstand storms yet rigid near the calyx to maintain control of the inverted posture. On the other hand, the very long stems could well have had another function. Stalked crinoids have the advantage of elevation, hence being able to filter from a region of the water not exploited by other organisms. A long-stalked crinoid could have gained lift from the current, provided that it was able to orient the parabolic fan at the correct angle (Fig. 9.45b). The whole system can be likened to a kite tethered by a cord. But unlike a kite the crinoid could probably regulate its own degree of lift by altering the curvature of the arms, the angle of the fan or the attitude of the pinnules. Given certain conditions, the crown could rise or sink in the water to exploit food-rich levels. In flexible-stemmed crinoids such as *Seirocrinus* the analogy of a kite holds good. Hence the interpretation of all such crinoids as pseudoplanktonic is not unequivocal; some of them could just as well have been fixed to the sea floor.

Formation of crinoidal limestones

Many Palaeozoic crinoids seem to have lived in 'crinoid gardens', i.e. in clumps or patches isolated from other such areas and in which diversity may be high or low. Other crinoids inhabited muddier water conditions.

After death, crinoids usually broke up through decay, disarticulation or the action of scavengers. The specific gravity of the various components first increases, owing to the removal of investing tissue, then decreases whilst the stereoplasm (tissue penetrating the stereom cavities) is replaced by sea water. Where isolated columnals are found they have been transported by rolling and may have come from some distance. Commonly great thicknesses of limestone are found, formed of short or long lengths of stem, isolated ossicles or calycal plates. In the Carboniferous limestone of England crinoidal debris of this kind can often form the bulk of the rock. These rocks are more or less autochthonous and represent the debris of crinoid gardens where the material was buried prior to complete disintegration.

When orientated, short stem lengths are found in abundance; they were rolled by the current into an area of accumulation, and the direction of the current that moved them can normally be told from their orientation. Unorientated debris found on the same bedding planes apparently became stuck and was not moved by the current. The rare cases when crown and stem are preserved represent almost immediate burial before decomposition or, as in the case of *Seirocrinus*, deposition in quiet and possibly anoxic waters.

SUBPHYLUM BLASTOZOA

Classes Diploporita and Rhombifera: cystoids

'Cystoids' are a heterogeneous group of Palaeozoic blastozoans all of which have respiratory 'pore-structures' traversing the plates of the theca.

Haplosphaeronis (Class Diploporita) (Fig. 9.46) is an Ordovician cystoid known mainly from northern Europe. It has an ovoid theca less than 30 mm in height with a flat base by which it was attached to the substratum. The theca has two circlets of polygonal plates: the lower circlet of seven plates (**LL**), the upper of seven **perioral plates (POO).** Above these lies the peristome, constructed of five perioral plates and having at its centre the slit-like mouth. From the mouth radiate five food grooves or ambulacra, each of which has several branches terminating in small studs (**ambulacral facets**) which were the points of

attachment of the food-gathering arms (brachioles). Brachioles are rarely preserved in cystoids but are very similar in structure and function to the arms of crinoids, though small and simple. They were probably spread out in a kind of horizontal collecting bowl, extracting food from the water and passing it to the mouth via the ambulacra.

A well-defined 'anal pyramid' of five or more triangular **anal plates** is set at the junction of two of the POO plates. The anal plates could hinge open along their bases and close again after evacuation. A small slit known as the **hydropore,** probably the entrance to the water-vascular system and characteristic of cystoids, is set near the peristome. All plates except these of the peristome are densely covered with twin perforations, surrounded by a raised rim and reminiscent of the pore-pairs of echinoids. These are known as **diplopores.** They probably led to an external 'papula', uncalcified and resembling an elongated tube-foot extended into the surrounding water for respiration.

Structural characteristics of cystoids

Some cystoids like *Haplosphaeronis* were fixed by their bases to the sea floor or to a shell; others had stems which have been occasionally preserved with the holdfast intact, but these stems were always short and never reached more than a few centimetres in length. Many of the Ordovician cystoids had thecae composed of numerous irregular plates (e.g. the diploporite *Echinosphaerites* and *Aristocystites* and the rhombiferan *Echinosphaerites*).

The food grooves in cystoids were protected by cover plates. These food grooves may be short as in *Sinocystis* (Fig. 9.46) or long, but only in Family Cheirocrinidae (Class Rhombifera) is there any documented trend to lengthen the ambulacra with time.

The curious Ordovician genus *Asteroblastus* has a star-shaped system of food grooves, very similar in appearance to the ambulacra of blastoids. Brachioles arose laterally from these, and the diplopores were restricted to the interambulacral regions alone. It has been suggested that blastoids descended from cystoids through *Asteroblastus*. This is unlikely, however; it is more probably a case of convergent evolution. Brachioles are rarely found in cystoids, and where they are known they are always short and slender. The **gonopore** (for the shedding of genital products)

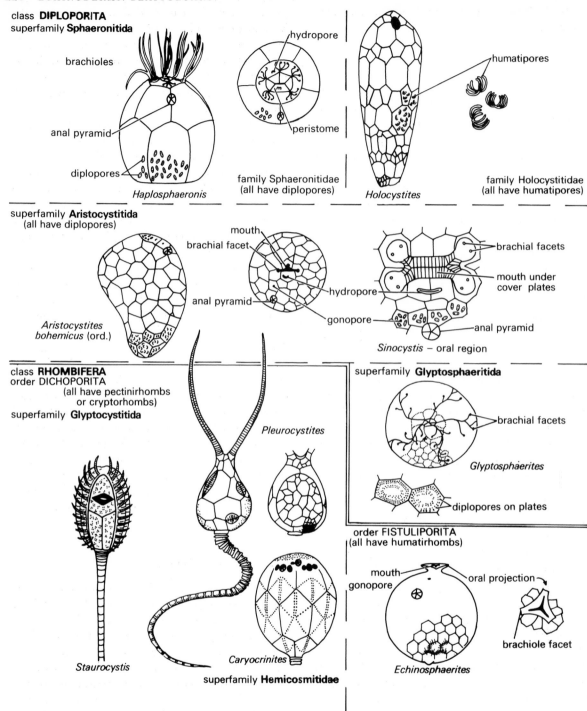

class **DIPLOPORITA**
superfamily **Sphaeronitida**

brachioles

anal pyramid

diplopores

Haplosphaeronis

hydropore

peristome

family Sphaeronitidae
(all have diplopores)

humatipores

Holocystites

family Holocystitidae
(all have humatipores)

superfamily **Aristocystitida**
(all have diplopores)

*Aristocystites
bohemicus* (ord.)

mouth
brachial facet

anal pyramid

hydropore

gonopore

brachial facets

mouth under
cover plates

anal pyramid

Sinocystis – oral region

class **RHOMBIFERA**
order DICHOPORITA
(all have pectinirhombs
or cryptorhombs)
superfamily **Glyptocystitida**

Pleurocystites

Staurocystis

Caryocrinites

superfamily **Hemicosmitidae**

superfamily **Glyptosphaeritida**

brachial facets

Glyptosphaerites

diplopores on plates

order FISTULIPORITA
(all have humatirhombs)

mouth
gonopore

oral projection

brachiole facet

Echinosphaerites

Figure 9.46 Cystoid morphology and classification according to Paul's system, showing the high taxonomic value of pore structures. (Diagrams mainly based on Kesling *et al.*, in '*Treatise*' Part (S))

and the hydropore are usually distinct but sometimes combined.

Pore-structures (Fig. 9.47)

The pore-structures of cystoids, which are of very many kinds, provide a firm basis for classification (Paul 1968, 1972). Diplopores, as in *Haplosphaeronis*, are pairs of perpendicular canals through the theca, each opening in an external depression (the **peripore**). Other kinds of pore-structure are based on U-shaped tubes of various kinds known as **thecal canals** which open to the external or internal surface through **thecal pores.** External pores are round and simple, sieve-like or slit-shaped; internal pores are always simple. Thecal canals are always tubular and may traverse the theca perpendicularly or tangentially (diplopores are paired thecal canals forming U-pairs in the soft parts).

Two quite distinct functional types of pore-structure can be distinguished. **Endothecal** pore-structures (**dichopores**) had canals that ran below the external surface of the theca and communicated with the surrounding sea water through external pores. If sea water was pumped through the canal, presumably by cilia, respiratory exchange could have been effected by gaseous diffusion through the thin calcified wall of the thecal canal which bathed the internal body fluids. Respiration was thus internal. By contrast, **exothecal** pore-structures (**fistulipores**) had internal pores and an external canal running outside the theca. Thus the body fluids would have to be brought through the canal and back down again within the theca, so that respiratory exchange was outside the theca. These two alternative methods of facilitating respiratory exchange have their advantages and disadvantages. Endothecal structures were liable to choking by foreign particles, but they may have slits, sieve plates or other devices to keep such foreign bodies out. A sieve usually protects the inhalant pore alone, enabling it to be identified. Exothecal structures did not have this problem but were liable to abrasion or breakage, and in any case they were much less efficient in terms of relative areas of surface exchange.

Diploporite cystoids such as *Haplosphaeronis* have numerous exothecal diplopores, as already described. These are simple paired tubes crossing the plate perpendicularly and leading to uncalcified external papulae. An alternative kind of structure is the **humatipore** (Fig. 9.47) in which the paired perpendicular tubes are linked by a complex of exothecal tubes. These are either flush with the surface or occupy raised humps. Diploporites like *Holocystites* (Fig. 9.46) have their plates replete with such humatipores.

Rhombiferans may have exo- or endothecal pore-structures always arranged in parallel sets with a rhomb-shaped contour. Each rhomb crosses a plate boundary. **Pectinirhombs** are highly organised units constructed of parallel dichopores. There are normally only two to four in any one cystoid. Each dichopore may open in a single slit (**conjunct**) or have the central part roofed over (**disjunct**) to form a separate entrance and exit. Dichopores may be joined laterally (**confluent**) forming a single folded membrane. Such pectinirhombs have a raised external rim and may be conjunct or disjunct. If disjunct, they have a thickened bar running along the line of the suture carrying a raised ridge on one side only. This has been explained as a mechanism for preventing the recycling of currents. The ciliary action of the cellular membrane lining the dichopores would produce a one-way current. Small particles dropping through to the base of the dichopore would be light enough to be disposed of by the ciliary current. Endothecal **cryptorhombs** are made of parallel dichopores without the raised external structure of pectinirhombs. They have sieve-like incurrent and simple excurrent pores. **Humatirhombs** are made of exothecal fistulipores likewise parallel and arranged in rhombs.

Classification of cystoids

Many taxonomic schemes have been proposed for this complex and difficult group. The scheme of Paul (1972) adopted here (Fig. 9.46) differs somewhat from that of Kesling in the *Treatise*. Paul has given high level taxonomic value to the pore-structures. In Class Diploporita there are only diplopores scattered over the thecal surface, many to a plate, other than in Family Holocystitidae where there are humatipores instead. In Class Rhombifera the members of Order Dichoporita all have pectinirhombs or crytorhombs, which are used further to divide the order into superfamilies. Humatirhombs characterise the other order, the Fistuliporata.

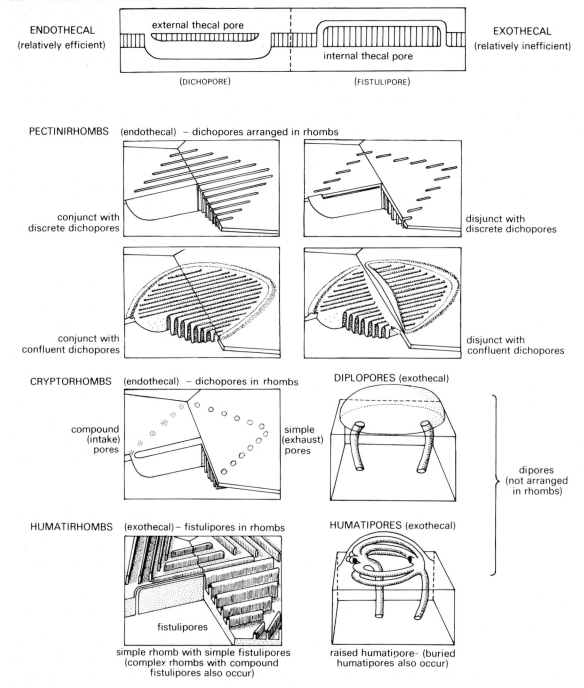

Figure 9.47 Different kinds of pore-structures in cystoids. (Based on Paul, 1968, 1972)

Cystoid ecology

Most cystoids were sessile and fed on small organic particles using their brachioles. These brachioles are seldom fossilised but do not normally seem to have been very large. Judging by the size of the ambulacral facets in the Holocystitidae, however, the brachioles may in this family have been of fair dimensions. Possibly the closely spaced brachioles of such genera as *Haplosphaeronis* formed a food trap like that of some modern Phoronidea.

Cystoids generally lived in shallow waters, and evidently diploporites were amongst the first colonisers of areas in which sediment deposition began again after a break. Cystoids had some tolerance of suspended sediment; conceivably exothecal pore-structures could have been adapted to sediment-laden water. Where conditions were particularly favourable cystoids lived together in clumps, and occasionally these assemblages were preserved in life position.

Some cystoid genera are strikingly modified. The rhombiferan *Pleurocystites* (Fig. 9.46) has an asymmetrical flattened theca which has large irregular plates with pectinirhombs on one side and small flexible plates on the other. The two brachioles are large and forward-projecting; the anus is located by the side of the stem, which is very flexible and tapering and lacks a holdfast. *Pleurocystites* probably lay on the sea floor with the flexible part of the theca down (Paul 1967). The stem morphology would allow undulating movements which would permit forward propulsion, thus the whole morphology of *Pleurocystites* seems to be specialised for a mobile benthic life. As such it bears a remarkable superficial resemblance to similarly specialised 'homalozoans' (e.g. *Cothurnocystis*) (Fig. 9.50b), which quite independently acquired similar characters through convergent evolution.

Why did cystoids need such elaborate pore-structures for respiration when the crinoids, which outlasted them, got on perfectly well without them? Paul (1976) has presented a reasonable hypothesis about this based on the known palaeogeographical distribution of cystoids. One of the earliest known cystoids, the Tremadocian *Macrocystella* (which was originally thought to be an eocrinoid), lacks any pore-structures though its plates are thin and have radial ridges organised in rhombs. It lives in cold circum-polar waters, which would have had a fairly high oxygen concentration. Pore-structures occurring in its descendants, in other cystoids, and indeed in all other echinoderms with elaborate respiratory structures, are found predominantly in shallow tropical seas. Nearly all Ordovician occurrences of pore bearers are between 30° N and S of the palaeoequator. In such warm shallow tropical seas oxygen tensions are very low during the night, and it is suggested that the pore-structures are well adapted for coping with this. Furthermore, from evidence available at present it seems that the more efficient pore-structures (pectinirhombs) occur in genera living within 15° N and S of the equator; the less efficient ones have a wider distribution.

Class Blastoidea (Figs 9.48, 9.49)

Blastoids (Sil.–Perm.) are an extinct class of small pentamerally symmetrical 'pelmatozoans' characterised by internal respiratory structures known as **hydrospires.** An individual blastoid had a short fixed stem surmounted by a calyx from which arose a crown of brachioles. The stem and brachioles are rarely preserved, and the majority of species are known from the calyx alone. The plating differs from that of crinozoans. In some genera there are extra and complex plates, but only the standard plating structure is given here. Some eighty genera are known.

Pentremites come from the Mississippian of North America. Nearly seventy species belong to this genus. *P. symmetricus* (Fig. 9.48) from the Lower Carboniferous of Illinois has a bud-shaped calyx with five petaloid 'ambulacra' running from the summit to two-thirds of the way to the base. Each ambulacrum is indented by a vertical median groove. There are three basal plates (BB). Above them are five radial plates (RR), deeply indented by the ambulacra, and surmounting the radials are five small rhomb-shaped interambulacral **deltoid plates** (Δ). Below each ambulacrum lies a long spear-shaped lancet plate, largely covered by the small plates of the food grooves.

At the summit there is a central star-shaped mouth, and surrounding it are five large **spiracles,** each set at the summit of a deltoid plate. One of these (the **anispiracle**) is larger than the others because it contained the internal anus, suggesting that the

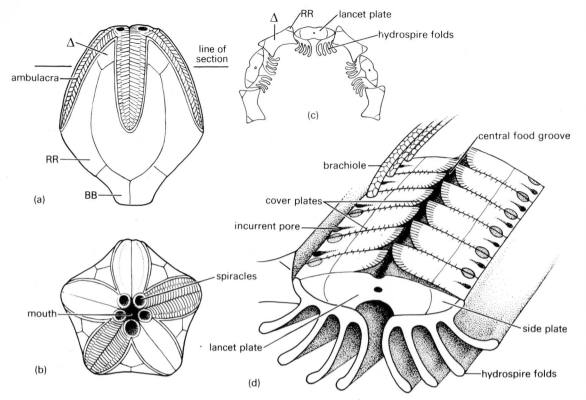

Figure 9.48 *Pentremites*, a Lower Carboniferous spiraculate blastoid: (a) lateral view; (b) adoral view; (c) transverse section with hydrospires; (d) oblique three-dimensional section showing brachiole and hydrospire system. ((a)–(c) x 2, (d) x 10)

spiracles were some kind of outlet system.

The ambulacral and subambulacral structure is complex and is best studied by making thin sections. These show that the lancet plate bears two lateral **side plates** associated with the elements of both the feeding and respiratory systems. Along each edge of an ambulacrum ran a single line of brachioles which, when broken off, left distinct facets. These brachioles led to food grooves with biserial cover plates crossing the ambulacrum and joining with the median food grooves which ran vertically to the mouth. Feeding was probably like that of crinoids, traces of the water-vascular system having recently been discovered. The hydropore was probably internal. Between adjacent brachioles is a narrow slit from which a pore leads down to the hydrospire system. Each of the paired hydrospires is simply a thin-walled rigid calcified tube, with a convoluted inner surface, lying below the ambulacrum and connected to the many **hy-drospire pores** by individual thin tubes. A pair of hydrospires from adjacent ambulacra connects with a single spiracle. The subdivision of the inner face of the hydrospire into four smaller parallel tubes results from the infolding of the calcified membrane of which it is constructed.

Diversity and function of hydrospires

Hydrospires may, as in *Orbitremites* (Carb.) (Fig. 9.49e), consist merely of straight undivided tubes. Usually, however, as in *Pentremites*, the hydrospires are convoluted (Fig. 9.48c, d), forming a **hydrospiraculum** which may have from two to seven folds. The hydrospire pores were probably lined with cilia, and their beat presumably created a unidirectional current drawing water in through the pores, passing it along the hydrospiraculum and

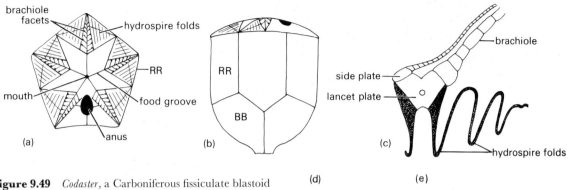

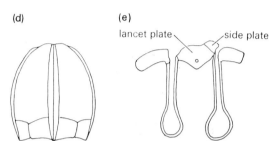

Figure 9.49 *Codaster*, a Carboniferous fissiculate blastoid with exposed hydrospire folds: (a) adoral view (x 3); (b) lateral view; (c) ambulacra with hydrospires in section. *Orbitremites*, a Lower Carboniferous spiraculate blastoid: (d) in lateral view (x 3); (e) hydrospires in section.

exhaling it through the spiracles. The folding would undoubtedly allow a greater surface for respiratory exchange.

How efficient were these structures? It is a fairly simple matter to calculate input/output flow rate by measuring cross-sectional areas of the hydrospire pores and the spiracles. In *Globoblastus*, where a complete study through ontogeny has been made (Macurda 1965b), the potential outflow rate in an adult could have been six times the inflow rate. Thus at an inflow velocity of 0·1 mm/sec the water volume could be changed completely in 100 sec. in an adult, 40 sec. in a juvenile. In *Globoblastus* differential growth gradients seem to relate to hydrodynamic and feeding efficiency by controlling its external form.

Classification and evolution of blastoids

Blastoids arose in the Silurian from an undetermined ancestor. Two orders are distinguished: the Fissiculata and Spiraculata. The Fissiculata have open hydrospire folds, whilst the Spiraculata (e.g. *Pentremites*, *Globoblastus*, *Orbitremites*) have spiracles.

Fissiculate blastoids are apparently more primitive. *Codaster* (Fig. 9.49a–c), for example, has a subconical theca with a flattened upper surface. The upper edges of the radial plates form a ring around the horizontal deltoid, radial and lancet plates. The anus is off-centre at the junction of two radials and a deltoid. Parallel with the edges of the lancet plates are

the upper edges of the hydrospire folds, graded in size away from the lancet so that the hydrospire system has a half-rhomb-shaped outline. There are only a few reduced hydrospire folds between the anus and the adjacent lancets. The side plates are large and almost cover the lancet plates, as shown in Figure 9.49c. It is presumed that the hydrospires functioned in a broadly similar manner to that of the spiraculate blastoids, but the system was probably less efficient.

The Fissiculata are widespread and lasted until the Permian. Such genera as *Orophocrinus* (e.g. Macurda 1965a) are well-known components of the Mississippian reef-knoll and carbonate-shelf fauna. Bizarre genera include *Astrocrinus* (L. Carb.) which has one aberrant ambulacrum and a displaced anus. Amongst several odd genera from the Permian of Timor is *Thaumatoblastus* which has immensely extended lancets and a rounded base.

Fissiculates could have given rise to spiraculates by the lateral growth of the lancet and side plates and the inward migration of the hydrospires, though there is no unequivocal evidence. All spiraculates have true hydrospire pores and spiracles. They have generally the characters of *Pentremites* but may be long and thin (e.g. *Troostocrinus*) or near-globular with the ambulacra extending right down to the concave base. Genera are distinguished by the plating structure and the number of hydrospire folds.

Ecology and distribution of blastoids

Blastoids were never an abundant component of any fauna except locally, where in thin bands they are sometimes so numerous as to make up the bulk of the rock. There are, for instance, well-known bands in the Visean of northern England where *Orbitremites* is so abundant as to permit mathematical study of variation and relative growth (Joysey 1959). These individuals were especially abundant in crinoid bank deposits capping reef knolls.

The few Silurian blastoid genera, both fissiculate and spiraculate, are entirely restricted to the North American continent, but by the Devonian blastoids had become worldwide. They reached a maximum in abundance and distribution in the Lower Carboniferous with about forty-five genera. Though they are rare in the higher Carboniferous, blastoids, often of peculiar form, are found in profusion in some very localised places in the Permian of the eastern hemisphere. By the end of the Permian they were extinct.

SUBPHYLUM HOMALOZOA (Fig. 9.50)

The calcite-plated asymmetrical fossils known variously as homalozoans, calcichordates, or by the older and more noncommital name 'carpoids', are perhaps the most bizarre and controversial of all known extinct invertebrates. They have generally been regarded as echinoderms, but at least one modern scholar has ascribed some of them to Subphylum Calcichordata of Phylum Chordata (which includes the vertebrates), suggesting that they are actually calcite-plated chordates with echinoderm affinities.

None of the diverse organisms included in the carpoids has radial symmetry; indeed most show no planes of symmetry at all. The Ordovician *Dendrocystites* (Class Homoiostelea) has a lobed asymmetrical flattened theca of small irregular plates, a single arm or brachiole, and a stalk of **heterostele** with a stout polyserial proximal part and a long tapering distal portion of elongated plates. The anus lies near the proximal part of the heterostele.

Cothurnocystis (Class Stylophora. Order Cornuta) is likewise laterally flattened. It has a boot-shaped theca, with a stalk (**aulacophore**) not unlike the heterostele of *Dendrocystites*. A marginal frame of stout plates surrounds the theca, extending into a 'tongue', 'heel strap' and 'toe' for the boot. On one of the flat surfaces there are projecting studs and a strengthening strut crossing the theca. Small rounded plates forming a flexible cover occupy the surfaces of the theca within the marginal plates. These are slightly different on the two sides. On the opposite side to the strut are some fifteen slits arranged in a curving arc, each with a hanging flap of small plates hinged above and fixed externally (Fig. 9.50c). Behind the 'tongue' is a pyramid of plates closing an orifice.

It is agreed by all authors that *Cothurnocystis* probably lay flat on the sea floor supported by the marginal spines and the studs, with the surface with the strut lying in proximity to the sea floor, the slits uppermost and the aulacophore fixed in the mud or free. But there is no agreement between Ubaghs (1971) and Jefferies (1968) as to the homologies of the various structures. Ubaghs has considered the slit-like pores to be the inhalant openings of respiratory organs, the large orifice behind the 'tongue' to be the anus, and the aulacophore to be a feeding organ. Jefferies (1968), on the other hand, has proposed vertebrate homologies for the various organs. He has suggested that the aulacophore is equivalent to the vertebrate tail, that the slit-like pores are exhalant and homologous with vertebrate gill slits, and that the large orifice is the mouth and entrance to the respiratory system. The presence of the hanging curtain of small plates outside the pores does in fact suggest that they were exhalant structures. The anus, according to Jefferies, is a small perforation near the origin of the stem.

Even more of a problem is the Ordovician genus *Mitrocystella* (Class Stylophora. Order Mitrata) (Fig. 9.50d) which has a large flattened, though asymmetrical, theca of irregular plates and a short stumpy down-curved stem or tail. Jefferies (1968) has described this form as resembling a 'large calcite-plated tadpole'. The theca has a marginal rim of large plates, whilst the slightly smaller plates of both sides of the theca are irregular. An opening at the opposite end to the stem leads to a central cavity internally divided by ridges. In the inner dorsal surface of the theca are impressed series of stellate and elongate grooves; these have been reconstructed by Jefferies into a complex series of structures reminiscent of the brain and cranial nerves of a fish. These and other

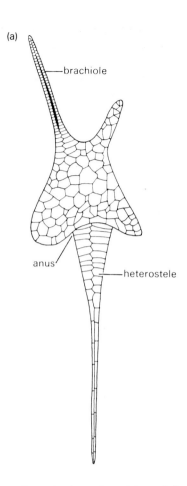

(a)

—brachiole

anus

—heterostele

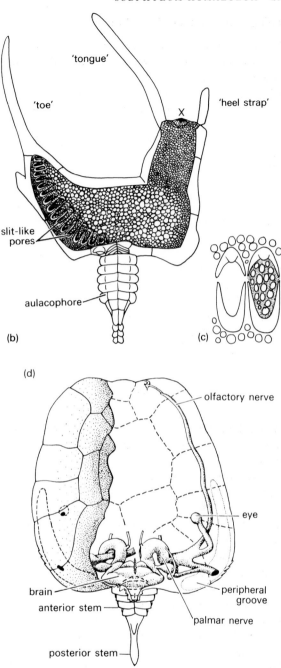

(b)

'tongue'

'toe'

'heel strap'

X

slit-like
pores

aulacophore—

(c)

(d)

olfactory nerve

eye

brain

anterior stem—

peripheral
groove

palmar nerve

posterior stem—

structures (e.g. the lateral position of the anus) led Jefferies to homologise the slit-like anterior orifice with a mouth, the stem with a vertebrate tail, and the ramifying canal system with the higher nervous complex within the vertebrate skull. Ubaghs (1971), on the other hand, has interpreted the fringed opening as a periproct and located the mouth where Jefferies has suggested the anus is. Furthermore, he has drawn attention to the identical structure of the 'carpoid' skeleton with that of other echinoderms and has not agreed that the impressions on the inside of the theca are in any way related to cranial nerves. Hence the issue remains controversial and unresolved and in the absence of soft parts is likely to remain so. It is known on embryological grounds that echinoderms and chordates are indeed related, and it is quite conceivable that the carpoids represent a distinct branch of an evolving chordate–echinoderm complex which split off at a critical time, in any case before the advent of radial symmetry. At present most workers

Figure 9.50 (a) *Dendrocystites* (Ord.) (x 1) (redrawn from Ubaghs in '*Treatise*' Part (S)); (b) *Cothurnocystis* (Ord.) in frontal view – the orifice X is interpreted by Jefferies as the mouth and by Ubaghs as the anus; (c) slit-like pores seen externally showing the 'hanging curtain' of covering plates (based on Jefferies, 1968); (d) *Mitrocystella* (Ord.) from above with right-hand plates removed showing nervous system as interpreted by Jefferies, 1968.

are inclined to regard 'carpoids' as true echinoderms, albeit with chordate affinities.

EVOLUTION

There is no geological evidence as to the ancestry of echinoderms. The first representatives appear already differentiated in the Lower Cambrian. The evidence that chordates and echinoderms are related is based upon embryology and comparative anatomy of living forms. Jefferies (1968), as mentioned, believes that 'carpoids'. were ancestral to the first chordates, though the evidence is incomplete and the whole issue controversial. But there is some evidence that the chordate–echinoderm group on the one hand, and the lophophorate (brachiopod–bryozoan) group on the other, were independently derived from a common worm-like ancestor, allied perhaps to the modern Sipunculida.

The nature of the ancestor has been sought by some with reference to the early larval stages of echinoderms. It has been proposed that the ancestral adult of both groups closely resembled the **'dipleurula'** larval stage of modern echinoderms. This small planktonic larva (Fig. 9.51) is cylindrical, with a ventral mouth and anus and symmetrical coelomic pouches on either side. There is also a ciliated band forming a loop and running down either side. According to the 'dipleurula theory', the ciliated band was present in the ancestral adult, together with its co-ordinating nerve plexus. In the branch that led to chordates this band fused in the midline to provide the rudiments of the dorsal nerve cord. In the ancestral blastozoan–crinozoan stock, on the other hand, torsion led to the twisting of the gut into its present position with a central, upwardly directed mouth and a lateral anus, whilst one of the coelomic pouches was lost.

Earliest echinoderms and their radiations (Fig. 9.51)

Echinoderms were already differentiated on their first appearance in the Lower Cambrian. They include the earliest edrioasteroid *(Stromatocystites)*, the first eocrinoids and a few species of a small echinozoan class, the Helicoplacoidea. Representatives of Class Eocrinoidea (Subphylum Blastozoa) were quite diverse and highly organised by the Middle Cambrian and persisted into the Silurian. In many respects they resemble the Crinoidea, but their plates close right up to the peristome and they do not have a tegmen. They have rather cystoid-like brachioles and frequently pores along the sutures, which are especially clear in the two Middle Cambrian genera *Gogia* and *Lichenoides*.

It has been suggested that eocrinoids were ancestral to all echinoderms, but this is very probably not the case as they are already far too advanced and specialised in the Cambrian.

The Helicoplacoidea (L. Cam.), first discovered in 1963 and known from a few localities only in western North America, are pear-shaped or fusiform echinoderms with a mouth at one end and possibly an anus at the other. A single ambulacrum with one branch runs spirally round the body to the mouth. The interambulacral region is likewise spirally plated. The whole test was evidently expansible and could be inflated from the inside. Where helicoplacoids fit in echinoderm phylogeny is quite uncertain. They do not bear any close relationship to the similarly fusiform holothurians and are probably best regarded as an independently derived group which arose from a primitive stock that had not acquired pentameral symmetry. But they may perhaps retain some trace of the organisation of the earliest echinoderms, and a resemblance to the hypothetical fusiform 'dipleurula' ancestor may not be entirely superficial.

There are other short-lived Cambrian echinoderm groups, so distinct that they too have the status of classes. These are the basket-like Lepidocystoidea, the radial Camptostromatoidea of the top Lower Cambrian, and the Ctenocystoidea of the Middle Cambrian. These are likewise very small classes lacking pentameral symmetry; they are known from a few isolated localities only. The first two are difficult to place in any subphylum. Their diversity testifies to an apparent radiation of echinoderms in the Lower Cambrian, but their paucity in species and numbers of known specimens makes it impossible to say much about their phylogenetic relationships. They may have all been derived independently from one or more pre-pentameral ancestors.

Cambrian echinoderms are not common. The great expansion of pentameral forms came in the Ordovician, though the non-pentameral 'carpoids' and the early non-pentameral cystoids reached their

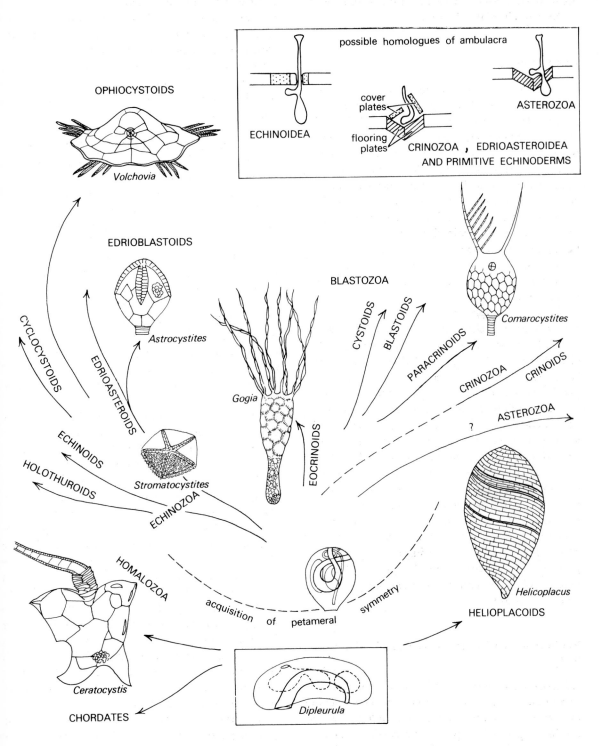

Figure 9.51 Possible relationships amongst Palaeozoic echinoderms. Insets: (below) a lateral view of a modern *Dipleurula;* (above) possible ambulacral homologues.

greatest diversity in the Ordovician also.

The Blastozoa, of which eocrinoids may have been the ancestral stock, rapidly diversified. There arose the cystoids and later the blastoids, whilst short-lived forms of bizarre appearance, the paracrinoids and parablastoids, made their appearance and soon became extinct. All of these retained the primitive ambulacral structure characteristic of early echinoderms: namely an upwardly facing mouth and food grooves with both cover and flooring plates, which in the Crinozoa lead to brachioles of similar construction.

The first echinozoans other than helicoplacoids were the edrioasteroids, which have ambulacra of broadly similar construction to crinozoans and, again, an upwardly directed mouth. In some ways they are not unlike crinoids without brachioles. In *Aulechinus* and other early echinoids the interambulacra are quite similar in plating to those in *Edrioaster*, and the ambulacrum has both cover and flooring plates. But the former are closed and perforated by a pore, and the mouth faces downwards. These echinoids could perhaps have been derived from an edrioasteroid stock by inversion, the larvae settling mouth downwards. But this is not a simple transformation, and it may be safer to infer that echinoids and edrioasteroids came from the same ancestral stock. The advantage of echinoid organisation, however, is the potential for mobility, allowing active searching for food rather than waiting passively for it to come along. The later evolution of echinoids involved the loss of the flooring plates, so that the radial water vessel is enclosed below the ambulacrum of fused cover plates. Possibly the enclosing plates below the radial water vessel of Palaeozoic genera could be homologous with the lost flooring plates. Little is known of the origin of holothurians, unless it is to be sought in the Ordovician *Eothuria* (Superorder Megalopodacea). Nor is anything understood of the relationships of other short-lived echinozoan classes that are rare as fossils: the Cyclocystoidea (M. Cam. – L. Dev.) which are discoidal with a thickened marginal ring and a flexible upper surface of radial plates, and the Ophiocystoidea (L. Ord. – M. Dev.) which are somewhat echinoid-like but have enormous plated tube-feet on the adoral surface.

Asterozoans may have come from a crinozoan stock, as discussed earlier; alternatively an edrioasteroid origin might be sought. The systematic position and relationships of 'carpoids' have already been dealt with.

The Ordovician radiation of echinoderms is very difficult to unravel, and the distinctness of the groups that are actually seen in the fossil record is such that links between the different groups are unclear, even in spite of some apparent homologies in structure. It is particularly difficult in the case of the small but well-defined classes whose representatives occur only in isolated pockets.

In summary, there was a Lower Cambrian radiation of echinoderms, including both radial and non-radial forms. The radial forms included the first blastozoans and the earliest edrioasteroids, which continued into the Ordovician. The earliest 'carpoids' were likewise Cambrian. The subsequent Ordovician radiation was a time of great and rapid diversification, especially of the pentameral groups. But the phylogeny and mutual relationships of the various groups is anything but clear. During subsequent periods the great range of higher taxa established in the Ordovician was drastically pruned. The non-radial forms were extinct by the Lower Devonian, and many of the pentameral groups became extinct in the Permian if not before. Evolution in the later Palaeozoic and afterwards occurred only in well-tried and adaptable taxa, which survive with great success today. The pattern of evolution within these groups varied greatly. In asterozoans, for instance, the peak of organisation was reached early, and there seem to have been no major innovations since the Palaeozoic. In echinoids, by contrast, many new types emerged throughout successive geological periods, and in certain lines of descent each evolutionary step acted as a springboard for the next.

Evolution of the tube-feet

The efficient use and differentiation of the tube-feet has clearly been very important to echinoderm evolution (Nichols 1974). Crinoids have the simplest and apparently most primitive system. In crinozoans there are no ampullae and tube-feet are extruded only by the contraction of the radial water vessel, which is the only pressure generator. Echinoids and dendrochirote holothurians have ampullae, which are the main pressure generators, but contractions of the radial water vessel are still important.

Aspidochirotes, on the other hand, can extend a relaxing tube-foot only when a neighbouring one contracts, sending water into the extending one.

Asteroids and ophiuroids have ampullae but to a varying extent inflate the tube-feet by contracting the lateral and radial water vessel. The tube-foot extensor system in extinct echinoderms has been inferred by analogy with that of modern forms.

Why pentamery?

Not all echinoderms have pentameral symmetry. Early cystoids are radially symmetrical but may have five symmetrical plates in the anal pyramid, whilst later cystoids acquire pentameral thecae. But other than in 'carpoids' and other early groups, pentameral symmetry is so characteristic that there must surely be a good reason for it. Nichols (1971) has suggested that the sutures binding echinoderm plates must have acted as lines of weakness. Only if there had been an odd number of plates would 'opposite' sutures have been offset so as to give greater strength against breakage. A circlet of three plates is too few, whilst five seems to be an ideal number for maximum resistance. A similar angular offset is needed to guard against accidental rupture along lines of pores, as in an echinoid, which has been called 'postage stamp weakness'. The offsetting of pore rows in multiples of five has been an ideal compromise between strength of test and maximum number of tube-feet.

On the other hand, damaged echinoid tests never usually break along the sutures, for the latter are reinforced with strong collagen fibres; hence the sutures are not necessarily the weaker part of the test. Another suggestion is that a pattern of five combines a nearly constant width presented in almost any direction in the horizontal plane with a small number of rays.

Convergent evolution and intermediate forms

A striking characteristic in the early echinoderms is for the same kind of structure to develop over and over again in distinct though unrelated stocks. An example is the folded membrane structure found in blastoids, cystoids and the crinoid family Porocrinoidea, in all of which it seems to have evolved independently. Paired pores have arisen independently in diploporite cystoids and in echinoids. Lanceolate ambulacra and a stem are present in blastoids and in the peculiar genus *Astrocystites*, which in other respects is more like an edrioasteroid. It has been proposed that a new class, the Edrioblastoidea, be established to accommodate this genus, but it may simply be better to regard this genus as an edrioasteroid convergent on blastoids.

Very many of the Lower Palaeozoic echinoderms are thus of a composite type which, as Regnell (1960) has pointed out, lie in the direct lineage of none of the later echinoderm groups but present features from several of them. Perhaps the most extreme case may lie in the convergence between the 'carpoid' *Cothurnocystis* and the cystoid *Pleurocystites*. Indeed the controversy over the relationships of these same 'carpoids' illuminates the fact that within the early echinoderm–chordate line only a few kinds of structures seem to have been able to evolve and carry out particular functions.

The 'intermediate forms' that have evolved, often with a puzzling combination of characters, have been the bane of echinoderm taxonomy. It is not easy to try to produce an objective phylogenetic scheme where such extreme genera are involved. Quite apart from the taxonomic question there are deeper evolutionary issues. Why, from the genetic point of view, should early echinoderms have produced similar structures, independently but often and so consistently? It is of course true that early members of a group tend to be rather generalised whilst later ones are frequently specialised, having suppressed or eliminated more generalised traits. Perhaps the genetic potential for, say, folded membrane structures was present in many Palaeozoic echinoderms, though suppressed in the majority and only realised in a few isolated groups. Whatever the ultimate answer, these perplexing early invertebrates, more than any other groups, seem to spotlight some complex and little-known aspects of evolutionary theory that have so far received but scant attention.

BIBLIOGRAPHY

Books, treatises, symposia

Moore, R. C. (ed.) 1966. '*Treatise*' Part (U). Echinodermata 3. Vols 1 & 2. (Asterozoa and Echinozoa)

Moore, R. C. (ed.) 1967. '*Treatise*' Part (S). Echinodermata 1. Vols 1 & 2. (Non-crinoid 'pelmatozoans')

Nichols, D. 1974. *Echinoderms*. London: Hutchinson (Invaluable text on morphology and evolution of living and fossil forms)

Nichols, D. 1975. *The uniqueness of echinoderms*. Oxford: Oxford University Press. (Short: descriptive functional approach)

Individual references

Bell, B. M. 1976. *A study of North American Edrioasteroidea*, 1–447. NY State Mus. Mem., no. 21. (Monographic interpretation of ambulacral structure)

Breimer, A. 1969. A contribution to the palaeoecology of Palaeozoic stalked crinoids. *Proc. K. Ned. Akad. Wet. B.* **72,** 139–50. (Application of knowledge of recent crinoid ecology to fossils)

Breimer, A. and C. D. Webster 1975. A further contribution to the palaeoecology of fossil stalked crinoids. *Proc. K. Ned. Akad. Wet. B.* **78,** 149–67. (As above)

Carter, R. M. 1967. On the biology and palaeontology of some predators of bivalved molluscs. *Palaeogeography, Palaeoclimatol. Palaeoecol.* **4,** 29–65. (Starfish habits and anti-starfish devices in bivalves)

Clarke, J. M. 1912. A remarkable occurrence of Devonic starfish. *Bull. NY State Mus.* **15,** 44–5. (Bivalve–starfish association)

Durham, J. W. 1966. Evolution among the Echinoidea. *Biol. Rev.* **41,** 368–91. (Classic study; illustrations of early echinoids)

Fay, R. O. 1961. Blastoid studies. *Univ. Kansas Palaeont. Contrib.* **3,** 1–147. (Monographic treatment)

Fell, H. B. 1963. Phylogeny of sea-stars. *Phil. Trans Roy. Soc. Lond. B* **246,** 381–435. (Possible crinoid origins of somasteroids)

Goldring, R. and D. G. Stevenson 1972. The depositional environment of three starfish beds. *N. Jb. Geol. Paläont. Mh.* **10,** 611–24. (Rapid burial essential for starfish preservation)

Hawkins, H. L. 1943. Evolution and habit among the Echinoidea: some facts and theories. *Quart. J. Geol. Soc. Lond.* **99,** 1–75. (Stimulating paper on echinoid phylogeny and adaptations)

Jackson, R. T. 1912. *Phylogeny of the Echini with a revision of Paleozoic species*, 1–443. Mem. Boston Soc. Nat. Hist., no. 7. (Large monograph of Palaeozoic echinoids, profusely illustrated)

Jefferies, R. P. S. 1968. The subphylum Calcichordata: primitive fossil chordates with echinoderm affinities. *Bull. Brit. Mus. Nat. Hist. (Geol.)* **16,** 243–339. (Proposes that carpoids are chordates)

Joysey, K. A. 1959. A study of variation and relative growth in the blastoid *Orbitremites*. *Phil. Trans Roy. Soc. Lond. B* **242,** 99–125. (Classic study of a large population from blastoid-rich horizons)

Kermack, K. A. 1954. A biometrical study of *Micraster coranguinum* and *M. (Isomicraster) senonensis*. *Phil. Trans Roy. Soc. Lond. B* **237,** 375–428. (Statistical evaluation of *Micraster* evolution)

Kesling, R. and C. R. C. Paul 1968. New species of Porocrinidae and brief remarks upon these unusual crinoids. *Contrib. Paleont. Univ. Michigan* **22,** 1–32. (Folded-membrane structures)

Kier, P. M. and R. E. Grant 1965. Echinoid distribution and habits, Key Largo Coral Reef Preserve, Florida. *Smithsonian Mis. Coll.* **149**(6), 1–69. (Life habits of several modern species)

Kier, P. M. Evolutionary trends in Palaeozoic echinoids. *J. Paleont.* **39,** 436–65. (Very important paper; numerous illustrations)

Kier, P. M. 1973. A new Silurian echinoid genus from Scotland. *Palaeontology* **16,** 651–63. (Description of *Aptilechinus*, with reconstruction)

Kier, P. M. 1974. Evolutionary trends and their functional significance in the post-Palaeozoic echinoids. *J. Paleont.* **48** (suppl.). (Paleont. Soc. Mem., no. 5, 1–95). (Major reference work)

Koch, D. L. and H. L. Strimple 1968. *A new Upper Devonian cystoid attached to a discontinuity surface;* 1–49. Iowa Geol. Surv. Rept Invest., no. 5. (Edrioasteroids and cystoids in life position)

Macurda, D. B. 1965a. The functional morphology and stratigraphic distribution of the Mississippian blastoid genus *Orophocrinus*. *J. Paleont.* **39,** 1045–96. (Morphology and dynamics)

Macurda, D. B. 1965b. Hydrodynamics of the Mississippian blastoid genus *Globoblastus*. *J. Paleont.* **39,** 1209–17. (Inhalant–exhalant flow through spiracles)

Macurda, D. B. and D. L. Meyer 1974. Feeding posture of modern stalked crinoids. *Nature* **247,** 394–6. (Rheophilic crinoids' life habits)

Moore, R. C. and L. R. Laudon 1943. *Evolution and classification of Palaeozoic crinoids*, 1–153. Spec. Pap. Geol. Soc. Amer., no. 46. (Major monograph on crinoid evolution)

Mortensen, T. H. 1928–52. *A monograph of the Echinoidea*, 5 vols. Copenhagen and Oxford: Reitzel and O.U.P. (The largest work ever published on echinoids, with full descriptions and photographic plates of all known Recent species)

Nichols, D. 1959. Changes in the chalk heart-urchin *Micraster* interpreted in relation to living forms. *Phil. Trans Roy. Soc. Lond. B* **242,** 347–437. (Classic validation of evolution in *Micraster*)

Nichols, D. 1974. The water-vascular system in living and fossil echinoderms. *Palaeontology* **15,** 519–38. (Structure and evolution)

Paul, C. R. C. 1967. The functional morphology and mode

of life of the cystoid *Pleurocystites* E. Billings 1854. *Symp. Zool. Soc. Lond.* **20,** 105–23. (Benthic crawling habit postulated)

Paul, C. R. C. 1968. Morphology and function of dichoporite pore structures in cystoids. *Palaeontology* **11,** 697–730. (Functional morphology and its bearing on classification)

Paul, C. R. C. 1972. Morphology and function of exothecal pore structures in cystoids. *Palaeontology* **15,** 1–28. (Functional morphology and its bearing on classification)

Paul, C. R. C. 1976. Palaeogeography of primitive echinoderms in the Ordovician. In *The Ordovician System*, M. G. Bassett (ed.), 553–74. (Proceedings of a Palaeont. Ass. symposium, Birmingham, September 1974.) Wales: Univ. of Wales Press. (Temperature control of distribution of echinoderms with folded-membrane structures)

Philip, G. M. and R. J. Foster 1971. Marsupiate Tertiary echinoids from south-eastern Australia and their zoogeographic significance. *Palaeontology* **14,** 666–95. (Description of marsupiate echinoids illustrated by stereo-pairs, and controls of distribution)

Regnell, G. 1960. 'Intermediate' forms in early Palaeozoic echinoderms. *Proc. 21st Int. Geol. Congress*, Copenhagen **22,** 71–80. (Taxonomic problems raised by some peculiar echinoderm groups)

Rowe, A. W. 1899. An analysis of the genus *Micraster*, as determined by rigid zonal collecting from the zone of *Rhynchonella cuviera* to that of *Micraster coranguinum. Quart. J. Geol. Soc. Lond.* **55,** 494–547. (Classic evolutionary study)

Seilacher, A., G. Drozozewski and R. Haude 1968. Form and function of the stem in a pseudoplanktonic crinoid *(Seirocrinus). Palaeontology* **11,** 275–82. (How did perfectly preserved crinoids come to lie in position on the sea floor?)

Sprinkle, H. J. 1976. Classification and phylogeny of pelmatozoan echinoderms. *Systematic Zool.* **25,** 83–91. (Basis of classification used here)

Stephenson, D. G. 1963. The spines and diffuse fascioles of the Cretaceous echinoid *Echinocorys scutata* Leske. *Palaeontology* **6,** 458–70. (Functional morphology of a shallow burrower)

Ubaghs, G. 1953. Crinoïdes. In *Traité de paleontologie III* (Onychophores, Arthropodes, Echinodermes, Stromochordes), J. Piveteau (ed.) 658–773. Paris: Masson & Cie. (The best summary, as the crinoid volume of the '*Treatise*' is not yet published)

Ubaghs, G. 1971. Diversité et specialisation des plus anciens echinodermes que l'on connaisse. *Biol. Rev.* **46,** 157–200. (Early radiations of echinoderms; disclaims that carpoids are chordates)

Wright, J. 1954. *A monograph of the British Carboniferous Crinoidea*, 1–190. Palaeontogr. Soc. monogr. (Fully illustrated monograph)

10 Graptolites

Abundant stick-like fossils are sometimes found in Lower Palaeozoic black shales or more rarely in other argillaceous rocks of the same age. These are the graptolites, so called from their resemblance to written marks on the shale (Greek: $\gamma\rho\alpha\varphi\varepsilon\iota\upsilon$ = to write; $\lambda\iota\theta o\varsigma$ = stone). They may be straight or curved, sometimes spiral in form, single-branched, bifid or many-branched. When preserved in the most common way, i.e. flattened in shale, one or both edges appear serrated like a tiny saw blade, but otherwise not much structural detail may be visible.

These fossils are of singular importance in establishing a stratigraphical time scale for the Lower Palaeozoic and have been of zonal value since the mid-nineteenth century. Though much of the stratigraphical work that has been done is based upon specimens preserved as flattened carbonised films (they are still identifiable even so), the anatomy of graptolites can only be properly understood with reference to the relatively rare specimens preserved in three dimensions.

STRUCTURE

'Graptolites' is a vernacular term for members of Class Graptolithina (Phylum Hemichordata). They are colonial marine invertebrates now believed to bear a distant affinity to vertebrates, though for a long time they were regarded as allied to corals, which have a much simpler grade of biological organisation. Within Class Graptolithina there are several orders, of which only two are of any real importance: the Graptoloidea (Arenig–Emsian) and the Dendroidea (Trem.–Carb.). Examples of really well-preserved genera of these two main orders will be described below, but it has to be made clear that graptolites are rarely preserved like this; in the normal course of preservation very much fine detail is lost. Even so, it is surprising how much remaining structure can sometimes be found even in crushed and carbonised specimens.

Order Graptoloidea

SAETOGRAPTUS CHIMAERA (Fig. 10.1)

This Ludlovian species (Urbanek 1958) is found only in glacial erratic boulders, originally derived from rocks under the Baltic Sea and now scattered over the North German and Polish plain; specimens can be isolated from the rock with acid.

As in all graptolites the 'skeleton' of *Saetograptus* consists of a series of hollow interlinked tubes, constructed of a thin sheet-like material known as **periderm.**

The first formed part of the graptoloid is the **sicula:** a conical tube with its aperture pointing downwards and terminating at its apex in a long thread-like **nema,** extending well beyond the upper limit of growth of the sicula. The sicula is divided into two parts with different ornamentation: an upper **prosicula,** and a lower **metasicula.** The prosicula has an ornament of longitudinal and spiral **striae;** the metasicula, like all other parts of the graptoloid, is ornamented with well-marked rings representing periodic, perhaps daily, growth increments and known as **fusellae. Fusellar tissue** consists of thin half-rings or complete rings of skeletal material stacked one above the other and uniting along zigzag **sutures.** In the sicula the suture on one side has the characteristic zigzag form shown in all graptolites; on the other side the fusellar half-rings are joined to a stout rod (the **virgella**) which projects below the aperture and may curve slightly under it.

From the sicula there grow up a number of cup-like **thecae** (singular: theca), the first-formed one of which begins at a **primary notch** in the growing edge of the sicula and on its virgellar side, before the development of the latter is complete. The **initial**

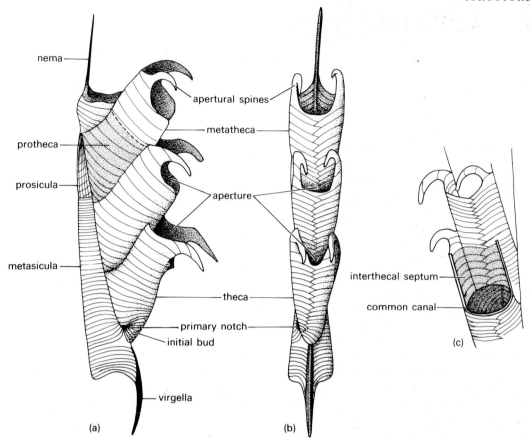

Figure 10.1 Morphology of *Saetograptus chimaera:* (a) lateral view; (b) frontal view; (c) structure of theca and common canal with part removed (x 40 approx.). (Modified from Urbanek, 1958)

bud grows up from the primary notch to form the first of the thecae, which in turn gives rise to successive thecae increasing in size away from the sicula until about the fifth or sixth theca, after which thecal dimensions remain constant. The thecal apertures point upwards but are obliquely inclined and provided with paired **apertural spines.**

The cavity of the first-formed theca connects with that of the sicula through a **foramen,** and all other thecae likewise link up with one another through a **common canal** close to the nema, for the partition walls of the thecae do not extend from the distal part of the nema. Where the nema is embedded in the wall of the graptoloid it is known as the **virgula.** The partition walls themselves are double-layered structures, the wall being secreted from both sides (Fig. 10.1c). Each theca has two parts: the **protheca,** proximal to the nema, through which the common

canal runs; and a **metatheca,** which is the external part divided by partition walls and possessing apertural spines. The growth lines of the thecae tend to be more widely spaced toward the aperture, presumably signifying a period of faster growth towards the end of thecal formation.

Since graptolites were colonial animals, with the inhabitants of the thecae and the sicula (whatever they may have been like) linked to one another through the common canal, it is presumed that food caught by one individual would be ingested and shared by the whole colony. This colony is known as a **rhabdosome** (an older and less correct name is **polypary**), and the branches are **stipes;** *Saetograptus* has only one stipe.

There is no direct evidence about what sort of animal was found in each theca. It is assumed that in each theca there was a **zooid** – a simple animal with

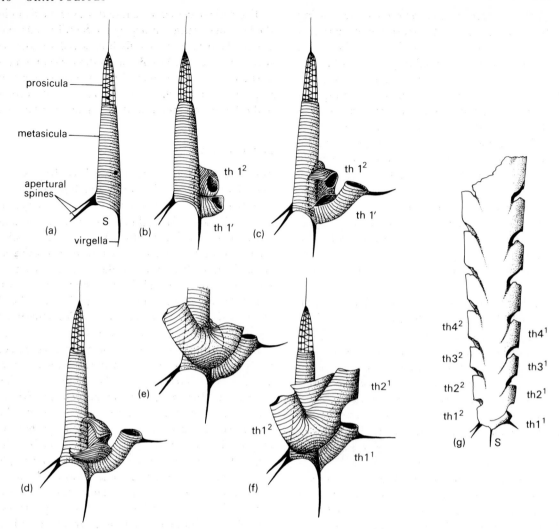

Figure 10.2   (a)–(f) Development (x 30 approx.) and (g)   (Modified from Bulman, 1944–7)
adult morphology (x 12 approx.) of *Diplograptus leptotheca*.

some sort of food-gathering apparatus, possibly a
lophophore – and that all the zooids were connected
internally and perhaps externally too by tissue. The
nature of the zooids will be considered after the
affinities of the graptolites are discussed in more detail
(p. 246).

DIPLOGRAPTUS LEPTOTHECA (Fig. 10.2)

Diplograptus is a Caradocian graptoloid with two
stipes in the rhabdosome. Both of these are united
back to back with the nema in between; the

rhabdosome is said to be **scandent** and **biserial.**
This graptoloid is structurally more complex than
Saetograptus, especially in the proximal end, i.e. the
sicula and the first-formed thecae which surround it.
Though the structure of the proximal end is difficult
to work out in mature specimens, individuals in all
growth stages have been found, and these can be
arranged in a developmental series showing exactly
how the different thecae originate and form (Bulman
1945–47).

  The sicula is broadly similar to that of *Saetograptus*
but has two apertural spines, and the initial bud

arises, not by the formation of a notch in the growing edge of the sicula, but by the resorption of a foramen high up on the metasicula before budding. This system is actually more normal in graptolites than the specialised mode of development in *Saetograptus*.

The thecae are numbered in pairs according to a conventional notation, so that the first-formed theca is th. 1¹, the second which originates from it is th. 1², the third 2¹, the fourth 2², etc. In a biserial graptoloid such as *Diplograptus* the thecae on one side are numbered: 1¹, 2¹, 3¹, 4¹ ... and on the other: 1², 2², 3², 4² ...

In *Diplograptus* th. 1¹ grows down almost vertically to about the level of the sicular aperture and then is sharply hooked upwards to terminate with an apertural spine at its lip. Th. 1² arises from the th. 1¹ near the sicular foramen, before the growth of th. 1¹ is complete.

Th. 1² is first directed upwards, then is twisted into recumbent S-shape, and finally grows across the front of the sicula becoming welded to it and forming a **crossing canal.** Curiously enough, the fusellae that form the crossing canal are laid down not only at the growing aperture of the th. 1², but also as a flange along the already formed tubular surface of th. 1¹. The two parts of th. 1² unite to form a tube open at both ends, so that when fully formed th. 1² terminates in an upward curve, whilst th. 2¹ arises on the opposite side arising from th. 1². Then th. 2² grows up from th. 2¹, forming another crossing canal. Likewise the later formed thecae all pass in front of the sicula or the nema, each time forming a crossing canal.

The mode of growth is marked by the fusellae (cf. Fig. 10.4) so that the developmental history is clear from them, but it is more useful to be able to see all the stages in growth from the developmental series here illustrated.

Order Dendroidea

The dendroids are the most ancient of all the graptolites and were apparently ancestral to the graptoloids. But their morphology is complex, and they have various organs not possessed by the descendent graptoloids. Normally dendroids are preserved, as are most graptoloids, as compressions in shale, but rare three-dimensional specimens can be isolated from the rock or studied by means of thin section, so their detailed morphology is quite well known.

The basic structure of dendroids was worked out in the late nineteenth century by the Swedish palaeontologist Carl Wiman, mainly from serial sections. It was later elaborated by Bulman (1945–47) and most particularly by Koszlowski (1948), who has described in great detail a rich and varied fauna of dendroids and other graptolites from the Tremadoc of Poland.

DENDROGRAPTUS (Fig. 10.3)

Dendrograptus, known from the three-dimensional material studied by Koszlowski, exhibits the representative dendroid structure admirably. The inverted conical sicula stood upright upon the sea floor with its apical base expanded into a holdfast. At a point about halfway up the sicula arises the **stolotheca,** which is equivalent to the prothecal series surrounding the common canal in *Saetograptus* and, with daughter stolothecae, forms a continuous closed chain all the way up the rhabdosome. From this there arise two kinds of thecae at successive and equally spaced nodal points. These are the large **autothecae** and the smaller narrower **bithecae,** which always come off the stolotheca at the same level. The bithecae in *Dendrograptus* maintain constant width, but each is normally looped over between the associated autotheca and the stolotheca, so that the aperture is on the opposite side of the stolotheca to the origin. In other dendroids the bithecae may be simple straight cylinders or be variously looped or coiled. The autothecae are very much larger and expanded upwards, each terminating in an outwardly inclined but often elaborately sculptured aperture with a median tongue.

The arrangement of the autothecae and bithecae follows the 'Wiman rule' of alternating triads; bithecae usually arise at alternate nodal points on opposite sides of the main stem and are carried distal to the autothecae. When the rhabdosome branches the stolotheca splits and the Wiman rule alternation continues on each daughter stolotheca. In some species the auto- and bithecae arise in sequence at angles of 90° rather than 180° from the one below.

Within each rhabdosome there is an internal tubular system running from the sicula all the way along the stolotheca and the base of each autotheca and bitheca. This **stolon system** is only found in well-preserved specimens, but Wiman detected it in his serial sections and believed it to have carried an

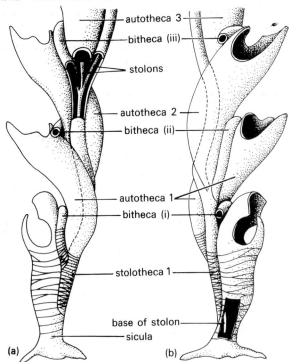

Figure 10.3 Morphology of the proximal end of a dendroid, *Dendrograptus communis:* (a) frontal view; (b) reverse view: The cut-away panels show the stolons (x 40 approx.). (Modified from Kozlowski, 1948)

internal apparatus such as a nervous system. Koszlowski confirmed this suggestion when studying modern pterobranchs, which are small tube-dwelling marine animals, for they are apparently the nearest living relatives of graptolites and have an equivalent stolon enclosing just such a nervous system as Wiman postulated. No stolon system is found in the Graptoloidea, but it is developed in some of the other orders, especially the Stolonoidea, where there are very many peculiar and irregularly branching stolons.

The fusellar tissue of dendroids is broadly similar to that of graptoloids. But there is also another kind of peridermal material: the **cortical tissue,** which is laid down in thin flat sheets covering and enveloping the earlier thecae. (It is usually present in graptoloids as well but is very thin and only clearly seen with the electron microscope. Dendroid cortical tissue is much thicker and is clearly visible.) Such cortical sheets form the holdfast which affixes the colony to the sea floor. Presumably cortical tissue was laid down by an

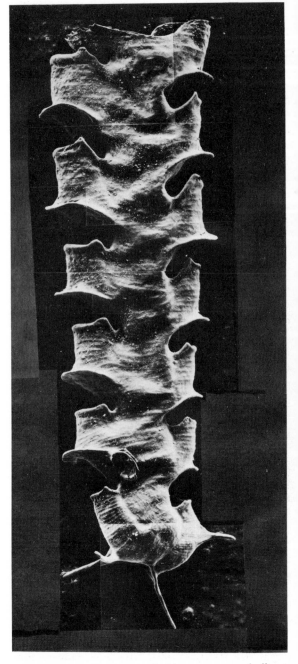

Figure 10.4 *Climacograptus inuiti* – a montage built up from SEM photomicrographs, showing banded fusellar tissue (total width at top approx. 1 mm). (Photograph by courtesy of P. Crowther)

extrathecal epithelium, though how this was done is uncertain.

Preservation and study of graptolites

The three genera described above are of some of the best-preserved graptolites known. They all come from fine argillaceous limestones, and though specimens may be very abundant locally such occurrences are rare. Three-dimensional specimens of such a kind must be freed from the rock so that they can be studied in the round. The rocks are first broken into fragments, treated first with hydrochloric acid and then with hydrofluoric acid, and after washing may be picked out and individually transferred to a concentrated nitric acid and potassium chlorate solution which renders them translucent. They are then dehydrated in a concentrating alcohol series, cleared in xylol and finally mounted on slides in Canada Balsam. Such specimens are then of a clear brown hue and virtually transparent.

With the light microscope, at relatively low magnifications, individual thecae and the sicula may be very clearly marked with fusellar rings. Excellent details may be seen with scanning electron micrography (Fig. 10.4). Such fusellae are occasionally seen even in flattened specimens, though in typical specimens of *Saetograptus chimaera* the carbonisation of the periderm has often gone too far to have conserved the microstructure and growth rings may be lost, yet the three-dimensional structure of the rhabdosome is still preserved. When carbonisation is complete the graptolite may be isolated from the rock, as with *S. chimaera*. But sometimes carbonisation is incomplete, and attempts to free the graptolite cause it to crumble when extracted. Then the only way to elucidate the structure is to make closely spaced serial sections through the specimen after embedding it in plaster or resin, cutting the sections normal to the long axis of the rhabdosome. If the structure is very complex, especially in the region of the sicula and the first few thecae (the proximal end), it is often useful to build up models in wax from these sections, by superimposing one layer after another. In practice the wax is cut to show the internal structure of the graptolite preserved as a core, for in this way relationships of the thecae are more clearly visible (Fig. 10.5).

Not infrequently the periderm is completely destroyed, but an internal mould of the whole rhabdosome may remain, retaining the three-dimensional organisation as a pyrite core. This is quite often the case with specimens preserved in black shale, the pyrite being derived from organic matter decaying in an anoxic environment. As with the wax model the periderm has gone, but the shapes and convolutions of the thecae remain in their original relationships. But the most common, and unfortunately the least revealing, of the various kinds of preservation is that where the rhabdosome has been completely flattened and preserved only as a compression: a carbon or aluminosilicate film on the rock. Even here, however, thecal shapes are usually retained so that the fossils can be readily identified, and it is from such specimens that most graptolite species and faunal sequences are known.

Ultrastructure and chemistry of the graptolite periderm

For a long time it was believed that the graptolite periderm was constructed of chitin or a related hydrocarbon. The translucent brown appearance of isolated graptoloids, the widespread use of chitin throughout the animal kingdom as a structural component, and the reported presence of glucosamine which is a known chitin breakdown product all seemed to suggest chitin. But more recent work has failed to detect glucosamine, though abundant amino acids have been identified, and it is now agreed that, whatever the original material of the graptolite periderm may have been, it was not chitin.

The ultrastructure of the graptolite periderm, which has now been intensively studied (Urbanek & Towe 1974, 1975), sheds some light upon this complex topic. Transmission electron micrography of exceptionally well-preserved graptoloid and dendroid material shows that the fusellar and cortical tissues are constructed of several kinds of fabric (Fig. 10.6). Furthermore, the existence of cortical tissue in graptoloids as well as dendroids has been confirmed, though in the latter it may be very thin. **Cortical fabric,** which makes up most of the cortical tissue, consists of closely packed parallel **fibrils** stacked in successive layers with alternating orientation. The manner of packing and the size of the cortical fibrils is strikingly reminiscent of the appearance of the protein collagen, which is an important structural component of many animal tissues, and this close

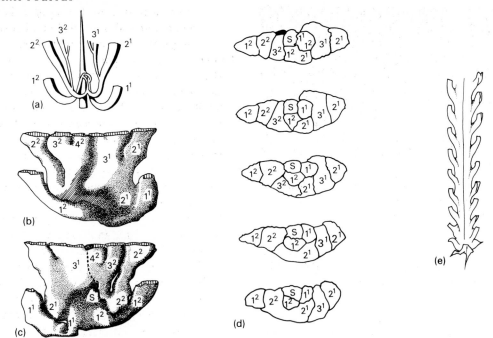

Figure 10.5 Morphology of *Glyptograptus austrodentatus americanus*, an early (Llanvirn) diplograptid: (a) simplified diagram showing relationships of thecae and sicula at the proximal end; (b), (c) frontal and reverse views of a wax model (x 30 approx.) built up from serial sections, some of which are illustrated in (d) (x 25 approx.); (e) adult rhabdosome (x 6) (redrawn from Bulman, 1963, *Palaeontology* **6,** 665–89).

resemblance indeed indicates that the original periderm material was collagen.

Fusellar fabric, though probably also collagenous in nature, is made of fibrils in a more open meshwork. The bulk of the fusellar tissue is made of this fabric. There are also at least two kinds of non-fibrillar sheet-like material. Such **sheet fabrics** are found delimiting particular layers within the cortical tissue, forming a thin external coat on the outside of the fusellae, deposited as a thin secondary sheet inside the thecae, and occurring inside the stolon sheath.

All these kinds of fabric may occur in either cortical or fusellar tissue, but one kind is usually predominant. Thus cortical tissue is largely cortical fabric, though sheet-like fabrics form layers within it as well as outer and inner linings. Fusellar tissue may also be sporadically present in patches.

Fusellar tissue is mainly fusellar fabric and may be entirely so, but in some dendroids laminae of cortical fabric may be present, and sheet fabrics are normally present.

Graptoloids may have, in addition to all the other fabrics, a peculiar **virgular fabric** which has been found, for instance, in the monograptid nema, and also in retiolitid graptoloids in which the periderm is reduced to a meshwork of girders. This fabric has long fibrils set parallel with one another in an electron-dense matrix, and the fibrils themselves have a radial internal structure which has not been found elsewhere.

The study of the nature and distribution of these structural fabrics, which is actively being investigated, suggests a wide variety of distributional and compositional patterns amongst the graptolites, especially in the Graptoloidea. This will probably have considerable value in taxonomy.

CLASSIFICATION

Class Graptolithina (colloquially graptolites) is divided into the two important orders: the Dendroidea and Graptoloidea (Fig. 10.10). But there

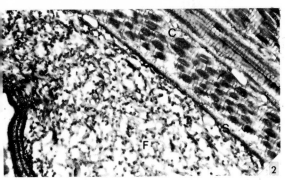

Figure 10.6 Ultrastructure of *Dictyonema* with fusellar tissue (F), cortical tissue (C), and sheet fabric (S) (x 14 000). (Photograph by courtesy of K. M. Towe)

are also some short-lived orders (Tuboidea, Stolonoidea, Camaroidea and Crustoidea) (Fig. 10.8) confined to the Tremadocian which will be referred to only in passing.

ORDER 1. DENDROIDEA (M. Camb.–Carb.): Many-branched graptolites with large numbers of small thecae, sometimes with connecting links (**dissepiments**) between stipes. Stipes with two kinds of thecae (autothecae and bithecae) opening off a continuous closed chain: the stolotheca. Lower part of rhabdosome has concentric sheets of periderm (cortical tissue) covering the standard fusellar tissue. Most genera were shrubby, upright and fixed to the sea floor, but *Dictyonema* was apparently pendent. Dendroids are classified into four families, largely based upon their external appearances, for their detailed structure is known only in a few genera:

FAMILY 1. DENDROGRAPTIDAE (M. Camb.–Carb.).

FAMILY 2. PTILOGRAPTIDAE (L. Ord. – U. Sil.).

FAMILY 3. ACANTHOGRAPTIDAE (U. Camb. – M. Dev.).

FAMILY 4. ANISOGRAPTIDAE (Trem.): This family has characters in many ways intermediate between those of dendroids and graptoloids. The anisograptid rhabdosomes may be **bilateral** (i.e. with two primary branches) as in *Clonograptus*, **triradiate** as in *Bryograptus*, or **quadriradiate** as in *Staurograptus*. The structure of the thecae is known only in a few anisograptids, since their preservation is normally poor. *Kiaerograptus* (Fig. 10.7a) and

some early species of *Bryograptus* have both autothecae and bithecae, but other species of *Bryograptus* only have autothecae.

ORDER 2. GRAPTOLOIDEA (L. Ord. – L. Dev.): Rhabdosomes have few stipes, up to eight in the early forms, but reducing in later genera to two and finally to one. Thecae are of one kind only, equivalent to the autothecae of dendroids, and may be arranged on one side of the stipe or on both. The terminology of different thecal shapes is given in Fig. 10.7d. Within the graptoloids there is much variation in thecal morphology, by contrast with that of dendroids. Sicula usually prominent. Graptoloids are classified in four suborders:

SUBORDER 1. DIDYMOGRAPTINA (Ord.): Early graptoloids with normally pendent to reclined stipes, though some biserial forms are known. Eight, four or two stipes present, with a simple proximal end structure. Includes FAMILIES DICHOGRAPTIDAE, DICRANOGRAPTIDAE and NEMAGRAPTIDAE amongst others.

SUBORDER 2. GLOSSOGRAPTINA (Ord.): A peculiar group of biserial graptoloids, in which the thecae curve so as to enclose the sicula in front and behind, so that the stipes are thus aligned side by side but facing in opposite directions. Often spiny. Include FAMILIES GLOSSOGRAPTIDAE and CRYPTOGRAPTIDAE. Interesting but numerically unimportant.

SUBORDER 3. DIPLOGRAPTINA (Ord.–Sil.): Biserial graptoloids with scandent stipes arranged back to back. Development like *Diplograptus*. The various families, mainly defined on thecal form, include FAMILIES DIPLOGRAPTIDAE, LASIOGRAPTIDAE, RETIOLITIDAE and DIMORPHOGRAPTIDAE.

SUBORDER 4. MONOGRAPTINA (Sil.): Scandent graptoloids with uniserial construction, as in *Saetograptus* and *Monograptus*. May have **cladia** (lateral branches) as in *Cyrtograptus*.

ORDER 3. TUBOIDEA (L. Ord.–Sil.) (Fig. 10.8a): Dendroid-like, with auto- and bithecae, but with irregular branching structure and a reduced stolotheca.

ORDER 4. CAMAROIDEA (Ord.) (Fig. 10.8b): Encrusting forms, with autothecae of peculiar shape, each having an inflated balloon-like base, with a vertical **collum** (chimney). Stolotheca and bithecae present.

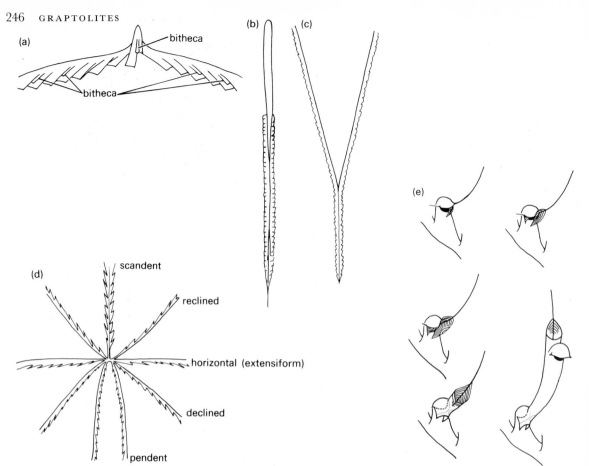

Figure 10.7 Various dendroid and graptoloid rhabdosomes: (a) *Kiaerograptus* (Tremadoc) with rare bithecae (x 6 approx.) (redrawn from Spjeldnaes, N., *Palaeontology* **6,** 121–31); (b) *Cystograptus vesiculosus* (Land.) with a trifid nemal vane (x 1·5); (c) *Dicranograptus ramosus* (Caradoc) (x 1); (d) terminology applied to the different shapes of rhabdosome in graptoloids; (e) *Cyrtograptus* (M. Sil.), stages in early cladial development (x 1·5). (Redrawn from Bulman in '*Treatise*' Part (V))

ORDER 5. CRUSTOIDEA (L. Ord. – U. Ord.) (Fig. 10.8c): Encrusting forms, with dendroid-like morphology, but the autothecae have modified apertures.

ORDER 6. STOLONOIDEA (Ord.) (Fig. 10.8d): Encrusting forms of irregular morphology; stolothecae have many irregular ramifying stolons; autothecae present but no bithecae.

BIOLOGICAL AFFINITIES

For a long time it was accepted that the affinities of the Graptolithina lay with Phylum Cnidaria and most probably with the hydrozoans. The publication of Koszlowski's work in 1948, however, made clear that this view was no longer acceptable. Koszlowski noted pronounced resemblances in structure between the graptolites and the small modern deep-sea animal *Rhabdopleura*: a colonial organism belonging to Phylum Hemichordata and with relatives known from scanty material in the Cretaceous and probably as far back as the Ordovician. The hemichordates are an 'advanced' phylum allied to the vertebrates; if Koszlowski's suggestions are correct, then the Graptolithina must be of a much higher grade of organisation than was originally imagined.

The Hemichordata comprise two classes: the modern Enteropneusta which are the 'acorn worms', such as *Balanoglossus;* and the Pterobranchia to which

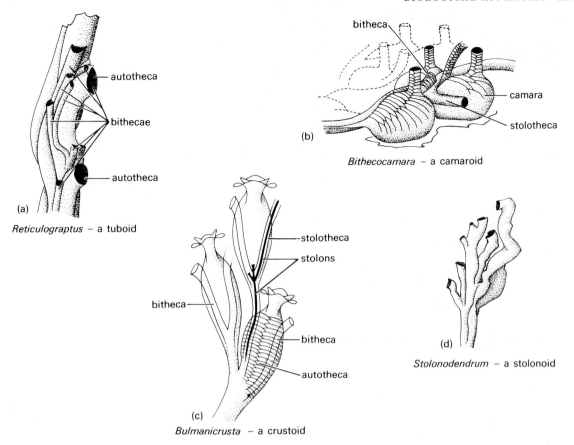

(a)

Reticulograptus – a tuboid

(b)

Bithecocamara – a camaroid

(c)

Bulmanicrusta – a crustoid

(d)

Stolonodendrum – a stolonoid

Figure 10.8   Minor graptolite orders. In *Stolonodendrum* the irregular tubes are stolons. The enclosing stolothecae are rarely preserved ((a)–(b) x 50, (c) x 30, (d) x 15). (Redrawn from Bulman in '*Treatise*' Part (V))

Rhabdopleura belongs. In both cases there are only a few Recent genera, and in the pterobranchs the only important ones are *Rhabdopleura* and *Cephalodiscus,* which is similar to *Rhabdopleura* in some ways but does not form true colonies.

Rhabdopleura (Fig. 10.9) is of very small size and is colonial with individual exoskeletal tubes, each housing a zooid and arranged in a creeping habit. Several tubes arise from a horizontal basal tube, the last one having a closed end with a terminal bud inside. The zooid which secretes and lives in each tube is small and has a pair of tentacular food-gathering arms known as the lophophore, though this is not necessarily equivalent to the lophophore of other invertebrates. Each individual zooid is supported by a contractile stalk which links it with a **pectocaulus** running through the colony as the stolon system does in a dendroid. Initially the pectocaulus is soft, but it

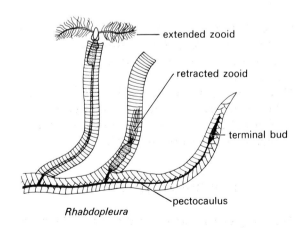

Rhabdopleura

Figure 10.9   The modern hemichordate *Rhabdopleura* (x 18) (Modified from Kozlowski, 1948)

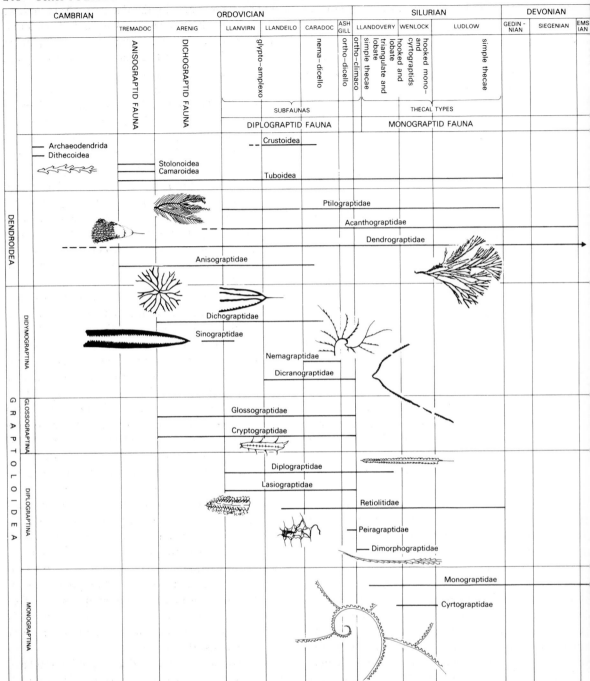

Figure 10.10 Evolution and faunal sequence in graptolites – a generalised picture. Genera illustrated (top to bottom)
Dithecodendrum, Ptilograptus, Dictyonema, Dendrograptus, Staurograptus, Tetragraptus, Didymograptus, Nemagraptus, Dicellograptus
Glossograptus, Diplograptus, Lasiograptus, Orthoretiolites, Dimorphograptus, Cyrtograptus.

later becomes rigid. The pectocaulus with its similarity to the dendroid stolon gives some indication of affinity, but perhaps the relationship of the two groups is most clearly suggested by the presence of fusellar tissue which forms the periderm of both graptolites and *Rhabdopleura*.

EVOLUTION

Evolution in the shape of graptolite rhabdosomes (Fig. 10.10)

The following summary of evolution in graptolites is concerned with gross morphology: the kind of details that can be seen in flattened specimens. Later some particular details will be added that can only be established through the examination of three-dimensional material.

The earliest graptolites known are Middle Cambrian and come from the Siberian Platform; they are small twig-like genera recently discovered and classified by Obut (1974) in two new orders, the Dithecoidea and Archaeodendrida, though more material needs to be forthcoming if this taxonomy is to be sustained. In the Tremadocian and later the evolution of graptolites is better known. The Upper Cambrian graptolites are all dendroids, along with representatives of the short-lived orders.

Dendroids were quite diverse in Upper Cambrian times, and though they were thereafter never a very abundant component of any fauna they continued until the Carboniferous, various genera having different time ranges. **Dendritic** or twig-like forms were perhaps the most common, though there were also some **pinnate** genera (e.g. *Ptilograptus*). The peak of dendroid complexity is reached in the Ordovician to Devonian genus *Koremagraptus*, an acanthograptid with a complex and anastomosing branch system. Each branch has many subparallel stolothecae all united together in columns, with their attendant autothecae and bithecae irregularly twisted up together in a thick stem. The vast majority of dendroids grew upright on the sea floor like small shrubs, fixed by holdfasts which, like their 'stems', were strengthened by cortical tissue. But one or two species of *Dictyonema* (a more rigidly engineered and cone-shaped genus) apparently departed from this ancestral habit and lived in an inverted position,

probably suspended by the thin nema from floating seaweed. Small **attachment discs** have been found attached to the end of the nema; hence a suspended mode of life seems likely though it was certainly not the norm for most graptolites.

It is generally accepted that *Dictyonema* was ancestral to the graptoloids, and the rather enigmatic anisograptids appear to have been an intermediate group. Most anisograptids are Tremadocian in age; there are records of a possible extension into the later Ordovician, but this is not certain. They may have two, three or four primary branches as in *Clonograptus*, *Bryograptus* and *Staurograptus* respectively; they are rarely preserved other than as compressions. But some Scandinavian genera are pyritised, and these are of particular interest for they show characters intermediate between those of dendroids and graptoloids at the microscopic level.

The Upper Tremadocian genera *Kiaerograptus* (Fig. 10.7a) and *Bryograptus* apparently have bithecae, but these are fewer in number than those of standard dendroids, and there is only one bitheca for every two autothecae; furthermore, they are reduced in size. Why the bithecae were dispensed with during the later evolution of graptolites is not clear. Presumably their function, which could have been reproductive, was taken over by autothecae; this might have happened through an intermediate kind of graptolite such as the Llandeilian *Calyxdendrum*, in which the bithecae, rather than having external apertures, open directly into the autothecae. The anisograptids became extinct by the end of Llandeilian time and their progeny, the Graptoloidea, became dominant. Continuous sections in the Tremadoc are few and far between; the best is in the Yukon Territory and has recently been described to show a continuous sequence of graptolite faunas. The fauna here is mainly of anisograptids and other dendroids, but there are also some early graptoloids (dichograptids): the forerunners of the main stock that dominated the lowermost Ordovician.

Dichograptids are the earliest graptolids and have no bithecae. They are usually symmetrical, branch dichotomously and may be many-branched or few-branched. A burst of adaptive radiation in the early Ordovician resulted in a profusion of types; eight-, four- or two-branched, with rhabdosomes pendent or declined, or even scandent, like *Phyllograptus* where the four stipes are arranged back to back and are in contact all the way up from the basal sicula, or *Tristichograptus* which has three scandent stipes. Other genera are *Tetragraptus*, with four declined or pendent stipes, and the common, zonally useful *Didymograptus*, in which the twin stipes may be pendent like a tuning fork or extended in line. Several of the early **multiramous** (many-branched) dichograptids had a web-like structure between the stipes at the proximal end, which increased the surface area and probably gave assistance in flotation.

From an unspecified two-stiped dichograptid ancestor there arose at some time during the Lower Ordovician the various stocks that came to dominate the Upper Ordovician (Fig. 10.12a–e). These never have more than two stipes. First there are the diplograptids, which are scandent forms. These have their origin well back in Llanvirnian time, their first representative being *Glyptograptus dentatus*. Thence they can be traced to the end of the Llandovery. The several diplograptid genera are distinguished one from another mainly on the characters of their thecae. Thus *Diplograptus* has simple straight thecae, *Climacograptus* has thecae of a square-cut appearance (Fig. 10.4), *Glyptograptus* has gently curving thecae, etc. All these graptoloids are more-or-less elliptical in cross-section and are easy enough to identify when flattened, provided that they are preserved with the 'long axis' of the ellipse parallel with the bedding plane, for then the thecae are clearly seen in lateral view. If, however, the long axis of the elliptical cross-section is perpendicular to the bedding then only the apertures show, and since they all look similar in this **scalariform view** they are not easy to identify.

Other Upper Ordovician genera include the V-shaped, stoutly built *Dicellograptus*, the more slender but otherwise similar *Leptograptus*, and the Y-shaped *Dicranograptus* (including *D. ramosus* (Fig. 10.7c) which can be more than 30 cm long). In the latter genus the first-formed part of the rhabdosome is scandent; in subsequent growth the branches diverged. The occurrence of diplo-, dicello- and dicranograptids in the Upper Ordovician gave rise to a belief that the diplograptids evolved from a dicellograptid ancestor by 'zipping themselves up' and that the condition in *Dicranograptus* represented an incomplete zipping process. Attractive though this suggestion appears it is unrealistic, for the first diplograptids appear well before the earliest dicello- and dicranograptids, and in any case the detailed structures of the

'proximal end' – the few first-formed thecae and the sicula, which are all taxonomically diagnostic – are quite different.

The dicello-, dicrano- and leptograptids became extinct before the Silurian, though the diplograptids continued until the end of the Llandovery as important members of the total graptolite fauna. One family of diplograptids, the Retiolitidae, which persisted until the end of the Silurian, has the periderm reduced to a meshwork of girders (**clathria**) with or without a reticulate net covering it. This presumably reduced the mass of the graptoloid and aided in its flotation. Judging by the persistence of the Retiolitidae (U. Ord. – U. Sil.), they must have been a most successful group.

The Silurian is dominated by monograptids: scandent forms with the thecae arranged along only one side of the stipe. They first appear just above the base of the Silurian. Monograptids come in various shapes and sizes; some are very long and straight, others short and stumpy, and there are various highly modified forms including curved or spirally coiled genera which, like *Cyrtograptus* (Fig. 10.7e), may have lateral arms (cladia). Various changes in thecal morphology, as will be explained in detail later, appear in succession during the Silurian. The last monograptids flourished in the early Devonian but are restricted to certain parts of the world only.

The picture of graptolite evolution presented here has been known for a long time and is well established. But it is based mainly upon compressions, and reference to three-dimensional material expands and illuminates it to a remarkable degree.

Evolution of the proximal end in graptoloids

The Graptoloidea differ from the Dendroidea in the absence of bithecae, the reduction of cortical tissue, the form of the sicula, the presence of a nema, the mode of rhabdosome branching and the structure of the thecae. They are also unique in the way in which the first few thecae develop from the sicula. In general terms, the early graptoloids have a rather simple kind of proximal end development; the Later Ordovician ones are much more complex, such as that of *Diplograptus* which may now be seen in evolutionary perspective. Various types of proximal end development are shown in Figure 10.11, illustrating the evolution of structure, but it must be remembered

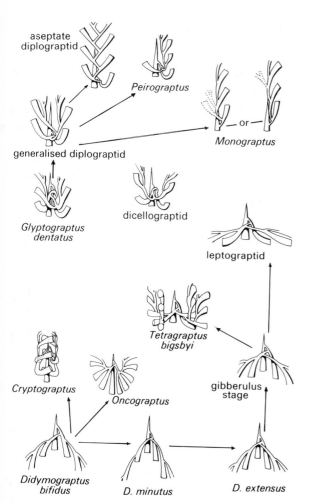

aseptate diplograptid

generalised diplograptid

Peirograptus

— or —

Monograptus

Glyptograptus dentatus

dicellograptid

leptograptid

Cryptograptus

Tetragraptus bigsbyi

Oncograptus

gibberulus stage

Didymograptus bifidus

D. minutus

D. extensus

Figure 10.11 Evolution in graptolite proximal end structures (redrawn from Bulman in '*Treatise*' Part (V)).

that the models represented here are greatly simplified from the actual structures encountered in real graptoloids. Most dichograptid proximal ends are of unmodified form. In these (as in *Didymograptus bifidus*) th. 1^2 grows out from th. 1^1 across the sicula and is welded to it; it forms a single crossing canal. But in later dichograptids (e.g. *Didymograptus extensus*) th. 2^1 is also welded to the face of the sicula, and so there are two crossing canals. In such a species as *Tetragraptus bigsbyi*, for all its apparent complexity there are still only two crossing canals, for th. 2^2 is not welded to the front of the sicula.

Leptograptus has usually two crossing canals, and though it may have three its development is hardly modified from that of the later dichograptids. The

dicello- and diplograptids have never less than three crossing canals. The structure of the proximal end in the scandent diplograptids differs markedly from that in their precursors, and the thecae are often twisted up together in a knot. Diplograptids may be **aseptate** (e.g. *Diplograptus*), or they may have a **median septum** separating the later thecae so that there are normally only three crossing canals rather than an indefinite number. Thus the later formed thecae (usually those succeeding th. 2^1) open on the same side of the median line as they arise.

In *Climacograptus*, for instance, th. 3^1 arises from th. 2^1, whereas in *Diplograptus* th. 3^1 arises from th. 2^2 and th. 3^2 from th. 3^1.

Similar in development to diplograptids in many ways are the dicellograptids, which likewise have three crossing canals with curiously twisted early thecae. Normally in *Dicellograptus* and its relatives the sicula reclines against one of the stipes.

There are some very modified modes of development in certain scandent graptoloid genera. *Cryptograptus*, for instance, has a thecal system quite different from that of the diplograptids; its most peculiar feature is the torsion of the thecae which results in the series th. 1^1, 2^1, 3^1 ... opening on opposite sides (frontal and reverse) of the sicula to th. 1^2, 2^2, 3^2 ... *Skiagraptus* has a modified version of this structure which is even more complex.

The origin of the scandent biserial graptoloids has been much disputed, but some evidence suggests a common origin of both diplograptid and glossograptid types from a broad leaf-like scandent dichograptid, such as *Apiograptus*, though the problem of how the crossing canals arose is far from clear.

Monograptids presumably arose from a diplograptid ancestor, but there are no direct links between them. As already shown, the monograptids' th. 1^1 arises from a notch low on the metasicula, and the serially arranged th. 2^1, 3^1, 4^1 ... grow upwards at an angle. But the change in proximal end structures from the complex and twisted system of the diplograptid precursors to the simple construction of the monograptid proximal end is very sudden, and there are no known intermediates.

The earliest known monograptids have been identified in the lowermost Llandovery monograptid genus *Atavograptus*, where the thecae are of simple construction and the rhabdosomes are straight or gently curved, but the proximal end is of typical monograptid construction. This gives no help with the problem of monograptid origins, but in the English Lake District in the lowermost zone of the Silurian there are some bedding planes that are covered with specimens of *Atavograptus ceryx* in which the population appears to be **dithyrial,** i.e. consisting of both uniserial and biserial individuals, the latter being similar in many ways to *Glyptograptus* (Rickards & Hutt 1970). Perhaps it is in such populations that monograptid origins are to be sought (Rickards 1974). Presumably selection acting on such populations resulted in a changed ecological balance, with the uniserial forms eventually becoming dominant.

It is worth noting that there were some 'false monograptids' in the early Llandovery as well as species truly of monograptid type. *Peiragraptus,* known only from specimens from the Llandovery of Anticosti Island, has incomplete diplograptid development with a solitary th. 1² opposite a range of uniserial thecae; in *Dimorphograptus* and allied genera the rhabdosome begins like that of a normal monograptid (though lacking some characteristic features) but distally becomes biserial.

These odd genera must be seen as the results of similar selection pressure acting on diplograptid stocks, resulting in the production of different groups with reduced numbers of thecae. Some of these, like the *Atavograptus* stock which gave rise to all the later monograptids, were eminently successful; in other cases the resultant groups did not succeed in producing any long term descendants.

Thecal structure and its evolution (Fig. 10.12)

Many graptoloids are defined at the generic and specific level upon the structure of the thecae. The dichograptids normally have straight and simple thecae. Similar thecae are found independently in some monograptids and in certain biserial graptoloid genera also. Characteristic genera that are descended from the dichograptids, such as the biserial *Glyptograptus* and *Orthograptus*, have gently curved thecae, whereas *Leptograptus* has exceedingly thin thecae which are more-or-less parallel with the common canal. In some of the Upper Ordovician groups a sharp-angled bend (**geniculum**) is developed in the ventral wall; this is usually associated with an incurved (**introtorted**) thecal aperture above, often facing directly into the base of the theca above it

and leaving little room for the zooid to come out. Such thecae are characteristic of dicrano- and dicellograptids, whilst the biserial climacograptids have geniculate thecae with a horizontal aperture and a straight outer thecal wall above the geniculum, parallel with the median line.

Within the monograptids there is often considerable thecal variation not only between genera but also within a single individual. In all monograptid species the thecae change shape along the rhabdosome, and the proximal thecae may be quite different from those of the distal end (Fig. 10.12f, g). If there is torsion in monograptid thecae it is usually an outward twisting (**retroversion**) as opposed to the introtorsion of the Upper Ordovician genera.

The basic form of the thecae in monograptids serves to define successive faunas which can be used stratigraphically (Figs 10.10, 10.12h–n). Lower Llandovery monograptids have straight or gently curving thecae. Above these come faunas with simple and **triangulate** thecae, whereas Upper Llandovery genera have mainly **lobate** and **hooked** thecae.

Triangulate thecae have tiny apertures and in lateral view are usually of triangular form. They reach their extreme development in the **isolate** thecae of *Rastrites* (Fig. 10.12 m) where each theca is exceptionally long and is isolated from its neighbours. Hooked thecae have the shape of an open hook, whilst lobate thecae are formed as short thick and compact hooks each inturned in a cowl-like manner to embrace its own aperture.

In the Wenlock the most abundant graptoloids have hooked thecae, and there are also a number of cyrtograptids; these are graptoloids coiled in a helical spiral form, with radially arranged straight or curved lateral branches (cladia). Cyrtograptids can often be very large and were apparently free-floating. In the Ludlow, diversity in thecal form reaches a peak. Many forms have reverted to a simple thecal type, but they exist side by side with genera in which the apertures are very modified (as in *Cucullograptus*) or positively asymmetrical. The graptoloids of the Lower Devonian are quite diverse in thecal form, with hooked or straight thecae having cowl-like apertural overhangs. Apparently graptoloids went on evolving to the last, and there is no clear reason why they should have become extinct.

In Silurian graptoloids especially it has been possible to make detailed microevolutionary studies

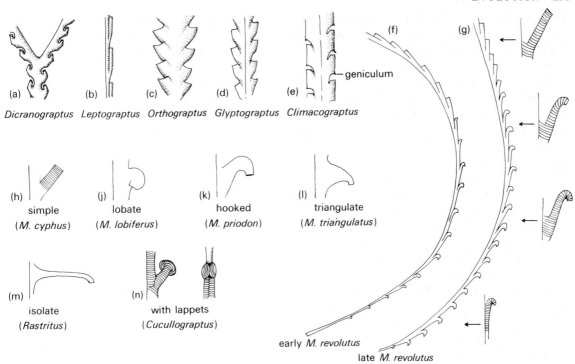

Figure 10.12 (a)–(e) Thecal morphology in some Ordovician graptoloid genera; (h)–(n) Silurian monograptids; (f) and (g) show change in thecal shape away from the proximal end of stratigraphically early and late *M. revolutus* (x 2). (Redrawn from Sudbury, 1958)

of the change in thecal shape with time. The various forms of monograptid thecae (hooked, lobate, isolate, etc.) appear to have been associated with definite evolutionary trends at particular times. Some of these trends have been analysed in detail (Sudbury 1958), and it has been shown that new thecal types are normally introduced at the proximal end, later spreading along the rhabdosome in descendants of the original type. Thus from *Monograptus triangulatus* there is a trend towards isolation of the thecae, with *Rastrites maximus* as an end point and *R. peregrinus* (which has the more isolate thecae confined to the proximal region) as an intermediate; in any population collected from a single bedding plane, individuals may vary greatly in their state of 'advancement'. Perhaps the best documented of these Silurian series is the trend from *M. argenteus*, where only the early thecae are hooked, towards *M. priodon*, in which all the thecae have the form of open hooks. In all of these the new type of theca came in at the proximal end and spread from it, but there are some cases where distal introduction was the rule. Urbanek

(1960) has postulated that certain growth-stimulating substances like the auxins of plants diffused from the sicula as the graptoloid grew, promoting differential curvature in the proximal thecae but not affecting the latter thecae, since their quantity decreased during ontogeny. In the descendants of these early forms the growth stimulators continued to be produced from the sicula until a later stage of development. Where new types were introduced distally, Urbanek assumed that a growth inhibitor, rather than a stimulator, diffused from the sicula. Whilst the operation of this system cannot be proved, it remains a valuable and interesting suggestion.

Cladia

Cyrtograptus, amongst other genera, produces lateral branches (cladia) at intervals. The development of these can be followed in well-preserved material (Fig. 10.7e). Each theca has a hood-like **lappet** with a pair of short lateral whiskers projecting from it. When a cladium develops from such a thecae, one of the two whiskers elongates and becomes a nema. New fusellar tissue grows along this as if from an initial bud. The aperture of the basal theca is constricted by the growth of the fusellar tissue but remains open. By the time the first theca of the new cladium has developed, the primary stipe has advanced by some seven or eight thecae. Since thecal form in *Cyrtograptus* changes on going along the rhabdosome, the characters of any one theca on the primary stipe can be matched exactly with one on the cladium; always there is a lag so that the first cladial theca is paralleled by a theca on the primary stipe seven or eight thecae ahead. Both ends of the rhabdosome, on the primary stipe and cladium, have been growing in the same way at the same time.

Though best known in *Cyrtograptus*, cladia are present in certain other genera also. *Diversograptus* has a sicular cladium, so that the straight main stipe grows in the opposite direction to the cladium. There may be accessory lateral cladia growing either from the main stipe or from the primary cladium. The related genus *Abiesgraptus* has a number of sicular cladia.

HOW DID GRAPTOLITES LIVE?

Most dendroids, together with stolonoids, tuboids, camaroids and crustoids, appear to have been bottom-living sessile organisms. They have no flotational structures, and the cortical tissue of dendroids such as *Dendrograptus* appears to have been used primarily for strengthening the proximal end of the rhabdosome into a root-like holdfast. Thus the dendroids can be envisaged as having grown upright on the sea floor with the sicular end down and the stipes outstretched as in a small shrub. But *Dictyonema* species such as *D. flabelliforme* are conical rather than shrubby and had little cortical tissue. The thread-like nema could not have supported a rhabdosome growing on the sea floor. Furthermore, there are sometimes found small tufts of thin fibres terminating the nema, which could have acted as fixing devices for the attachment of the dendroid to floating seaweed. This may be suggested by the presence of much carbon in the black shales in which many graptolites are found. Presumably *Dictyonema* evolved from the standard sessile dendroids by inversion of the

rhabdosome, sicular end up, though an 'upright' orientation seems to have been restored in the scandent graptoloids of the later Ordovician and Silurian.

A nemally attached mode of life is likely for *Dictyonema* and probably some other graptoloids as well, but the majority were probably free-floating, forming the main preserved part of the Lower Palaeozoic plankton. In this undoubtedly there were other organisms, some of which such as the small epiplanktonic brachiopods and occasional small trilobites, are sometimes found preserved in grapto-litic shales. Possible adaptations for floating are numerous; modern organisms achieve flotation either by having gas or fat bubbles within their tissues or by having the tissues isotonic with the surrounding sea water. In addition, supplementary flotational mech-anisms have been described in the graptoloids (Bulman 1964). Some dichograptids have flat **expanded webs** surrounding the proximal end, which could have increased their surface area. Many scandent graptoloids have a float-like structure of vertical **radiating vanes** arranged around the nema, which probably gave the floating graptoloid stability. Less often there may be a **vane** or plate at the proximal end, the function of this being less clear. Certain spirally coiled graptoloids (*Cyrtograptus* and *Monograptus turriculatus* amongst others) have been considered as floating forms, which would be rotated round and round, spiralling upwards in the water under the influence of currents and thus buoyed up. The presence of a flat proximal membrane in *Cyrtograptus* may have been an extra flotational aid.

The suggestions given above are the standard view on the life of graptoloids, but recently the hypothesis of graptolite automobility has been advanced (Kirk 1969, 1972). It has been suggested that the graptolites were mobile, not just passively drifting, but actively swimming with the sicula upwards while feeding currents were drawn by the thecae towards the ciliary-ornamented zooids. The reduction in stipe number, the increasing symmetry and the change in inclination of the stipes are all seen as part of an evolutionary response to this new mode of life, and though the whole matter is very controversial and has been much debated it remains an interesting hypothesis. At the same time, since the nature of the extrathecal tissue in graptoloids is as yet not understood, this vital gap in our knowledge may well prevent any further development of our understanding of graptoloid ecology.

Whether or not graptoloids were actually mobile, they seem to have been a primary component of the plankton of the Palaeozoic. Their common association with black shale (graptolitic facies) with its high carbon content and presence of syngenetic pyrite suggests that these planktonic organisms were preserved in conditions where no benthos flourished; then they would be able to be preserved undisturbed by bioturbating animals. Since they are found in other kinds of sediment they must have been widespread, but their association with a particular argillaceous facies is largely a matter of preservation failure elsewhere.

FAUNAL PROVINCES

The distribution of graptolites on a global scale, at least in the Ordovician, was not always uniform (Skevington 1974). In the early Arenig it seems that most graptolite genera had a cosmopolitan distribution, but from the later Arenig onwards until the end of the Llandeilo there were two well-defined graptolite faunal provinces – the 'Atlantic' or 'European' province and the 'Pacific' province – whose faunas are not closely similar. The area of the Atlantic province includes England and Wales, south-eastern Ireland, most of Europe, North Africa and, rather oddly, Peru and Bolivia. On the other hand, the graptolites of the Pacific province are found in North America, Argentina, all of Australia and Australasia and (at first sight paradoxically) in Scotland, north-western Ireland and western Norway. The graptolites collected from any part of one of these provinces bear a fairly close resemblance to each other at the generic level.

Thus amongst other distinctive factors pendent (tuning fork) *Didymograptus* species are very common in the Atlantic but not in the Pacific province, whilst biserial scandent graptoloids common to both provide a good basis for stratigraphical correlation.

By the later Ordovician and throughout the Silurian there was a higher degree of cosmopolitanism, and the two faunal provinces were replaced by a single one.

What could have been the controls of provincialism? With free-floating organisms, geographical barriers are likely to be less important than they are with shelf-living benthos, unless they cross temperature zones. Such more-or-less latitudinal zones defining regions of differing temperature (tropical, subtropical, warm-temperature, cold-temperature, boreal, polar) are the most important single control of the global distribution of organisms today, and presumably they were of the Lower Palaeozoic plankton also.

The Pacific faunas of the Lower Ordovician were apparently circum-tropical and were confined entirely between the latitudes of 30° N and S of the equator. Atlantic faunas, on the other hand, lay south of the 30° S latitude except for the European region where Atlantic faunas are found nearer to the equator. It has been proposed that the isotherms more or less followed palaeolatitudes, except where a tongue of colder water projected northwards in the European region (as with today's Peru Current), carrying with it its own indigenous fauna.

Such a pattern could well account for Lower Ordovician graptolite distribution, with temperature as the primary control, but why then did the faunas of the later Ordovician become more uniform? One reason is that Families Dichograptidae and Sinograptidae which had been important in defining the two provinces had by this time become extinct, but it is most significant that almost no later Ordovician graptolites are found outside the 30° N and S latitudes; they are confined within the region formerly occupied by the circumtropical Pacific province. This reduction of the geographical distribution of graptolites in the later Ordovician may well have been under climatic control. High latitudinal realms no longer supported graptolites, possibly because the waters were too cold, and the graptolites of the tropical zones were unable to adapt to cold conditions.

By the latest Ordovician (late Ashgill) only some five or six species of graptolites remained: a decimated stock living at a time of widespread glacial conditions, from which came the earliest monograptids. During the milder climates of the Silurian their descendants were able to expand and diversify and to spread uniformly and widely across the world once more, to the regions that their ancestors had occupied in the earlier Ordovician.

STRATIGRAPHICAL USE

To geologists, graptolites are primarily of interest as stratigraphical indicators. The sequence of graptolite faunas has been used for the subdivision of Ordovician and Silurian rocks since the time of Lapworth, whose work in southern Scotland in unravelling the complexities of geological structure is classic. The main problems in the stratigraphical use of graptolites are:

(a) Graptolites are usually preserved as compressions, and there may be problems in precise identification.

(b) The time range of many graptoloid species is rather long, and it is not possible with graptoloids to give the kind of stratigraphical precision that is possible with (for instance) ammonites.

(c) Graptolites are normally confined to black or grey shaly facies by preservational factors, but often great thicknesses of Ordovician and Silurian strata, which might be expected to contain graptolites, do not in fact have any fossils at all. Furthermore, it is relatively rare to find graptolites in coarser-grained rocks away from graptolitic facies, and there are consequently singular problems in correlating between areas of different facies.

Usually geologists look for areas where, because of shoreline oscillations, there are exposed sequences of alternating graptolitic and shelly facies, so that a particular graptolite assemblage is time-bracketed within a trilobite–brachiopod assemblage and hence the two may be directly correlated. Alternatively, the discovery of mixed graptolitic and shelly faunas in the same argillaceous sediment is most helpful. Even so, there are successions where the shelly faunas are indigenous to the region or have very long ranges and where in the absence of graptolites precise correlations are not possible.

Individual graptolite zones are based and defined on the time ranges of particular short-ranged species, though the named species is normally only one of a number in the total fauna. In the European succession two Tremadocian, eleven Ordovician, thirty-two Silurian and three Devonian zones have been erected and are in common use; in Australia, however, the Tremadocian and Ordovician are represented by twenty-five zones of which only some of the Upper Ordovician ones contain graptolites identical to those in the European zones. Some widespread species, such as *Glyptograptus teretiusculus*, *Nemagraptus gracilis* and *Dicellograptus complanatus*, are normally considered as representing time-equivalent horizons, but there may be some diachroneity.

Though such graptolites may be very widespread, there are many others which are geographically restricted and useful for strictly local correlation alone. Because of this difficulty several palaeontologists have tried to define faunal units which are larger than zones but of more-or-less worldwide application. These successive **graptolite faunas** (Fig. 10.10) are easily identified and in any case epitomise, as Bulman (1958) has shown, the geological history of the group.

The following sequence of graptolite faunas through time is from Bulman (1958). The various faunas and subfaunas here noted are defined on the first appearance of new graptoloid types.

The Tremadocian is characterised by an **anisograptid fauna** consisting mainly of *Dictyonema* species and anisograptids. Since provinciality was well marked at this time there are some problems in intercontinental correlation. The time equivalence of different *Dictyonema* species in various parts of the world is now becoming reasonably clear, even though these species may be mutually exclusive.

Very few graptolites of Upper Tremadocian age are known, other than one fauna from the Yukon Territory and another from Norway in which *Bryograptus*, *Kiaerograptus* and possibly a *Didymograptus* have been described. But other than this doubtful record there are no true graptoloids known from Tremadocian rocks.

The succeeding **dichograptid fauna** of the Arenig is marked by the incoming of the earliest real graptoloids. *Tetragraptus* is found right at the base and is usually associated with didymograptids as well as with remaining anisograptids. The extensiform didymograptids of the Arenig are replaced in the Llanvirn by tuning fork species of *Didymograptus* (though by the Llanvirn the succeeding diplograptid fauna became established), and the genus continues into the Caradoc before becoming extinct. Provinciality is well marked in the Lower Ordovician, and the endemic Pacific genera *Oncograptus*, *Cardiograptus* and *Isograptus* have been found stratigraphically useful in beds of equivalent age in western North America and Australia.

A few biserial graptoloids are found at the very top of the Arenig, but the real flowering of the biserial genera is from the Llanvirn until well into the Silurian. The **diplograptid fauna** spans the period from the Llanvirn to the lowermost Silurian, before the incoming of the **monograptid fauna.** The diplograptid fauna is divided into four subfaunas. The *Glyptograptus–Amplexograptus* subfauna of the Llanvirn and Llandeilo contains many tuning fork graptoloids in addition to the genera from which the fauna takes its name; evidently the main period of differentiation of the biserial graptoloids took place at this time, even though they are not numerically abundant. Lower Caradocian beds contain

a *Nemagraptus–Dicellograptus* subfauna in which *Dicranograptus* is also found and which contains the final dichograptids. The *Orthograptus–Dicellograptus* subfauna replaces it in the Upper Caradoc and Ashgill. In the later Ashgill this subfauna is somewhat impoverished as is the *Orthograptus–Climacograptus* subfauna of the lowermost Silurian, which immediately predates the arrival of the monograptid fauna.

This fauna extends throughout the whole of the Silurian (other than in the two lowermost zones) and into the Emsian. Though it has not been divided into subfaunas, different monograptid types, characterised mainly by their thecal construction, are found successively and can be used stratigraphically.

BIBLIOGRAPHY

Books, treatises, symposia

Teichert, C. (ed.) 1970. *'Treatise'* Part (V) (Revised). Graptolithina. (Updates Bulman's original Treatise of 1955)

Rickards, R. B., D. E. Jackson and C. P. Hughes (eds) 1974. *Graptolite studies in honour of O. M. B. Bulman*, 1–260. Spec. pap. in palaeont., no. 13. (20 original papers and bibliography of Bulman's works)

Individual references

Bulman, O. M. B. 1945–47. *A monograph of the Caradoc (Balclatchie) graptolites from limestones in Laggan Burn, Ayrshire*. Palaeontogr. Soc. monogr., (i) 1945, 1–42; (ii) 1946, 43–58; 1947, 59–78. (Detailed morphology and ontogeny; profusely illustrated)

Bulman, O. M. B. 1958. The sequence of graptolite faunas. *Palaeontology* 1, 159–73. (Stratigraphical use)

Bulman, O. M. B. 1964. Lower Palaeozoic plankton. *Quart. J. Geol. Soc. Lond.* 119, 401–18. (The graptolite biocoenosis as the preservable plankton of its time)

Elles, G. L. and E. M. R. Wood 1922. *A monograph of British graptolites* (with synoptic supplement by I. Strachan, 1971), 1–539. Palaeontogr. Soc. monogr. (Descriptions and stratigraphical use of all British species then known; standard work)

Kirk, N. 1969. Some thoughts on the ecology, mode of life, and evolution of the Graptolithina. *Proc. Geol. Soc. Lond.* 1659, 273–92.

Kirk, N. 1972. More thoughts on the automobility of the graptolites. *Quart. J. Geol. Soc. Lond.* 128, 127–33. (A different interpretation from that of Bulman)

Koszlowski, R. 1948. Les graptolithes et quelques nouveaux groupes d'animaux du Tremadoc de la Pologne. *Palaeont. Polonica* 3, 1–235. (Classic, profusely illustrated

study of three-dimensional dendroids and other groups)

Obut, A. M. 1974. New graptolites from the Middle Cambrian of the Siberian Platform. In Rickards *et al.*, op. cit., 9–13. (Oldest known graptolites)

Rickards, R. B. 1974. A new monograptid genus and the origins of the main monograptid genera. In Rickards *et al.*, op. cit., 141–7. (Monograptids may have originated from dithyrial population)

Rickards, R. B. and J. E. Hutt 1970. The earliest monograptid. *Proc. Geol. Soc. Lond.* 1663, 115–19. (Monograptid origins)

Skevington, D. 1974. Controls influencing the composition and distribution of Ordovician graptolite faunal provinces. In Rickards *et al.*, op. cit., 59–73. (Temperature controls; detailed map)

Sudbury, M. 1958. Triangulate monograptids from the *Monograptus gregarius* zone (Lower Llandovery) of the Rheidol Gorge (Cardiganshire). *Phil. Trans Roy. Soc. Lond. B* 241, 485–555. (Evolutionary lineages)

Urbanek, A. 1958. Monograptidae from erratic boulders in Poland. *Palaeont. Polonica* 9, 1–105. (Fine structural detail; many illustrations)

Urbanek, A. 1960. An attempt at biological interpretation of evolutionary changes in graptolite colonies. *Acta Palaeont. Polonica* 5, 127–234. (Develops theory of chemical growth-regulation in graptolites)

Urbanek, A. and K. M. Towe 1974. Ultrastructural studies on graptolites. 1: The periderm and its derivatives in the Dendroidea and in *Mastigograptus. Smithsonian Contrib. Paleobiol.* 20, 1–48.

Urbanek, A. and K. M. Towe 1975. Ultrastructural studies on graptolites. 2: The periderm and its derivatives in the Graptoloidea. *Smithsonian Contrib. Paleobiol.* 22, 1–24. (Transmission electron microscope studies of well-preserved cuticle, e.g. Fig. 10.6)

11 Arthropods

Arthropods have, as their most distinctive characteristics, a hard outer coat or **exoskeleton** and **jointed appendages** which they use for movement and feeding. They include insects, crustaceans and spiders, as well as extinct trilobites and eurypterids (the giant Palaeozoic water-scorpions). Arthropods are perhaps the most successful and diverse of all invertebrates, and since their tough exoskeleton offers considerable potential for fossilisation (especially if mineralised) their geological record is good.

Arthropods are segmented coelomate metazoans, as are the annelid worms with which they possibly share a common ancestry. In annelids the only skeleton is a hydrostatic one provided by the coelomic fluid, which does not give a rigid anchorage for muscles. The arthropods, however, because of their firm exoskeleton, and also since many have an internal skeleton (**endoskeleton**) as well, possess a rigid base for attachment of the **internal muscles** that move the limbs; hence they have the potential for rapid locomotion. Many, furthermore, have hard jaw structures (**mandibles**) which can grind, crush or bite. Thus arthropods, in terms of movement and feeding, have an over-riding superiority over the annelid worms and have been able to invade many different environments which remain closed to the latter.

CLASSIFICATION AND GENERAL MORPHOLOGY

Diversity of arthropod types

Arthropods take their name from the jointed appendages that are a constant feature of their organisation (Greek: $\alpha\rho\theta\rho o\varsigma$ = joint, $\pi o\delta o\varsigma$ =foot). These same appendages, which include the jaw structures, have become very differentiated, and their number, arrangement and morphology are often of critical importance in taxonomy. Many different taxonomic schemes have been erected for the arthropods, and there is still controversy over whether the arthropods arose from single or from multiple ancestors. Nevertheless the view of Manton (summarised in Manton 1973, 1977) that arthropods are polyphyletic is generally accepted. Manton's studies suggest that 'arthropodisation' occurred at least three times, leading to three or more distinct phyla. The 'Phylum Arthropoda' is therefore an unreal entity consisting of heterogeneous elements, and the taxonomic rank of some of the subgroups is still in flux.

Manton's system of classification covers both recent and fossil arthropods, though the taxonomic status of trilobites, which are the most abundant fossil arthropods, is not yet fully resolved.

PHYLUM UNIRAMIA (Cam.–Rec.): Jaws bite with the tip of a whole limb. Trunk lacks **biramous** (two-branched) appendages; the limbs are **uniramous.**
  CLASS 1. ONYCHOPHORA (Cam.–Rec.): These have a non-rigid segmented body, with unjointed legs that can be inflated with blood for rhythmic walking. Jaws short, ventrally directed, with blade-like terminal claws. e.g. *Peripatus* (Recent terrestrial), *Aysheia* (Cambrian marine).
  CLASS 2. MYRIAPODA (Carb.–Rec.): Myriapods and centipedes; entirely terrestrial. Have jointed mandibles biting in the transverse plane. Some giant fossil representatives, such as the terrestrial Upper Carboniferous *Arthropleura*.
  CLASS 3. HEXAPODA (Dev.–Rec.): Insects, aerial and terrestrial. The most diverse and numerous of all terrestrial animal species. Mandibles primitively roll and grind, but in some cases strong secondary transverse biting is possible.
PHYLUM CRUSTACEA (Cam.–Rec.): Dominantly marine arthropods with many fossil representatives (e.g. Fig. 11.1). A very varied phylum with highly

Figure 11.1 *Eryon propinquus*, a Jurassic crab from the shallow-water Solnhofen Limestone, Germany (x 0·5). The only living eryonids are deep-water species.

differentiated limb morphology. The less advanced aquatic crustacea pass food forwards from behind towards the |mouth| along a median food groove, using the limbs or by feeding currents generated by the limbs. In advanced aquatic crustaceans the food is lifted up from the substratum by specialised appendages. The biting jaws (mandibles) are formed from **gnathobases:** internal extensions of the appendages which meet in the median plane and roll or grind together as the limb moves backwards and forwards. The outer part of the leg may disappear.

The earliest crustaceans were Subclass Phyllocarida (Cam.–Rec.), which have a large bivalved **carapace** almost covering the body (e.g. Jones & Woodward 1888–99, Rolfe 1962). These, abundant in certain facies in the earlier Palaeozoic, were almost certainly ancestral to the more advanced shrimp-like or lobster-like forms, the earliest of which are Devonian in age and which have become extremely important since then.

Phylum Chelicerata (Cam.–Rec.): A large arthropod group in which all representatives have the body divided into two parts: the **prosoma** (fused head and thorax) and the **opisthosoma** (abdomen). The jaws can primitively bite together in the transverse plane, using a biting movement quite unlike the secondary transverse biting movement of the crustaceans. The presence of a pair of **chelicerae** (pincers) in front of the mouth is characteristic.

Class 1. Merostomata (Cam.–Rec.): Aquatic chelicerates, with two important subclasses well represented as fossils:

Subclass 1. Xiphosura (Cam.–Rec.): Includes the modern horseshoe crab *Limulus* and its fossil representatives (Order Xiphosurida) and the Cambrian Order Aglaspida.

Subclass 2. Eurypterida (M. Ord.–Perm.): A group of large fresh-water and marine water-scorpions (similar in appearance to terrestrial scorpions, but not closely related).

Class 2. Arachnida (Sil.–Rec.): All are eight-legged; some are marine, but most are terrestrial. Includes spiders, harvestmen and scorpions. Fossil spiders are occasionally found in Upper Carboniferous coal-bearing sequences, and many examples are preserved in late Tertiary amber, which was exuded as a resin from the bark of conifers and trapped spiders and insects. Pycnogonida (sea-spiders) have small bodies with four pairs of stout or spindly walking legs. They are all marine and especially common in modern Antarctic waters where there are over 100 species including giant forms. They have been regarded as an independent subphylum, but Manton's studies show that they are more likely to have derived from an early group of marine arachnids.

Whilst the taxonomic status of trilobites and similar forms is yet uncertain, there may be a case for elevating them to the rank of an independent phylum. For the moment the older classification of the *Treatise* is retained:

?Subphylum Trilobitomorpha (Cam.–Perm.): Marine arthropods known only as fossils.

Class 1. Trilobita (Cam.–Perm.): The earliest known arthropods, with hard three-lobed exoskeletons and serially repeated biramous limbs all the way down the body.

CLASS 2. TRILOBITOIDEA (M. Cam.–Dev.): A rather heterogeneous¦and probably not a natural group, supposed to have trilobite-like limbs all of the same kind, and known almost entirely from one horizon: the Middle Cambrian Burgess Shale of British Columbia. Such forms may have been abundant through time but had such thin organic shells that they were preserved only in exceptional condition (see Ch. 12).

Considering the great diversity of arthropods, living and fossil, and in view of their full treatment by Manton (1977), discussion of arthropods here is limited to only trilobites and chelicerates.

Features of arthropod organisation

The exoskeletal **cuticle** has undoubtedly been one of the principal factors in the success of the arthropods. It gives a physical and chemical barrier between the animal and its environment, yet allows a degree of osmotic and temperature regulation. Furthermore, it supplies good protection against predators and a firm base for the attachment of the internal muscles that move the limbs. It also provides a satisfactory location for various kinds of sense organs, linked to the nervous system through fine **tubular canals** in the cuticle and strategically positioned for environmental monitoring.

But though the advantages of having a cuticle are clear, there are also problems which the inhabitant of a hard outer casing has to contend with. Of these, articulation of the joints, respiration and growth are perhaps the most acute.

All arthropods have segmented bodies. The exoskeleton of most modern ones consists of hard **sclerites,** each usually consisting of a dorsal **tergite** and a ventral **sternite,** making a ring for each segment. These may form a rigid cylinder or, alternatively, the exoskeletal rings may be able to move against one another by means of internal muscles running from one segment to the next. The joint between each segmental ring is protected by flexible unmineralised cuticle attached to the junction between segments. Similarly, each 'leg' (Fig. 11.2) is a chain of hard cylinders connected by short links of soft and flexible material and powered by internal muscles. These muscles may be attached to internal knobs (**apodemes),** often formed by simple infolding of the exoskeleton, or to an endoskeleton. The muscles operate according to the normal antagonistic system, so that when one set of muscles contracts to move the limb in a particular direction the opposite set will simultaneously relax (as with the biceps and triceps muscles of the human arm). When the opposing set contracts, the limb moves the other way.

Respiration in aquatic arthropods normally takes place through **gills.** These are usually lamellate organs with a very thin cuticle, extending from the bases of the appendages into the surrounding water. In very small arthropods diffusion over the whole body surface suffices, and there are no gills, but in larger ones there are gills which can sometimes be preserved fossil. Sometimes gills are located in specialised internal chambers. Modified gills in internal chambers are also found in some terrestrial arthropods, such as the **lung books** of arachnids. Many terrestrial arthropods breathe by **tracheae,** which are branched, spirally thickened tubes bringing air direct from the outside to the tissues. Such tracheal respiration is possible only in animals of relatively small dimensions, which is perhaps a critical factor in controlling the upper size limit of insects.

Growth is perhaps the most difficult physiological problem for arthropods, since the rigid exoskeleton encasing them cannot enlarge once it is formed. It has to be shed or moulted at intervals while a new and larger exoskeleton forms. This moulting process is known as **ecdysis.** It is a limiting system in arthropods and not an entirely perfect one, for some 80–90 per cent of arthropod mortality occurs during moulting. Before the exoskeleton is cast, a new soft cuticle, somewhat larger than the existing hard shell, is formed below the old one. At this stage it is elastic, soft and wrinkled. Whilst this is being formed, the lower part of the old cuticle is partially dissolved from below by corrosive fluid poured out from cutaneous glands below the new cuticle. Just before the old exoskeleton is cast, the animal stops feeding but takes up much oxygen and water. As it swells it makes spasmodic movements of the body to shake off the hard shell, and finally it is able to withdraw itself completely. The old skeleton may have special lines of weakness which facilitate its splitting; this was the

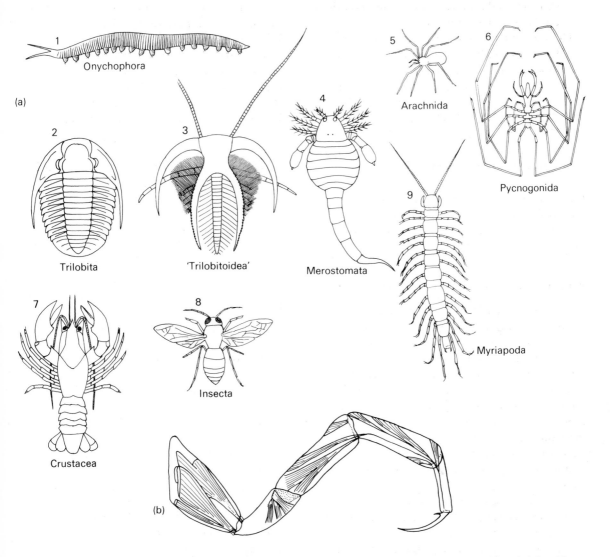

Figure 11.2 (a) Diversity of arthropod types with representative examples of major groups; (b) limb joints and internal musculature in arthropod leg. (Redrawn from V. B. Wigglesworth, 1965 (6th edn) *The principles of insect physiology*. (London: Methuen))'

case in trilobites where the lines of weakness are known as **sutures.** The final swelling of the body to its full size takes place through the uptake of water after the casting process is complete. The animal is now soft-shelled|and|cannot|move|much;|cuticle|hardening and further secretion takes place when this stage is over. The initial cuticle is paper-thin, and the acquisition of the full cuticle thickness may take some time.

During moulting the animal is very vulnerable, both to predators and to the possibility of tearing. Ecdysis is furthermore a wasteful process as much organic material is lost each time. But because of ecdysis we have a good record of the moult stages of many fossil arthropods, and by arranging the cast shells in a size series it is possible to elucidate the various transformations that the fossil arthropod went through from the larval to the adult stage. In trilobites especially this has proved to be of the greatest interest.

?SUBPHYLUM TRILOBITOMORPHA

Class Trilobita

Trilobites are the earliest of all known arthropods. Their first representatives are found in rocks of lower Cambrian age just above the earliest non-trilobite Cambrian fauna. The last died out in the late Permian. Throughout their 350 million years of geological history, they preserved a remarkable constancy of form, though with many variations. All trilobites were marine. Well over 1500 genera are known, and there are several thousand species of which many, especially those of Cambrian and Ordovician rocks, have great stratigraphical value.

General morphology of trilobites

Certain characteristics distinguish trilobites from other arthropods. In all trilobites the head (**cephalon**) is a single plate, made up of several fused segments. There are usually sense organs on the head, and there are also certain lines of weakness, known as **cephalic sutures,** which look like cracks on the surface and apparently facilitated ecdysis. The body (**thorax**) consists of a number of **thoracic segments** hinged to one another and allowing some capacity for enrollment of the body, whilst the tail (**pygidium**), also segmented, is fused into a single plate like the head. These three primary divisions of the body are set at right angles to the unique three-lobed longitudinal division from which the class takes its name.

The limbs (or appendages) which were attached to the lower (ventral) surface are rarely preserved. When they are found they are seen to have a surprisingly structural uniformity within an individual trilobite. Except for the flexible uniramous **antennae** they are all two-branched (biramous), and their structure is virtually identical all the way down the body.

These special features are not shared by other arthropods, and trilobites form a class (or possibly a higher rank) of their own and apparently left no descendants.

ACASTE DOWNINGIAE (Fig. 11.3)

The Silurian *Acaste downingiae,* which is common in the carbonate facies of the borders of Wales and England, shows most of the characteristic features of an 'advanced' trilobite. The body has the shape of a laterally flattened ellipse. The head (cephalon) is somewhat larger than the tail (pygidium), and the central part of the body (thorax) has eleven articulated segments.

The cephalon is quite strongly vaulted with a pentagonal to semicircular shape and transverse posterior edges. In the centre of the cephalon is a raised central hump (the **glabella**), bounded at the sides by diverging **axial furrows** and reaching to the **anterior border.** The glabella is indented by three short and more-or-less transverse pairs of furrows (the **glabellar furrows**) and is closed off at the back of the cephalon by an arched **occipital ring.** Stretching out sideways from the occipital ring are the **posterior border furrows,** which delimit a thin strip of the cephalon at the posterior edge; this is the **posterior border.** Yet another indentation (the **lateral border furrow**) runs parallel with the semicircular lateral border, and in this trilobite it joins the narrower **anterior border furrow** round the front of the glabella.

Placed laterally to the glabella are the eyes, which here are large and crescentic. The lenses are borne on a visual surface and are arranged in a system of hexagonal close packing. There are about 100 lenses forming a compound eye like those of insects and crustaceans, but they are unusually large and separate from one another. This is characteristic of **schizochroal** eyes, which are confined to the particular suborder (Phacopina) to which *Acaste* belongs. Above the visual surface lies the flat **palpebral lobe,** separated by a crescentic **palpebral furrow** from the small raised **palpebral area.** The anterior edge of the eye is very close to the glabella, the posterior one further from it.

A thin though very distinct lineation (the **facial suture**) runs between the palpebral lobe and the visual surface; this suture extends forwards to entour the front of the glabella, being continuous with the facial suture on the other side of the head. Posteriorly, the facial suture turns out transversely on either side to terminate well in front of the **genal angle,** which is the most postero-lateral point of the cephalon. This sort of suture is said to be **proparian.** The region lateral to the glabella though within the suture is the **fixigena;** outside it is the **librigena.** In many trilobites the anterior branch of the facial suture cuts across the antero-lateral border, so that there are two

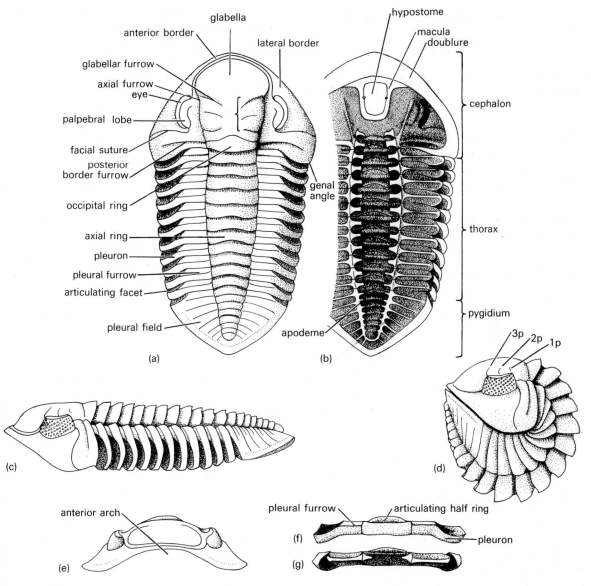

Figure 11.3 Morphology of *Acaste downingiae* Silurian, England: (a) dorsal surface; (b) ventral surface; (c) side view in natural life attitude; (d) enrolled specimen; (e) frontal view showing anterior arch; (f) detached thoracic segment in dorsal and (g) ventral view (x 2). (For additional terminology see text)

distinct sutures rather than a single continuous one. In such cases, which include the majority of Cambrian species, the cephalon may disintegrate before burial into three components: the two librigenae and the central **cranidium,** which is the glabella plus fixigenae. This disintegration does not usually happen in *Acaste* because the suture is continuous.

The lower surface of the cephalon shows that the antero-lateral borders are continued ventrally as a narrow flange (**doublure),** whose inner edge is concentric with the edge of the cephalon and directly below the antero-lateral border furrow. It does not extend, however, all the way to the occipital ring, being phased out along the posterior border. Two pairs of pronounced knobs (the **apodemes)** project ventrally from the posterior two glabellar furrows. These seem to be associated with the attachment and articulation of the ventral appendages. Identical apodemes are present on the thoracic segments, and evidently there were legs under the head as well as under the thorax, though such appendages are not preserved in *Acaste*.

A large plate with a central 'blister' is attached to the rear edge of the anterior cephalic doublure. This is the **hypostome,** which lies below the glabella. The mouth apparently lay directly behind the hypostome, and from it the oesophagus ran forward to the stomach which lay in the space between glabella and hypostome. The gut evidently ran posteriorly from the stomach along the thoracic axis to end under the pygidium. The shape and position of the hypostome is very variable amongst trilobites, and many genera can be identified on the basis of detached hypostomes found in the rock, even if the rest of the exoskeleton is missing. A couple of small swellings (the **maculae**) lie towards the rear of the hypostome. It has been suggested (Lindström 1901) that these are ventral eyes such as are possessed by various modern arthropods, but their fine structure is normally indistinct, and this may not have been so.

The thoracic segments are eleven in number in *Acaste*. They are all identical in form, though the posterior ones are slightly smaller. In each there is an arched **axial ring** identical in form to the occipital ring of the cephalon and defined laterally by paired **axial furrows.** The axial ring is bounded anteriorly by a groove (the **articulating furrow**), in front of which projects a semicircular **articulating half-ring** which in life fits neatly under the axial ring in front. The paired **pleura** (singular **pleuron**) project horizontally from the axial ring, though their outer extremities are sharply turned down. Each pleuron is indented by an oblique **pleural furrow.** There is a short doublure on the outer edge of each pleuron. The anterior edge of the downturned distal part of each pleuron is a flat and truncated **articulating facet.** When the trilobite rolled up in a ball for protection the paired facets slid below the rear edge of the preceding pleura, whilst the axial region, now expanded into a hoop, was protected by the series of articulating half-rings which were then exposed. The pleura are separated by **interpleural furrows.**

The pygidium is a flat plate of fused segments resembling those of the thorax. It has a similar half-

ring at the front and is articulated with the rear thoracic segments in the same way as the thoracic segments are all linked. The pygidial axis has a series of furrows equivalent to articulating furrows, becoming more closely spaced and fainter towards the rear. The lateral parts of the pygidium (**pleural fields**) are sculpted by two kinds of indentation: one series equivalent to the edges of the thoracic segments (interpleural furrows), the other to the thoracic pleural furrows. Often the latter are more strongly pronounced in the trilobite pygidium. The pygidial doublure has about the same width as the cephalon.

Since the trilobite cuticle or 'shell material' is fairly thin, each furrow (indentation) on the dorsal surface is marked as a ridge on the ventral side and vice versa. The exception is the fine **granulation** on the outer surface of *Acaste* which is not reflected ventrally. The small granules are remains of sense organs, probably the sites of small hairs susceptible to vibrations in the water.

Acaste probably spent most of its time near the sea floor in an outstretched attitude. The reconstruction given here is based upon several assumptions; for example, only when the cephalon is in the orientation illustrated with the anterior border raised to form an **anterior arch** will the visual field of the eye be horizontal. But it seems to be a reasonable reconstruction, since the axis and the tips of the anterior pleurae are parallel; the trilobite could easily rest or crawl upon the sea floor in such an orientation. The uplifted front of the cephalon (anterior arch) below which the hypostome projected, and the **posterior arch** formed by the progressively shortened posterior thoracic segments and tail uplifted from the sea floor, would allow the free passage of water below, aerating the feathery gills.

Acaste specimens, like certain other trilobites, are often found enrolled with the cephalic and pygidial doublures in contact. The outline of the two borders is identical, and the opposing surfaces when in contact are mirror images of one another. How was such enrollment achieved? It is probable that two longitudinal sets of paired internal muscles with antagonistic action were responsible for holding the body in an extended posture or rolling it up. One set (the **flexors**) ran all the way along the body, joining up the apodemes. There was thus a continuous pair of parallel lines of longitudinal muscle. The other sets of muscles were the **extensors.** Muscle scars on the ventral surface show that they joined the underside of each articulating half-ring to the lower surface of the preceding axial ring. When the flexors contracted the extensors simultaneously relaxed, in the normal pattern of antagonistic musculature found in invertebrates and vertebrates alike. Contraction of the flexor muscles would shorten the distance between the apodemes, and since the pleural edges all formed serial parallel hinges the thorax and pygidium could only move downwards; thus the tail was swung into the position under the head, the legs presumably having been lifted out of the way first. Contraction of the extensor muscles would bring the body back into the extended position. Hence the whole body of *Acaste* is finely adapted both for perfect spheroidal enrollment and active life in a functional mobile outstretched attitude.

Detailed morphology of trilobites

Class Trilobita is characterised on the one hand by a confining evolutionary conservatism and on the other by a remarkable plasticity within the limits dictated by the defined pattern of organisation. The morphology of the earliest and latest trilobites does not depart very radically from that of *Acaste*, and much of the observed range in morphology can be related directly to the biological functions of its various components. Hence it is appropriate to consider the morphological range in functional terms before proceeding with the classification.

Cuticle (Fig. 11.4)

Arthropodan cuticles normally consist of several layers, and the cuticle of trilobites is no exception. Many arthropod cuticles are constructed of chitin (a hydrocarbon allied to cellulose) and sometimes reinforced by mineral. In trilobites, however, the cuticle consists largely of calcite arranged in microcrystalline needles orientated normal or near-normal to the outer surface and set in an organic base whose nature has yet to be determined (Dalingwater 1973b, Teigler & Towe 1975). Chitin has not yet been detected. This cuticle consists of two layers: a relatively thin outer layer with large calcite crystallites having their *c*-axes normal to the surface, and a much thicker inner or principal layer of microcrystalline structure. The inner layer is laminated, the

individual laminae being concentrated in three zones; the outer and inner zones have closely spaced laminae, whilst in the central zone they are much more widely spaced. The cuticle is well supplied with **sensillae:** small structures interpreted as of sensory function. Thus many kinds of tubular canals traverse the cuticle, leading from the inside to the surface; the majority of these are straight or helically coiled **pore canals,** often with trumpet-shaped outer ends. These are most closely packed where the cuticle is highly curved. They were probably mainly sensory, carrying small hairs (**setae**) externally, each connected to the central nervous system by a nerve running up the canal. By analogy with modern arthropods, most of these would have been sensitive to vibrations or chemical change in the water.

Sometimes the pore canals are associated with a system of parallel ridges, closely spaced and forming a regular 'fingerprint' ornament over all or part of the surface. When examined, this system resolves itself as a series of **terrace ridges** (Fig. 11.4b) giving a serially repeated dip-and-scarp topography with the sensory pore canals opening at the base of the scarps (Miller 1975). Flume experiments have shown that this could have functioned as a current-monitoring system, sensitive to change in water current direction. Such terrace ridges are normally concentrated on the doublure, but they may also occur on the upper surface of the trilobite. In certain orders (e.g. the Proetida) terrace ridges are frequently developed, in others hardly at all.

Many trilobites have cuticular **tubercles** (Fig. 11.4c–g) as their dominant surface sculpture. In thin section these appear to be of various kinds (Dalingwater 1973, Miller 1976). Some are merely **domes** with a space below; others are true tubercles enclosing an internal space connected by pore canals to the outer and inner surfaces of the cuticles. Some kinds of true tubercles seem to have been the sites of large numbers of grouped pore canals or sensory organs of other function. There are also **pseudo-tubercles** which do not have the discrete appearance of tubercles proper. They tend to be concentrated in particular areas of the exoskeleton, especially on the glabella and in areas where the doublure is likely to come in contact with the sea floor.

In the example illustrated here, *Phacops rana* (Fig. 11.4f), different kinds of structures – terrace lines, domes, tubercles and pore canal openings of different

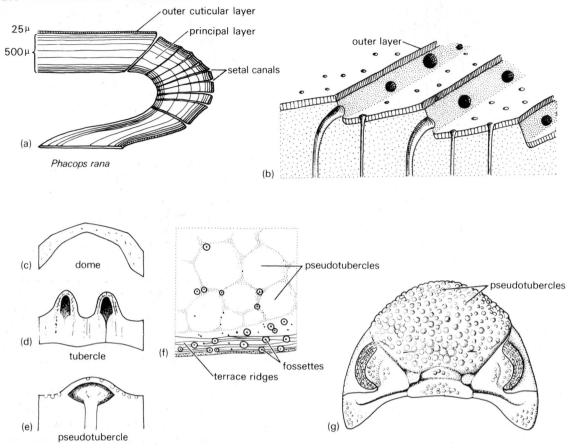

Figure 11.4   Trilobite cuticle: (a) section through cuticle of *Phacops rana* (Dev.); (b) section through terrace ridges of *P. rana* showing outer cuticular layer absent over the 'scarp slope' and two kinds of pore canals opening on 'scarp' and dip slopes respectively; (c) 'dome' in section; (d) tubercle; (e) pseudotubercle; (f) anterior border and part of cephalon of *P. rana;* (g) cuticular sculpturing of *P. rana* (x 1·5). ((c)–(f) Redrawn from Miller 1976)

sizes – have been mapped. Different parts of the cuticle were supplied with different combinations of sensillae acting as environmental sensors. Most cuticular structures seem to have been sensory, though specific functions can only be ever postulated for a given organ by analogy with sensors in modern arthropods; since exact analogies seldom occur, this procedure can be hazardous.

Some other kinds of cuticular 'ornamentation' were not apparently sensory, such as the **caecae** commonly encountered in Cambrian trilobites. The cuticle of many Cambrian trilobite genera was thin and relatively flat, and it is frequently impressed with a series of radial and ramifying ridges forming an elaborate and symmetrical pattern over the cephalon and sometimes the rest of the body. These caecae

seem to be moulded to the form of a tubular system originally lying below; either a circulatory apparatus, or the diverticula of a complex subcephalic set of digestive organs (Öpik 1960). These ridges have been referred to as **alimentary prosopon** and seem to have been connected with the oesophagus (Fig. 11.5a, b). Nutrients digested in the gut could presumably be passed directly to the other parts of the body by means of this system. But it is very rare to find alimentary prosopon in post-Cambrian trilobites, in which the shell is generally much thicker. Presumably the internal organs, of which the prosopon is the external impression, are still there but located further below the cuticle and no longer partially within it. Likewise, the transverse **ocular ridges** that are normal in Cambrian trilobites, linking the eye with the glabella

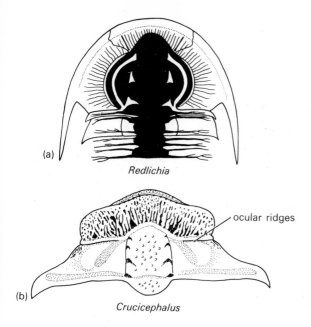

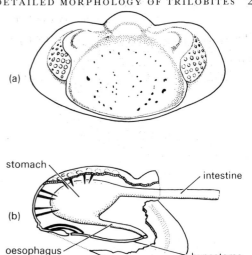

Figure 11.5 Cuticular sculpturing: (a) *Redlichia* (L. Cam.), alimentary prosopon (in black) (redrawn from Opik, 1961); (b) *Crucicephalus* (U. Cam.), showing peculiar cuticular sculpturing and ocular ridges (x 6). (Redrawn from Shergold, 1971, *Bull. Bur. Min. Res.* **112**)

Figure 11.6 (a) Frontal view of *Phacops rana* (Dev.) with cuticle removed, showing impressions, possibly muscular, on the underlying matrix (x 1·5) (redrawn from Eldredge, 1972a); (b) reconstruction of cephalon in lateral view showing possible arrangement of internal organs with stomach supported by muscles (black) and mouth opening behind hypostome (modified from Eldredge, 1972a).

(and incidentally usually carrying two prosopon caecae), are rarely present in post-Cambrian trilobites. The presence of such caecae usually suggests that an unknown trilobite is of Cambrian age, and the palaeontological patterns have been found useful in taxonomy.

Cephalon

Most trilobite cephala have the form of semicircular plates with well-defined structures, as represented in *Acaste*. There are some, however, in which the furrows limiting the different structures are all but effaced, so that the various parts are very indistinctly defined, e.g. *Trimerus* (Fig. 11.7f). Particular organs – the glabella, facial sutures, eyes, hypostome and certain specialised characters – merit further attention.

Glabella

The shape, size and structure of the glabella are widely variable. Glabellae may or may not reach the anterior border; in some cases they may be greatly swollen, have lateral lobes, be entire or indented with up to five pairs of glabellar furrows, as in the examples

shown in Figure 11.7. The stomach probably lay below the glabella and above the hypostome, and its size, and hence the size of the glabella, was probably related to the trilobite's diet. Some symmetrical indentations on the glabella of certain trilobites (e.g. *Chasmops*, where they are arranged in a V-shaped pattern, and *Phacops* (Fig. 11.6a) in which they form a pair of subcircular patches) have been said to be the scars of **suspensory ligaments** or muscles holding the stomach in place and allowing it to expand and contract. A reconstruction by Eldredge (1972a) (Fig. 11.6b) has shown how these might have been arranged.

Cephalic sutures (Fig. 11.7)

The cephalic sutures, which are unique amongst arthropods, include the facial sutures and **ventral cephalic sutures** which are sometimes present. The facial sutures are of three main kinds: **proparian** (Fig. 11.7h, j) where the posterior branch passes in front of the genal angle or spine; **opisthoparian** (Fig. 11.7b, c, d, g, k) in which it cuts the posterior border anterior to the genal angle, and **marginal** (Fig. 11.10a–c, g) where it runs along the edge and is

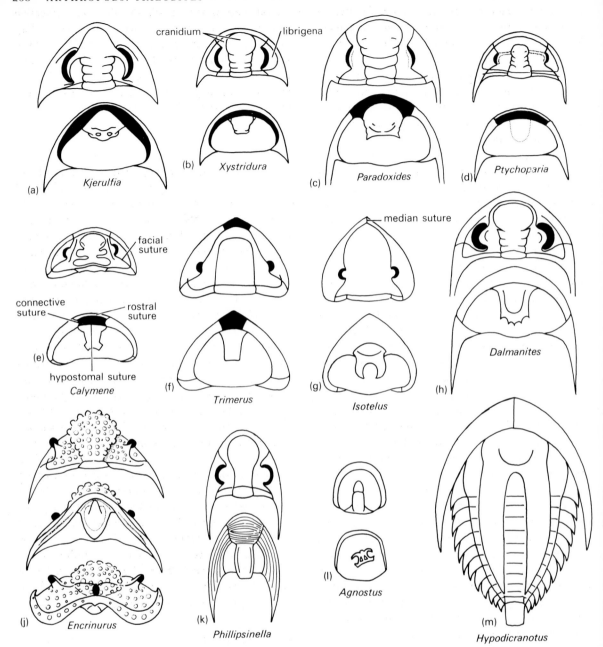

Figure 11.7 Various kinds of cephala in dorsal and ventral view, showing in morphological series an apparent reduction of the rostral plate (in black): (a) *Kjerulfia* (L. Cam.); (b) *Xystridura* (U. Cam.); (c) *Paradoxides* (M. Cam.); (d) *Ptychoparia* (U. Cam.); (e) *Calymene* (Ord.–Sil.); (f) *Trimerus* (Sil.); (g) *Isotelus* (Ord.); (h) *Dalmanites* (Sil.); (j) *Encrinurus* (Sil.); (k) *Phillipsinella* (Ord.); (l) *Agnostus* (with small winged, detached hypostome); (m) *Hypodicranotus* (Ord.). Sutures illustrated are facial, rostral, connective, hypostomal, median. Metaparian sutures in (a), proparian in (h) and (j), gonatoparian in (e) and (f), opisthoparian in (b), (c), (d), (g) and (k). (Mainly redrawn from Stubblefield, 1936; (m) redrawn from Whittington in 'Treatise' Part (0); (l) based on Robison, 1972, *Lethaia*, **5**, 239–48.)

not visible on the dorsal surface. In one family, the Calymenidae, the facial suture is **gonatoparian** (Fig. 11.7e, f) and runs directly through the genal angle.

These main sutural systems were believed to define natural groupings within the trilobites throughout much of the nineteenth century and the first part of the twentieth. They were originally used as a primary basis of classification, erecting orders: the Proparia (proparian and gonatoparian genera), the Opisthoparia (opisthoparian genera) and the Hypoparia (marginal-sutured genera). This classification eventually came under severe criticism and was finally abandoned when it was appreciated that the Proparia and Hypoparia were both composed of very heterogeneous elements which had no close natural relationships. Stubblefield (1936), for instance, has pointed out that Hypoparia could not be a natural grouping, for marginal sutures had evidently been derived independently in several groups of trilobites, the sutures becoming more marginal as the eyes were lost. There may be no primitively eyeless trilobites, and all of the blind trilobites (with the possible exception of the Cambrian agnostids) were derived from sighted ancestors.

Recent classifications have used the whole complex of trilobite characters to divide the trilobites into natural groupings, especially those of the axial region, and not merely a single character, however important it may appear to be.

On the ventral side of the cephalon there may be several other sutures, especially where the facial sutures cross over the anterior border and continue across the doublure. *Calymene* (Fig. 11.7e) is an example showing the maximum number of possible sutures. Here the facial sutures are continued and join with the lateral **connective sutures** with which they may be homologous. An elongated **rostral plate** is isolated by these and by the anterior **rostral suture** and posterior **hypostomal suture.** This rostral plate is not present in all trilobites, and its absence or reduction has given rise to much phylogenetic speculation. The morphological series here illustrated (based on Stubblefield 1936) suggests that it was originally present as an important ventral structural component in such early trilobites as the olenellids and that in many later lines of descent it was reduced or lost by different evolutionary pathways. In *Trimerus* (Fig. 11.7f) and *Encrinurus* (Fig. 11.7j) the

rostral plate is very small, and the two connective sutures are close together, whilst in *Isotelus* (Fig. 11.7g) the connective sutures are represented by a single **median suture.** In *Acaste* and its relatives there is no trace of a median suture, and the two lateral facial sutures unite around the front of the glabella so that the librigenae are fused.

Hypostome

The hypostome may be large or small, short or long. Usually it lies directly below the most convex part of the glabella. Rarely the hypostome projects backwards beyond the cephalon. Thus the Ordovician *Hypodicranotus* (Fig. 11.7m) has a very long hypostome extending almost to the pygidium and prolonged into a pair of long blades with a median space between. Hypostomes usually have a pair of lateral wings which may turn up inside the cephalon to rest close against its inner wall, though not actually in contact. Frequently the hypostomal suture is straight and could probably hinge against the doublure and be moved up and down. *Encrinurus* (Fig. 11.7j), in which the lateral wings are pronounced though slender and delicately formed, has a large three-lobed hypostome projecting forwards below a pronounced anterior arch. It is connected to the doublure by a V-shaped hypostomal suture, and is presumably immovable. Likewise in the case of *Isotelus* (Fig. 11.7g) and its relatives no movement was possible, since the anterior border of the hypostome is strongly curved and is let into the rear of the doublure. In the case of *Phacops* the hypostome is attached laterally, though not centrally to the doublure (Miller 1976). There is a narrow median gap between hypostome and doublure round which the terrace ridges run. Probably this allowed a greater flexibility than would otherwise have been possible, presumably when feeding.

Eyes (Figs 11.8, 11.9)

The eyes of trilobites are the most ancient visual system known, and indeed they are the earliest of all well developed sensory systems. Their evolution can be followed through some 350 million years of geological time (Clarkson 1975). Trilobite eyes are compound, and like the lateral eyes of modern crustaceans and insects they were composed of radially arranged visual units pointing in different

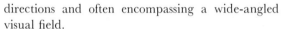

Figure 11.8   (a) Holochroal eye of *Paralejurus brongniarti* (Dev.), Bohemia, Dvorce–Prokop Limestone (x 8); (b) Schizochroal eye of *Phacops rana* (Dev.), Silica Shale, Ohio (x 18); (c) *Acernaspis (Eskaspis) sufferta* (Sil.), Pentland Hills near Edinburgh, Scotland (x 2·5).

directions and often encompassing a wide-angled visual field.

Though compound eyes are typical of arthropods, they have evolved from separate beginnings in a number of arthropod stocks; and though the eyes of trilobites are analogous to those of modern arthropods, they are not necessarily homologous. In most modern arthropods the visual units are the **ommatidia** (Fig. 11.9f), each being a cylinder of cells with the photosensitive elements (**rhabdom**) located deep within it. Each ommatidium is capped by a **corneal lens,** underlying which is a subsidiary dioptric apparatus (the **crystalline cone**). The lens and cone together focus light onto the rhabdom. The rhabdom consists of a cylinder of stacked plates, each made of parallel **microvilli.** Alternate plates have their blocks of tubules arranged at right angles to one another. These tubules are the site of the photoreceptive pigments whose chemical alteration by light triggers an electrical discharge in the **ommatidial nerves.** The nerve impulses are processed in a complex **optic ganglion** deep below the ommatidia, and some kind of integrated image is produced from the mosaic effect of light coming down individual and separate ommatidia. How 'good' the arthropod eye is in contrast with its vertebrate counterpart is very hard to assess, for the two are different kinds of eyes performing basically different functions.

The eyes of trilobites may also have had sublensar ommatidia, though this may not have been the case in all. Little evidence of internal structure survives. The lenses alone are preserved because they, like the cuticle, were constructed of calcite.

Most trilobite eyes are **holochroal** (Fig. 11.8a) having many round or polygonal lenses whose edges are all in contact and which are covered by a single **corneal membrane:** the equivalent of the outer cuticular layer with which it is laterally continuous. Holochroal eyes are the most ancient kind of eye in trilobites; they are found in the earliest Cambrian trilobites and persist until the final extinction of trilobites in the late Permian.

The eyes of Cambrian trilobites are poorly known, because in the majority the whole visual surface is encircled by an **ocular suture,** so that the visual surface dropped out after death or during moulting. This system was abandoned in most post-Cambrian groups, probably through paedomorphosis, and the visual surface was thereafter welded to the librigena: the only suture then running along the top of the eye as the **palpebral suture,** which forms part of the dorsal facial suture. The post-Cambrian radiation of holochroal trilobite eyes was substantial. The visual surface may be small or large, even hypertrophied in some groups, and encompasses a variable angular range.

Lenses in Cambrian trilobites, where found, are all thin and biconvex ((Fig. 11.9e), but in some of the Ordovician groups, especially those with thick cuticles (Fig. 11.9d), the 'lenses' have the form of long prisms with a flattish outer surface and a hemispherical inner end; such lenses have, however, a similar focal length to those that are biconvex. Each lens is a single crystal of calcite, which being highly biref-

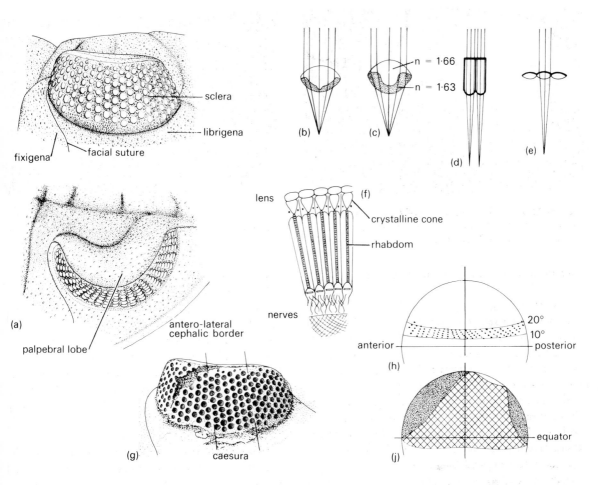

Figure 11.9 Trilobite eyes: (a) *Acaste downingiae,* lateral and dorsal views of schizochroal eye (x 7·5) (redrawn from Clarkson, 1966, *Palaeontology* **9,** 1–29). (b), (c) lenses of *Dalmanitina socialis* (Ord.) and *Crozonaspis struvei* with (shaded) intralensar bowl, conforming to ideal correcting lenses of Des Cartes and Huygens respectively. Models with a small refractive index difference between bowl and upper unit focus light sharply (based on Clarkson and Levi-Setti, 1975). (d) Passage of light rays through prismatic lenses of *Asaphus* (L. Ord.); (e) *Sphaerophthalmus* (U. Cam.) an olenid;

(f) internal organisation of a modern arthropod eye (redrawn from Snodgrass in '*Treatise*' Part (O)); (g) *Ormathops* eye (L. Ord.) preserved as an internal mould, and showing irregular disposition of identical lenses (x 7·5) (redrawn from Clarkson, 1971, *Bull. Rech. Geol. Min.,* 51–64); (h) visual field of *Acaste downingiae* eye with individual lens-axis bearings plotted on a Lambert equal area net (redrawn from Clarkson, 1966); (j) panoramic visual field of *Bojoscutellum,* only bearings of peripheral lenses plotted (redrawn from Clarkson, 1975).

ringent does not at first sight seem to be a very suitable material for a dioptric apparatus. But all the lenses are arranged with their *c*-axes normal to the visual surface, so light travelling parallel with the axis is not broken into two rays but continues unaltered to the rhabdom. Oblique rays may have been screened out by pigment cylinders below the lens, as in modern arthropod eyes, to which holochroal eyes seem to bear

a fair structural and functional resemblance.

On the other hand, **schizochroal** eyes (Figs. 11.8b, 11.9a–c, g), which are confined to Suborder Phacopina (Ord.–Dev.), have no known modern counterparts; they are a unique visual system unlike anything else in the animal kingdom. In schizochroal eyes, such as those of *Acaste,* the lenses are large and separated from each other by an interstitial material (**sclera**) of the same structure as the rest of the cuticle. Each lens has its own corneal covering, which plunges through the sclera and internally probably terminates in a sublensar conical structure, rarely preserved.

The structure of the lenses is most interesting. Whereas holochroal eyes have lenses ranging in diameter from 30 to 200 μm, though most are less than 100 μm, schizochroal lenses are usually much larger, being in the range 120–750 μm. They are usually steeply biconvex and of compound structure. Within each lens is a bowl-like unit separated from the upper part of the lens by a wavy surface which has been shown in various genera to be similar in shape to the surfaces of aplanatic correcting lenses, as designed by Huyghens (Fig. 11.7c) and Descartes (Fig. 11.7b). Experimental models (Clarkson & Levi-Setti 1975) have shown that a slight difference in refractive index between the oriented calcite of the upper unit and the bowl would operate together with the correcting surface to produce a sharp anastigmatic focus. Further correcting devices are seen in some phacopids in which the fibrous calcite needles, of which the upper part of the lens is composed, diverge fanwise towards the top of the lens, so as to be near-normal to the surface of the lens and hence to minimise the effects of birefringence. A central core found in some lenses may also have had some kind of correcting function (Campbell 1975a).

Schizochroal lenses, in spite of being made of calcite, were corrected against astigmatism and possibly against undue birefringent effects from oblique rays also. But why were they so large and of such high optical quality? Possibly their possessors were nocturnal, and the large lenses would have enabled them to see in the dim light. A more recent suggestion (Stockton & Cowen 1977) is that schizochroal eyes were capable of using adjacent lenses for stereoscopic vision through 360°, a system unique amongst arthropods.

Schizochroal eyes originated from holochroal ones, probably by paedomorphosis, for larval eyes of holochroal-eyed trilobites are in many ways like tiny schizochroal eyes. The earliest schizochroal genus *(Ormathops)* (Fig. 11.9g) had a rather haphazard and irregular system of lens packing, which arose from the geometrical contraints of packing lenses of identical size on a curving visual surface (Clarkson 1975). In later derivatives of *Ormathops* regular packing was achieved by graduating the lens size. But in spite of all

that is known about the structure of this most ancient visual complex and about the visual fields of various kinds of eyes (Fig. 11.9h, j), we are still a long way from knowing how the eyes of trilobites really worked.

Cephalic fringes (Fig. 11.10)

In Family Trinucleidae (Ord.) and Family Harpidae (Suborder Harpina) (Ord.–Dev.) the antero-lateral cephalic border is developed into an extensive pitted **fringe,** the suture has become marginal and the eyes are very reduced or absent. Two kinds of fringe were independently evolved and are dissimilar in appearance. In a few other trilobite groups the anterior border may be pitted though never in the same way nor so extensively; yet other trilobites possess expanded anterior flanges though without the pits.

Cephalic morphology and fringe of trinucleids (Fig. 11.10a–f). The cephalic features of the Ordovician Family Trinucleidae are unique. In trinucleids the glabella is usually highly convex, has shallow glabellar furrows and is bounded by broad axial furrows. Lateral to the glabella are the quadrant-shaped **genal lobes,** which may range from gently swollen to greatly inflated and are devoid of eyes and sutures. In *Tretaspis* and *Reedolithus*, there are small nodes set on the genal lobes. Each of these is a dome-shaped thin area of the test, though lens-like structures have been reported at the summit of the dome (Stormer 1930). Rarely, ocular ridges may link the eyes with the glabella. In all trinucleids the **genal spines** are very long and terminate well behind the short body, which has only six thoracic segments and a short triangular pygidium.

The fringe is a bilamellar structure in which the two lamellar counterparts are separated by a marginal suture, which becomes dorsal only where the genal spines join the cephalon. The upper lamella is thus contiguous with the genal lobes, whilst the lower has the genal spines attached. A series of funnel-shaped **fringe pits** indent both lamellae, each dorsal pit located directly above an inverted ventral counterpart, so that the whole structure is really a hollow pillar. The floor of each pit (ceiling in the case of the ventral one) is formed by a **terminal disc** or 'nozzle' with a minute **central perforation,** and at the suture which runs between them the two counterpart terminal discs are juxtaposed. The space

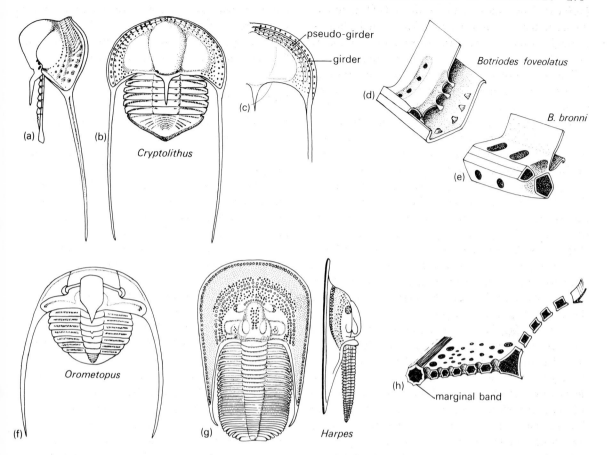

Cryptolithus

pseudo-girder

girder

Botriodes foveolatus

B. bronni

Orometopus

Harpes

marginal band

Figure 11.10 Cephalic fringes: (a), (b), (c) *Cryptolithus* (Ord.) in lateral, dorsal and ventral views (x 2) ((a), (b) redrawn from Whittington in '*Treatise*' Part (O), and (c) redrawn from Hughes, Ingham and Addison, 1975); (d), (e) reconstruction of the brim in *Botriodes foveolatus* and *B. bronni* (modified from Størmer, 1930); (f) *Orometopus* (Tremadocian) a possible trinucleid ancestor (x 4·5); (g) *Harpes* (Dev.) in dorsal and lateral view (x 1) ((f), (g) redrawn from Whittington in '*Treatise*' Part (O); (h) reconstructed section through brim and cheek showing perforations and thickened marginal band.

between the upper and lower lamellae opens to the lower side of the cephalon by a pair or series of ventral perforations. There must have been communication between the fringe cavity and the body cavity; probably part of the digestive or circulatory system was housed therein.

Some of the early trinucleids have an almost flat fringe with poorly ordered and irregular fringe pits. In later stocks, however, the fringe pits became ordered into a symmetrical pattern of ordered arcs, concentric with the antero-lateral border, intersecting with radial rows. Some small irregular 'F-pits' are found in small patches lateral to the genal lobes in the genera *Marrolithus* and *Cryptolithus*. The homologies of the arcs in various trinucleid genera, which is essential in taxonomy and identification, can be assessed with reference to a thickened ventral ridge – the **girder** (Fig. 11.10c) – which is concentric with the border and is now known to be homologous in all trinucleids. Arcs external to the girder are numbered E1, E2, ...; internal arcs are I1, I2, ... The development of **pseudogirders** concentric with the main girder in some genera has confused taxonomy in the past, but as the homologies are now clear the identification of species in this most stratigraphically important group, though time-consuming, no longer presents the problems it once held (Hughes *et al.* 1975).

How did trinucleids live, and what was the fringe for? Ventral appendages known in *Cryptolithus* have very long filamentous gills and stout walking legs. Trace fossils made by trinucleids show that these trilobites could use their legs to excavate shallow burrows, in which imprints of the fringe, genal spines and appendage scratch marks are often visible. Evidently trinucleids sat in their burrows with the head down (Campbell 1975b). Some series of superposed burrows made by the one animal whilst shifting position show that during such movement the cephalon always faced in the same general direction. It was probably rheotactically oriented to the current in the only stable position; lateral currents would have overturned it. Hence the fringe may be interpreted as a sensory organ, each pit having sensory hairs at its base which were responsive to changes in current direction and so enabled the animal to keep its head to the current. This may not, however, have been the sole function. Other suggestions made include the possibility that the fringe was some kind of filter, but this is unlikely because the **pronounced anterior arch** would let in currents below and because the tiny holes in the terminal discs (or nozzles) are not likely to have acted as filters. Our understanding of the trinucleid fringe and its functions must proceed, as in all cases of this kind, by extremely detailed morphological analysis coupled with a stratigraphical perspective of the range and development through time of the various modifications of the basic structural theme.

Harpid fringe (Fig. 11.10g, h). The fringe of trilobites of Suborder Harpina, unlike that of trinucleids, is flat and prolonged backwards into a pair of curving horns rather than genal spines. Furthermore, the distribution of pits on its surface is highly irregular, and the pits themselves are rather small and found on the genal regions as well as on the fringe. Some genera have a small triangular anterior arch, but generally the fringe is flat all the way round. The fringe is bilaminar with two closely juxtaposed laminae, and the pits, which in section have the form of opposed vertical funnels, perforate right through it. There is a pronounced marginal band, supplied with many sensillae, round the external periphery of the fringe. *Harpes* and its allies could enroll, having a

pronounced flexure of the thorax in the region of the first few thoracic segments (**discoidal enrollment**). It may have been a sedentary animal, using the brim to spread the weight of the body like a snowshoe. The pits undoubtedly lightened the body, but such an elaborate structure may have had more than one function.

Enrollment and coaptative structures

The enrollment system exhibited by *Acaste* is very common in post-Cambrian trilobites and is known as **spheroidal enrollment.** More rarely, trilobites may roll up by tucking the pygidium and last few thoracic segments under the cephalon, this being **double enrollment;** and in harpids and trinucleids there is **discoidal enrollment** in which only the first few thoracic segments bend, the rest being held as a flat laminar plate.

Cambrian trilobites, with the exception of the isopygous agnostids, could not enroll; they were prevented from doing more than curl up in a half-sphere because the distal free edges of the pleurae then came in contact. Many Ordovician and later groups, however, developed an elaborate enrollment system, articulating facets being particularly well developed in Order Phacopida to which *Acaste* belongs. In addition, within this order there are many different kinds of 'tooth and socket' structures, seemingly of quite independent origin, which are found on the cephalic and pygidial doublure and help to 'lock' the rolled-up sphere together. These are known as **coaptative structures** (Clarkson & Henry 1973), and though particularly well developed in the Phacopina they are present also in trinucleids and asaphids.

Amongst phacopids *Acaste* has no real tooth-and-socket structures, though the cephalon and pygidium are the same shape and the opposing doublures fit together neatly. In the related Upper Ordovician *Kloucekia*, a single median projection in the cephalic doublure interlocks with a corresponding excavation in the pygidial doublure. *Morgatia* (Fig. 11.11g) has a series of lateral sockets for the reception of the pleural tips of the thoracic segments and the first pygidial segment. In *Phacops* a deep marginal groove (the **vincular furrow**) is often present on the cephalic

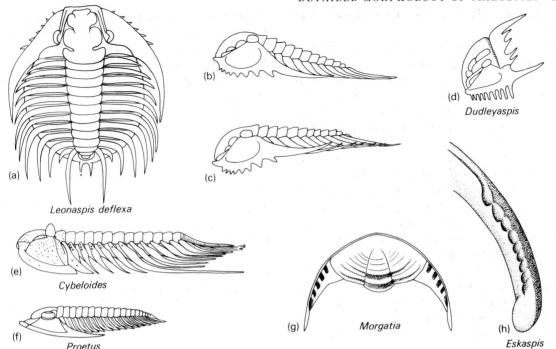

(a) Leonaspis deflexa

(b)

(c)

(d) Dudleyaspis

(e) Cybeloides

(f) Proetus

(g) Morgatia

(h) Eskaspis

Figure 11.11 (a) *Leonaspsis deflexa* (Sil.), dorsal view (x 2·5): (b), (c) the same in two alternative life attitudes for resting and swimming or browsing; (d) *Dudleyaspis* (Sil.), cephalon in side view, attitude equivalent to (c). ((a)–(d) redrawn from Clarkson, 1969) (e) *Cybeloides* (Ord.) in life attitude, lateral view, showing the body supported on long macropleural spines on the sixth thoracic segment; (f) *Proteus* in life attitude, lateral view, showing high carriage of the body; (g) *Morgatia* cephalon from below showing press-studs (vincular notches) for the reception of thoracic pleurae and with the pygidium in place (x 4) (after Henry and Nion, 1970, *Lethaia* **3,** 214–24); (h) *Acernaspis (Eskaspis)* cephalic doublure with vincular furrow and vincular notches (x 7·5) (after Clarkson, Eldredge and Henry, 1977, *Palaeontology* **20,** 119–42).

doublure. The edge of the pygidium fits into this, and the thoracic segments come to rest in extensions of the vincular furrow divided into individual lateral pockets. Such lateral pockets may still be present even when the median part of the vincular furrow is absent, as in *Eskaspis* (Fig. 11.11h). Such morphological differences as these are of some taxonomic value. Other examples of quite different mechanisms are shown by *Encrinurus*, and in the cheirurid *Placoparia* another stratigraphically documented series shows a progressive increase in the number of depressions on the dorsal surface of the cephalic anterior border. When enrolled, the pygidial spines rested in these like clutching fingers (Henry & Clarkson 1975).

Some other trilobite groups (e.g. the Proetida and Agnostida) appear to have been well adapted for enrollment, but only the Phacopida reached the summit of enrollment ability.

The trilobite thorax

The axis of the thorax in most trilobites is about as broad as the occipital ring. These axial rings with their half-rings and apodemes are of fairly standard construction throughout the trilobites, though there are differences in the length of the apodemes, the half-rings are sometimes absent, and surface sculpture and spinosity are variable. Certain phacopid genera from the Devonian of South America – a region long isolated so that a rich endemic fauna of trilobites developed from *Acaste*-like ancestors – repeatedly developed dorsal spines of a particular erect construction, sometimes singly, sometimes one for each segment of the thorax and on the occipital ring and pygidium as well.

The pleura tend to be rather flat in some Cambrian genera, though not in all, but they are often arched or bent down distally in the later groups and are normally marked with a pleural furrow or less often with a ridge. Usually the proximal parts of the pleura have parallel edges against which the edges of adjacent pleura can hinge; the distal parts do not touch and are free. The junction between the free and hinged parts is the **fulcrum,** at which the pleura are often sharply downturned. Cambrian genera have a simple articulating hinge structure in which contiguous pleural edges alone are used for articulation. Specialised hinge structures and articulating pleural facets did not appear until the Upper Cambrian; many of these seem to be associated with enrollment ability. The pleural doublure may be wide or narrow and is often ornamented with terrace ridges. Also to be found in post-Cambrian stocks, notably the Family Asaphidae, are small protuberances on the inner faces of the thoracic doublures, which presumably acted as stops preventing overgliding of the pleura as they came to rest during enrollment. These are the **organs of Pander;** they are not to be confused with **panderian openings** which, though similarly located, probably were the orifices of some kind of segmented and possibly excretory organs.

The most highly modified pleura of all are to be found in the Family Cheiruridae (Ord.–Dev.) and Order Odontopleurida (Ord.–Dev.) (Fig. 11.11a). In the former the distal ends may be curiously pointed, but the contiguous parallel edges are equipped with interlocking ledges which presumably facilitated highly efficient hinge articulation. The Odontopleurida are a very spiny group of trilobites, and each pleuron terminates in two spines. The anterior pleural spine is normally vertical; the posterior one is horizontal and usually short. But in modified odontopleurids (e.g. *Ceratocephala*) the vertical spines are very long and, being equipped with many secondary spines, virtually close off a sub-thoracic 'box' from the external environment: a unique feature whose function is quite unknown.

The trilobite pygidium

The pygidium is a fused plate which in the Lower Cambrian olenellids may consist of just one segment, but in later genera it consists of as many as thirty segments. It articulates with the last thoracic segment by means of an articulating half-ring, and if the thoracic segments have pleural facets so does the pygidium. With the exception of the Agnostida most Cambrian trilobites have small (**micropygous**) pygidia, whilst post-Cambrian genera tend to have **heteropygous** (smaller than the cephalon) or **isopygous** (equal-sized) pygidia. Rarely pygidia can be **macropygous** (larger than the cephalon), as in the Lichida. In *Encrinurus* the number of axial segments is supernumerary, probably through the division of existing axial rings.

In all pygidia the pleural furrows are homologous with the axial furrows, whilst the interpleural furrows are equivalent to the edges of the pleura; the ontogenetic reasons for this are discussed later.

Appendages of trilobites

The limbs are known only in a few species of trilobites as these delicate structures did not preserve well. Appendages were first described in 1876 and have been reported, often in a very fragmentary state, in about twenty species. They are well known and have been fully described in *Olenoides* (Whittington 1975) and *Kootenia* from the Middle Cambrian Burgess Shale of British Colombia, also in the pyritised Ordovician *Cryptolithus* by Raymond (1920) and from enrolled specimens of the Ordovician *Ceraurus* by Størmer (1939). The pyritised Lower Ordovician *Triarthrus* and Devonian *Phacops* have yielded much detail through dissection and X-radiography. In all these the structure of trilobite appendages shows remarkable constancy in that there is always first a pair of uniramous antennae located on either side of the hypostome, followed by paired biramous (two-branched) appendages, under the cephalon and one pair for each segment all the way down the body and in the pygidium also. These biramous appendages are almost identical to each other in all but size. The constancy of morphology here contrasts with that of the various crustacean groups, where there is wide variation even within the one genus. Yet within the trilobites whose appendages are known there is indeed some diversity, as will be shown.

About fifteen specimens of *Olenoides serratus* (Figs 11.12, 11.13a–c) with limbs are known from the Burgess Shale, and the appendages are preserved as a fine dark film, usually in place but sometimes detached. They are flattened and have a reflective surface so that many details can be captured on film, given light of suitable inclination. The antennae are

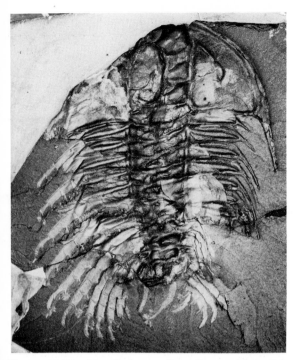

Figure 11.12 *Olenoides serratus* from Burgess Shale (M. Cam.), British Columbia U.S. National Museum no. 58589 (x 1·7). (Photograph by courtesy of Prof. H. B. Whittington)

composed of many jointed rings, as are the **cerci,** which are posterior equivalents of the antennae. Like the antennae they almost certainly had a sensory function; they are provided with minute setal hairs. Behind the antennae there are three pairs of biramous appendages under the cephalon, seven on the thorax and between four and six on the pygidium. Each biramous appendage was apparently attached to an apodeme, and even where the appendages are not preserved the number of cephalic apodemes will indicate the relative number of cephalic appendages. The basal joint of the appendage (**coxa**) gives rise to two branches: one the **walking leg** and the other the **gill branch.** Such terminology presupposes definite functions, and probably the gill was used for functions other than purely for respiration. But other terminology presupposes homologies, and though the term **telopodite** is often used for the walking leg, the correct terminology for the gill is not agreed; for the moment the term 'gill branch' is retained, though **outer ramus** is a good alternative.

The coxa is about half the length of the axial ring with which it is associated, and its inner margin is supplied with long sharp spines, straight and curved, between which are numerous shorter spines. A pre-coxa or initial segment other than the coxa has been described, but recent work does not seem to substantiate its presence. The walking leg is continued in the direct line from the coxa and consists of six jointed segments (**podomeres**). The proximal podomeres are very spiny on their inner surfaces, the more distal ones less so but with fine setae, and the last podomere has three terminal spines. From the coxa there also projects the gill branch, lying above the leg branch and directed posteriorly. The individual gill branches overlap markedly. Each has two lobes: a long **proximal lobe** (three times as long as it is broad) and a shorter ovoid **distal lobe.** The former has a posterior fringe of long parallel filaments diminishing in size towards the distal lobe, whilst the latter has no filaments and only a fringe of setae. An anterior marginal rim originally described on the gill branch is now shown to have been a preservational artifact.

Precisely how the coxa fitted onto the apodeme is uncertain; presumably the whole appendage could rotate in a horizontal plane, but it is not known whether it was capable of any dorso-ventral or transverse movement.

Before considering how the appendages functioned it seems appropriate to consider the range in form of known appendages. *Kootenia* appendages are fairly similar to those of *Olenoides* but are known only from one specimen. *Triarthrus* is a Lower Ordovician olenid. Where found in the 'Utica Slate' (actually a black shale) of New York State the appendages are pyritised and extend well beyond the body (Fig. 11.13d, e). The antennae converge in front of the hypostome and diverge again away from the cephalon. The three cephalic biramous appendages are small, lying within the cephalic border; the thoracic ones, especially the anterior, project far outside. There is a large coxa with internally directed spines, and the walking leg, though less spiny, otherwise resembles that of *Olenoides*. The gill branch is different for it has no real proximal lobe; the filaments are joined directly to a flexible segmented anterior rod, and the distal lobe is reduced and tiny. Beecher's work and that of his student Raymond (1920) were done by careful cleaning of specimens. More recently Cisné (1975), using X-radiography, has shown, in addition to details of limb morphology,

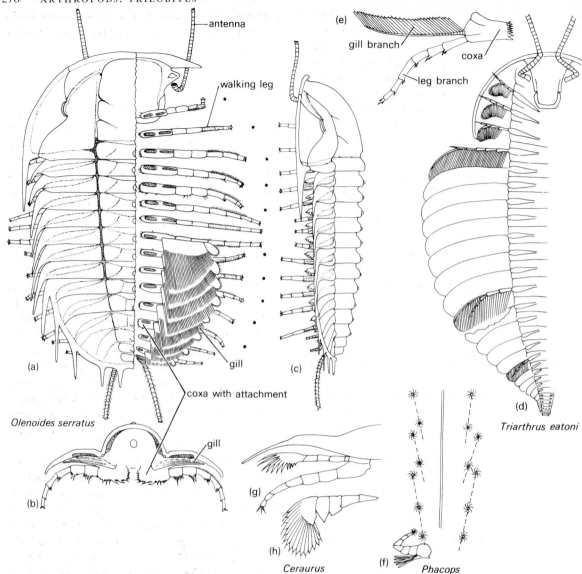

Figure 11.13 Trilobite appendages. *Olenoides serratus* (M. Cam.), showing reconstructed appendages: (a) dorsal view with part of the exoskeleton removed showing the appendages in place and the gill branch lying above the leg; (b) section; (c) lateral view (redrawn from Whittington, 1975). *Triarthrus eatoni* (L. Ord.): (d) reconstruction of the appendages in ventral view exhibiting the small post-pygidium and diminutive cephalic appendages; (e) a detached appendage (d–e redrawn from Cisne, 1975). (f) *Phacops* (Dev.) leg with a terminal 'brush' believed to have made stellate markings on either side of a central groove (based on Seilacher, 1962). *Ceraurus* (Ord.): (g) limb in frontal view; (h) gill branch in dorsal view (based on Størmer in 'Treatise' Part (O)).

a tiny segmented **post-pygidium** projecting from the rear of the pygidium proper, with a terminal anus, an apparent internal system of guts and diverticulae confirming what is known about the digestive system of other trilobites and, most extraordinary of all, an internal endoskeleton: a series of longitudinal rods and cross-bars to which box truss muscles were attached. Cisné has reported that this internal muscular system approximates that of some living primitive crustaceans (e.g. *Hutchinsoniella*).

The use of soft X-rays has also been useful in interpreting the appendage structure of Devonian

Phacops and *Asteropyge* specimens from the North German Hünsrückschiefer (Stuermer & Bergstrom 1973). These specimens of *Phacops* have stout walking legs with the usual six podomeres. There are three pairs of biramous appendages below the cephalon, the last two having strongly developed coxae prolonged internally and armed with spines. These could have functioned as gnathobases equivalent to those in some modern arthropods, and by rolling movements in a horizontal plane they may have been able to work as crushing jaws. There is one pair of appendages for each segment of the thorax and there are several small sets in the pygidium also. The main difference between the appendages of *Phacops* and those of other trilobites is that the gill branch filaments emerge directly from the coxa and are transverse and subparallel with each other. The whole array of gill branch filaments forms a kind of slatted curtain lying dorsal to the walking legs, which may have had both a respiratory and a filtering function.

The Ordovician cheirurid *Ceraurus* (Fig. 11.13g, h) has peculiar appendages, as shown by Størmer (1939) who made serial sections through enrolled specimens collected from a limestone of Trenton (U. Ord.) age. The branch structures were worked out from a wax model built up layer by layer from these sections, as has been done with the graptolites (q.v.). The walking leg is relatively unmodified, but the gill branch has five segments increasing in size distally, the last of which forms a flat paddle-like lobe fringed with filaments. These filaments are of normal construction, i.e. elongated thin plates arranged subparallel with one another and having each a terminal bristle.

The diversity of form in the gill branch implies some differences in function. It is generally accepted that the gill branch did in fact function for respiratory exchange, though Bergstrom (1973) has pointed out that since they are relatively hard structures they could have functioned merely as aerators, to agitate the water whilst the animal was moving and so to feed more oxygen to the gills, which he imagined as soft structures attached to the ventral membrane. Probably the paddle-like gill branches of *Ceraurus* and *Olenoides* could have facilitated swimming, but there is no direct evidence as to how well trilobites could swim, and the question remains open. There is, however, good evidence as to how the walking legs

operated, and the detailed study of these has yielded a surprising amount of information about trilobite behaviour.

Trilobite tracks and trails

Early in the history of palaeontology elongated trail-like markings were described from many different sedimentary rocks of Palaeozoic age. The commonest kind, first called *Cruziana* (Fig. 11.14a) and later referred to by several authors as 'bilobites', are ribbon-like markings with paired oblique chevron structures. They were first believed to be of algal origin. But Nathorst (1881) called attention to the fact that 'bilobites' always occurred at the interfaces of sandy and shaly layers in the rock. This, he presumed, was more likely to have been the result of a moving animal crawling over or rather ploughing through a firm muddy sea floor, leaving marks which had been filled with sand soon after, the difference in sediment type being responsible for their preservation. Not infrequently *Cruziana* markings are cut across by other and presumably later trails made by the same kind of animals, confirming the mobile animal hypothesis. Other kinds of markings existing on shale/sandstone interfaces could also have been made by moving animals, and the most likely contemporaries were trilobites.

There is little direct evidence that these and other kinds of markings were actually made by trilobites since trace fossils of any kind are rarely found with body fossils. Calcified shells are not normally retained by the permeable sandstones in which the trace fossils occur. But a few specimens of intact trilobites have been found in the Ordovician of Cincinnati (Osgood 1970), where *Calymene* specimens with intact cuticles have been found in their resting excavations (called *Rusophycus*). Other criteria for attributing particular trace fossils to trilobite movements are indirect, though compelling. Thus Cambrian and Ordovician *Cruziana* are numerous and diverse, whilst the generally poorer Siluro-Devonian *Cruziana* fauna reflects the decline of the trilobites. And though trilobites are not found in *Cruziana*-bearing sandstones, they may occur in intercalated beds of the same sedimentary sequence, as in the Ordovician of Poland and the Devonian of Germany. Most *Cruziana* and other similar markings were clearly made by 'legs' of uniform and undifferentiated structure, such

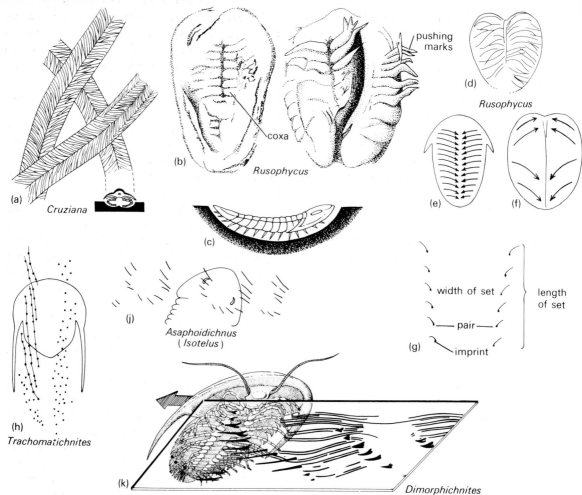

Figure 11.14 Trilobites as trailmakers. (a) intersecting *Cruziana* trails showing how a 'ploughing' trilobite could have made them by pushing backwards with its legs; the lateral striations on one trail could result from the gill branches moving across the surfaces. (b) *Rusophycus* from the Cambrian of Poland, showing resting marks of coxa, and pushing marks made by appendages as the trilobite pulled itself sideways out of its burrow (redrawn from Orlowski, Radwanski and Roniewicz, 1970). (c) Trilobite in a shallow burrow. (d) Leg marks preserved in a *Rusophycus*. (e) Directions of movement of the legs. (f) A burrow made by anterior and posterior legs moving in different directions. ((c)–(f) Redrawn from Seilacher, 1972) (g) Terminology of a set of walking marks. (h) *Trachomatichnites*, a trilobite (trinicleid?) walking trail in which individual sets can be distinguished. (j) *Asaphoidichnus* sideways movement marks made by a large (asaphid?) trilobite, possibly *Isotelus*. ((g)–(j) redrawn from Osgood, 1970) (k) Reconstructed ventral view showing how *Dimorphichnites* was made by a sideways-crawling trilobite (redrawn from Seilacher, 1955).

as are possessed by trilobites, and sometimes the 'prod marks' made by walking legs can be matched with the actual known structure of the distal part of the leg. Furthermore, the common bilobed and normally ovoid trace fossils known as *Rusophycus* are just the right size to have been made by trilobites, and sometimes indentations caused by the impression of a cephalon, genal spines or lateral parts of the body resting in the sediment are preserved in association with the *Rusophycus* marks (Fig. 11.14b).

Walking movement in arthropods

In all crawling arthropods the movement of the limbs relative to one another takes place in the same general way. Leg movements on each side of the body are synchronised in regular waves of forward movement, known as **metachronal rhythm.** Each leg comes forwards in turn, is placed on the ground and pushes backwards, helping to support and move the body in the process. At any one time a leg will be slightly behind the one in front of it in its forward movement, so that in a long-bodied arthropod (e.g. a centipede) successive waves of motion may be seen travelling up the body from the tail to the head. As the waves of movement sweep forwards, every tenth leg or so is in approximately the same position and at the same angle. Even in short-bodied arthropods (e.g. wood-lice) the pattern of movement is clear, and the waves of movement always travel forwards along the body. This was evidently the case in trilobites, and the walking marks especially may be interpreted on this basis.

Different kinds of trilobite trails

The various sorts of trace fossils believed to have been made by trilobites are classified according to the binomial system of nomenclature. The retention of Linnaean practice for such **form genera** and **form species** is unavoidable, since it is very rare that a particular trace fossil can be directly related to a known producer, and even when it can both the trail and the body fossil retain their own names. Of the various form genera that have been attributed to the life activities of trilobites some of the most important are:

(a) *Protichnites, Trachomatichnites:* walking or striding trails with individual leg impressions (Fig. 11.14h);
(b) *Diplichnites, Petalichnites, Asaphoidichnus:* as above but usually oblique (Fig. 11.14j);
(c) *Cruziana:* bilobed chevron-marked trails which are the traces of crawling, ploughing, shovelling or burrowing movements (Fig. 11.14a);
(d) *Rusophycus:* bilobed ovoid traces which are probably resting nests, burrows or surface excavations (Fig. 11.14b, c–f).

These movement trails would be generally classified as Repichnia (crawling traces) and *Rusophycus* as Cubichnia (resting traces), according to the taxonomic scheme of Seilacher. A single trilobite could well have made different kinds of marking, depending on exactly what it was doing.

Though trilobites often walked straight ahead, they frequently progressed obliquely or even sideways like a crab. In the case of forward movement, as represented by the tracks of *Protichnites* and *Trachomatichnites*, the pygidial appendages were the first to touch the ground, followed by those of the last thoracic segment, and so on until the wave of movement reached the head; meanwhile new waves of movement were progressing forwards from the hinder end. The result of such successive waves of movement is a series of sets of paired markings impressed like open-ended (truncated) V-shapes (Fig. 11.14g) (Seilacher 1955, 1964, Osgood 1970). Since the trilobite widens anteriorly from the pygidium the V is directed forwards (though some xiphosurids moving forwards had a series of reversed Vs). Where the sides of the V do not curve in again at the front it seems that the head appendages cannot have been used in walking.

Trilobites moving directly forwards made a track consisting of a series of superimposed sets of V-markings, so that it is not easy to sort out the number of imprints per set, though a good example is shown in Figure 11.14h. When trilobites walked with their movement oblique to the body axis (as in *Diplichnites*), the imprints are closely crowded and interfering on one side but quite separate on the other, and so they can more easily be counted; there is then some chance that the trail maker can be identified, as there should be some kind of parity between imprint number and the number of walking legs. *Diplichnites* might have been made by a trilobite walking obliquely as its normal mode of progression, or it could have resulted from the animal trying to keep steady on a slippery mud surface whilst a lateral current was flowing.

In these various walking trails the marks of the walking leg often retain the detailed imprint of the terminal digit which made them. In the Devonian Hünsrückschiefer of Germany many specimens of *Phacops* are found, some with their appendages intact (Fig. 11.13f) (Seilacher 1962). These terminate in a distal tuft of bristles, encircling the terminal claw, which if extended like an umbrella would fit ideally with the stellate impressions arranged in paired series on either side of a median groove, such as have been found at the silt/sandstone interfaces of beds in the same sequence. This is one of the few cases of unequivocal matching. In the Ordovician of Cincinnati the large asaphid *Isotelus* is present in a sequence that also contains the ichnofossil *Asaphoidichnus*, which consists of paired sets of trifid imprints (Fig. 11.13j). This trace fossil is large enough to have been made by *Isotelus*, and indeed it may give a clue as to the structure of the appendages of this trilobite, which are otherwise unknown.

Evidence of strongly oblique movement is known only in rare cases. The well-known *Dimorphichnites* (Fig. 11.14k) from the Cambrian of the Salt Range of Pakistan is one of these. In each 'unit' of *Dimorphichnites* there are two elements: on one side a set of round and deeply impressed imprints, on the other a set of lightly impressed and slightly sigmoidal raking marks. Seilacher's (1955) interpretation of these is that a trilobite (probably an olenellid) moving crabwise from left to right would press its right-hand legs into the substrate and, using these for support, would heave its body sideways by contracting its leg muscles. Then it would drag its left hand legs across the mud, only using them for minimal support, and by withdrawing its right hand legs one after the other and replacing them some distance to the right would then be in a position to repeat the procedure. A number of complete movements of this kind can be clearly seen on the lower surface of a sandstone unit. In one case a Lower Cambrian olenellid, moving from the left by such a mode of progression, encountered the track of another olenellid, turned round again and moved back again parallel with its own set of tracks. Then it jumped over its own tracks and moved away at right angles to its first track by slightly oblique forward movements. Seilacher interpreted these as grazing tracks since the trilobite obviously avoided previously made tracks. Similar tracks have been found in the Ordovician, apparently made by trinucleids.

Cruziana is a very widely distributed trace fossil, and up to thirty different form species have been described, though these can often be attributed to somewhat different burrowing techniques by the same animal or to the inclination of the body of the animal on different gradients. But the basic *Cruziana* is simple: an elongated bilobed trail, each lobe being convex downwards and usually striated with oblique, closely spaced ridges and grooves forming a herringbone pattern (Fig. 11.14a). The trilobite must have ploughed its way along the upper surface of the sediment incising these markings as it went; in the best-preserved specimens there was probably a thin layer of sand already overlying the plastic mud that took the impressions, protecting it from current erosion. More sand later filled the markings. The two lobes, with their oblique structures, were probably made by the walking legs moving inwards towards the median line and backwards. The oblique scratches seem to have been made by divided terminal claws, though as they are generally superimposed one upon another, the marks of individual legs are often difficult to distinguish.

Variants of the basic structure are often found. Sometimes lateral furrowed borders are present, probably the scratch marks of gills or pleurae. Certain *Cruziana* species have an extra pair of lobes within the outer furrows, perhaps the brush marks of the gill branches as the trilobite ploughed forwards. *Cruziana* has also been found with apparent pygidial scratch marks as well. One species shows a herringbone pattern clearly subdivided into sets with from three to nine parallel scratches; each successive set overlaps and truncates the preceding one in a manner suggestive of alternate stronger and weaker pushes.

Rusophycus (Fig. 11.14b–f), interpreted as trilobite burrows or temporary resting traces, is the name given to short ovoid markings which are normally bilobed and deepest in the centre. These may be smooth (coffee bean marks) or have herringbone ridges as in *Cruziana*. They usually occur singly, implying that a swimming trilobite had landed, excavated a temporary 'nest' and then swum away. They may also terminate *Cruziana* trails, as if the trilobite had crawled, rested and then swum off, and they have been found in association with striding trails. The smooth *Rusophycus* was probably formed beneath the body of a resting trilobite rheotactically directed upstream and hollowed out and smoothed by current scour. Some deeply excavated hollows are best interpreted as nests in which the trilobites lay for quite long periods, and it is these that have been found in the Ordovician of Cincinnati still with trilobites in them, though only a very few examples have actually been discovered. Some interesting examples described by Seilacher were clearly produced by the anterior appendages shovelling forwards and the posterior ones working backwards; hence *Rusophycus* has a bidirectional structure. Other examples were apparently excavated by using the anterior border itself as a shovel and scrabbling with the posterior appendages. Such modifications of normal practice allowed the formation of deeper burrows than the normal shallow resting traces. The trilobite could easily get out of its nest by pushing with the walking legs, but one intriguing example from Poland (Orlowski *et al.* 1970) (Fig. 11.14b) shows imprints of the legs outside the burrow, though only on one side as evidently the animal had levered itself out sideways before swimming away. This and other examples show the grooves where the coxae had rested.

Life attitudes and habits of trilobites

Acaste (Fig. 11.2d) shows a standard pattern of construction in side view. This attitude is very common, and even in such a spiny trilobite as *Cybeloides* (Fig. 11.11e) the spatial and angular relationships of the various parts of the body are similar. *Cybeloides*, however, has both genal spines and macropleural spines on the sixth thoracic segment. These would hold the body in an outstretched attitude on the sea floor with the posterior thoracic segments and the pygidium rising away from it, though some relaxation of the muscles would lower the 'tailgate'. Besides being used as props the spines could have been used in burying the trilobite, by a series of alternate flexion and relaxation movements.

Spinosity in trilobites was once regarded as an adaptation for a planktonic mode of life, the spines inhibiting the animal's sinking. But most trilobites are far too large for the spines to have been of any real value in this context, and it is more likely that they were, as well as being protective, used for spreading the weight of the trilobite when resting on the sea bottom. Thus long genal spines, as in *Proetus* (Fig. 11.11f), bore the weight of the whole exoskeleton, being the most ventral part of the animal; the thorax and pygidium were carried slightly higher than these and not in contact with the sea floor.

Perhaps the most striking case where spinosity and life attitudes are correlated is in the Odontopleurida (Fig. 11.15). Odontopleurids are always spiny and have not only lateral but also ventrally directed spines. Such genera as *Acidaspis* and *Dudleyaspis* (Fig. 11.11d) have a fringe of modified **denticles** extending along the antero-lateral border on either side of the glabella. Since they terminate in a flat plane they seem to have been adapted for supporting the cephalon in a particular attitude, presumably whilst

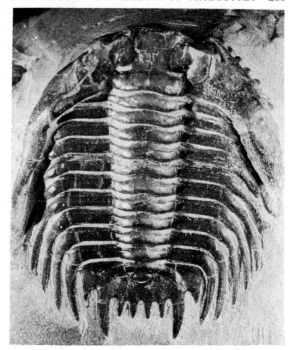

Figure 11.15 *Leonaspis coronata* (Sil.), Wenlock Limestone, Dudley, England. Complete specimen in dorsal view, cephalon slightly crushed (x 4).

feeding, for the short hypostome would then be in close proximity to the sea floor and so would the mouth which lay just behind it. There is no anterior arch in these modified genera or in *Ceratocephala*, which could support itself in a similar attitude though it does not have the denticles (Whittington 1956).

The root stock genera of the odontopleurids, *Leonaspis* (Fig. 11.15a–c) and *Primaspis*, have the anterior denticles lying in a curve rather than in a plane, and it has been postulated (Clarkson 1969) that these trilobites could take up two alternative life attitudes, both equally functional but for different purposes. In one the trilobite lies on the sea floor or crawls over it with an open anterior arch, with the cephalon supported by the genal spines and posterior denticles and with the thorax and pygidium, which slope posteriorly, resting on their spines (Fig. 11.11b). In the other attitude the cephalon is tilted forwards like that of *Dudleyaspis*, resting on the anterior denticles, whilst the body is held out horizontally (Fig. 11.15d). The first attitude seems to have been a resting position, the second an active or 'browsing' stance.

Dudleyaspis was a presumed derivative of the *Leonaspis* stock permanently specialised for life in the active attitude, for the posterior thoracic spines and the pygidial spines are bent down so as to terminate in the same plane as the anterior denticles. In *Ceratocephala* it is the anterior thoracic spines that are modified; they are large and thick and terminate in the same plane as the anterior border of the cephalon, showing how specialisation for the active attitude has been independently evolved.

The relatively unspecialised *Leonaspis* stock, with its dual mode of life, was very versatile, and the stratigraphical range of this one genus has been estimated as 170 million years. But its derivatives (*Dudleyaspis*, *Acidaspis* and others) were of relatively short duration, illustrating the general rule that the highly specialised taxa did not usually last very long.

Ecdysis and ontogeny of trilobites (Fig. 11.16)

Like all other arthropods, trilobites had to moult periodically in order to grow. The cast shells (**exuviae**) were thrown off as the animal ecdysed, being commonly disarticulated and broken along the cephalic sutures during the moulting process. Meanwhile the soft shell which had formed below the old exoskeleton was inflated to a larger size and hardened. Occasionally the moulted exuviae remain intact or only slightly displaced into a particular and fairly constant relationship (Henningsmoen 1975). Sometimes, especially in very fine argillaceous sediments, a gradational size series can be collected which gives a partial or complete record of ontogenetic development from the earliest larval stages (Whittington 1957). The first ontogenetic series known were described by Barrande in 1852; one of these, the Cambrian *Sao hirsuta*, is illustrated here and shows the general system of growth.

The earliest stage is a **protaspis**: a small disc some 0·75 mm in diameter. This is simply a cambered disc, open ventrally, which carries a segmented central lobe, later to become the glabella. The eyes at this stage are tiny and are located on the anterior margin; later they migrate inwards bringing the facial suture with them. The hypostome is not known in *Sao*, but in such genera as *Asaphus* and *Gravicalymene* (Fig. 11.16d), in which it has been described, it is extremely spiny, suggesting a functional change as the hypostomal spinosity diminishes. Many protaspides are

generally spiny, and the spines may later disappear totally in the adult. Olenellid trilobites, for instance (Palmer 1957), have protaspides with a pair of long blade-like spines, whose adult equivalent is merely a pair of tiny knobs lying within the genal spines, which are quite separate organs developing later.

As the protaspides grow by further moults a transverse furrow develops, separating the larval cephalon from the presumptive pygidium; the two can freely articulate against one another. The **meraspid** stage begins when the pygidium becomes free. The thoracic segments then form in a zone of growth along the front of the pygidium. They are actually part of the pygidium for a while, and then they are released in turn and liberated from its anterior part. The pygidium is thus known as a **transitory pygidium** as it does not acquire its permanent identity until the last thoracic segment has been released. Meraspides are numbered in degrees – 0, 1, 2, 3, . . . – according to how many thoracic segments have been freed from the anterior edge of the transitory pygidium. The process is perhaps best seen in the development of *Shumardia pusilla* (Fig. 11.16a) from the Tremadocian of Salop, England, since the macropleural spines on the fourth thoracic segment are a good marker. These are liberated at meraspid degree four whilst the fifth and sixth segments are still being formed at the front of the transitory pygidium. Meanwhile the cephalon is acquiring adult proportions.

When the adult number of thoracic segments has been reached the trilobite is now a **holapsis**, though it may have to pass through many more moults before it is of fully adult proportions. Rarely, as in *Aulacopleura* which had up to twenty-five thoracic segments, new segments are added until a very late stage in development.

Quite apart from their intrinsic interest, ontogenies are potentially useful in suggesting phylogenetic relationships between established taxa. If two distinct stocks have fairly similar protaspides, differing in characters from those of other stocks, they may share a common ancestry even if the adults look quite different. Thus when the ontogeny of corynexochid trilobite *Bathyuriscus fimbriatus* was elucidated (Robison 1967), it became clear that there were similarities between its protaspides and the protaspides of many ptychopariids. Furthermore, adult corynexochids show many characters that are found

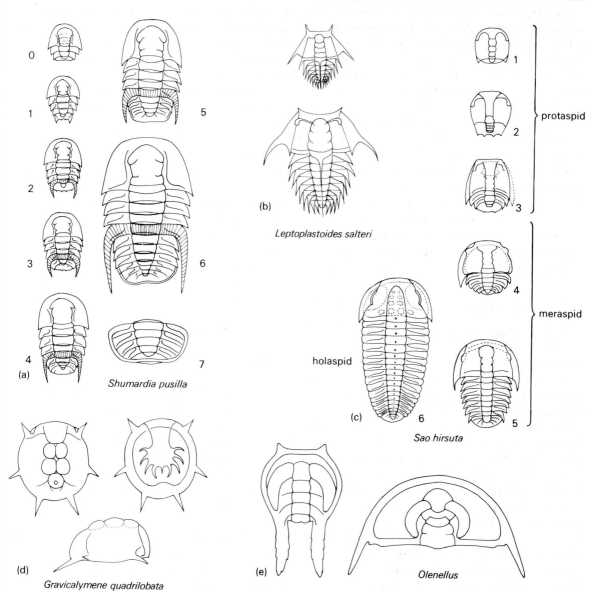

Figure 11.16 Trilobite ontogeny. (a) *Shumardia pusilla* (Tremadocian), ontogenetic series from degree 0 meraspis to holaspid; macropleural segment shaded; 0–5 are the isolated pygidium meraspids, 6 is a holaspid; the isolated pygidium 7 shows changes in segmentation after the full complement of thoracic segments had been reached, 0 (x 30), 1–6 (x 20), 7 (x 14) (redrawn from Stubblefield in '*Treatise*' Part (O)). (b) *Leptoplastoides salteri* (Tremadocian), early and late meraspids showing cephalic spine reduction (x 14) redrawn from Raw in '*Treatise*' Part (O). (c) *Sao hirsuta* (M. Cam.), Bohemia: protaspid (1)–(3) x 14, meraspid (4, 5) x 10, degrees 0 and 6) and holaspid (6) x 1·5 (after Barrande in '*Treatise*' Part (O)).](d) *Gravicalymene quadrilobata* (L. Dev.), Australia: dorsal, ventral and lateral view of protaspid, showing spiny hypostome (x 20) (redrawn from Chatterton, 1971, *Palaeontographica* **137**, 1–108). (e) *Olenellus*, protaspid (x 20) compared with adult cephalon (x 1); the larval spines are almost completely reduced in the adult and are not homologous with the genal spines (redrawn from Palmer, 1957).

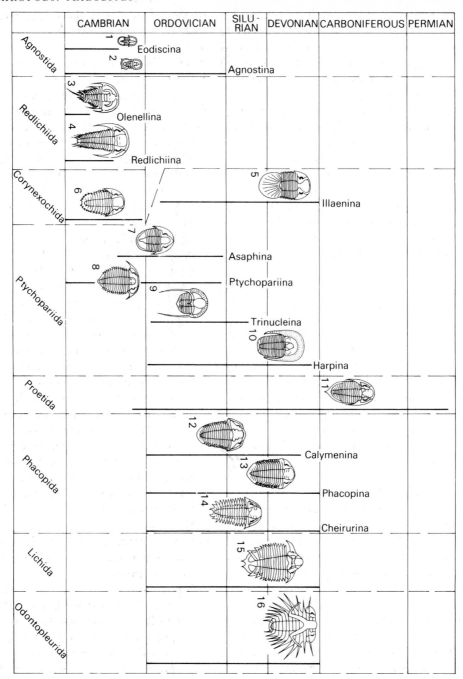

Figure 11.17 Time ranges of orders (e.g. Phacopida) and sub-orders (e.g. Calymenina) in trilobites. Genera illustrated are: (1) *Eodiscus* (M. Camb.); (2) *Homagnostus* (U. Cam.); (3) *Paedumias* (L. Cam.); (4) *Paradoxides* (M. Cam.); (5) *Scutellum* (L. Dev.); (6) *Kootenia* (M. Cam.); (7) *Asaphellus* (Trem.); (8) *Olenus* (U. Cam.); (9) *Cryptolithus* (Ord.); (10) *Harpes* (M. Dev.); (11) *Weberides* (Miss.); (12) *Calymene* (Sil.–Dev.); (13) *Acaste* (Sil.); (14) *Cheirurus* (U. Ord.–Sil.); (15) *Trochurus* (U. Ord.–M. Sil.); (16) *Dicranurus* (Dev.).

in the protaspid and meraspid stages of ptychopariids, and so it has been possible to establish a real phyletic relationship between the two groups, as well as showing that major differences in the holaspid morphology of the two groups have resulted mainly from differential growth rates.

Ontogenies in trilobites also point to the importance of paedomorphosis as an important factor in evolution, especially where new kinds of organisation seem to have arisen very rapidly and without preserved intermediates. Thus the suture of trilobites can be traced through ontogeny as initially proparian, though it may later become opisthoparian. Since there is some evidence that phacopid schizochroal eyes arose paedomorphically, it is not unlikely that the proparian suture of the Phacopina is likewise paedomorphic.

Classification of trilobites (Fig. 11.17)

The classification given below is a modified version of that in the '*Treatise*'. It is based upon the whole complex of axial and other characters, without any one or more being singled out as being of over-riding importance. Virtually all the characters used by the earlier systematists are herein incorporated, but in a more realistic perspective. In addition, such characters as resemblances at early ontogenetic stages become increasingly important. Yet this scheme is not wholly satisfactory; for instance, the large Order Ptychopariida needs to be split into smaller units which hopefully would represent natural groupings. Attempts have been made to produce other taxonomic schemes based upon different characters; a recent system by Bergstrom (1973) emphasises enrollment patterns, which he believes to be functionally quite dissimilar and hence phylogenetically distinct. Yet though these might well prove to be of importance, it may be many years before an agreed system emerges. For the moment the scheme in which the Class Trilobita is divided into eight orders appears to be reasonable in the light of existing knowledge.

CLASS TRILOBITA (Cam.–Perm.): Marine arthropods with largely calcitic exoskeleton divided into three longitudinal lobes and with a distinct cephalon, articulated thorax and pygidium. Cephalon with primitively furrowed glabella, facial sutures, compound eyes (sometimes lost) and doublure with hypostome. From two to forty thoracic segments, each with axis and pleura. Pygidium variable in size and shape. Spinosity variable. Ventral appendages, apart from uniramous antennae, are one pair per segment, biramous and all of the same kind, decreasing in size posteriorly.

ORDER 1. AGNOSTIDA (L. Cam. – U. Ord.): Small trilobites with subequal cephalon and pygidium. Thoracic segments number only two (SUBORDER AGNOSTINA) or three (SUBORDER EODISCINA). Hypostome rarely found, small and with ribbon-like wings. Eyes and sutures absent except in the eodiscid FAMILY PAGETIIDAE (L.–M. Cam.). e.g. *Agnostus, Eodiscus, Pagetia*.

ORDER 2. REDLICHIIDAE (L.–M. Cam.): An early group of trilobites with a large semicircular cephalon having strong genal spines, numerous and usually spiny thoracic segments and a tiny pygidium. Eyes large; hypostome fused to the rostral plate. Sutures may be **metaparian** (ankylosed) in SUBORDER OLENELLINA, opisthoparian in SUBORDER REDLICHIINA. e.g. *Olenellus, Redlichia, Paradoxides*.

ORDER 3. CORYNEXOCHIDA (L.–U. Cam.): A rather heterogeneous group of Cambrian trilobites. Glabella of varied form but usually parallel-sided or expanding anteriorly; sutures opisthoparian with subparallel anterior branches; rostral plate fused with hypostome. Thorax with seven to eight segments. Usually macropygous. e.g. *Olenoides, Zacanthoides, Bathyuriscus*.

ORDER 4. PHYCHOPARIIDA (L. Cam.–U. Dev.): The largest trilobite order, including many modified groups.

SUBORDER 1. PTYCHOPARIINA (L. Cam. – U. Ord.): Simple and usually forwardly tapering glabella not reaching front of cephalon, straight simple glabellar furrows, hypostome articulating along hypostomal suture. Thorax generally large. Pygidium small. FAMILY OLENIDAE is of great stratigraphical value. e.g. *Olenus, Ptychoparia, Leptoplastus, Triarthrus, Asaphiscus, Elrathia*.

SUBORDER 2. ASAPHINA (U. M. Cam. – L. Ord.): Large smooth isopygous opisthoparian trilobites in which a median suture may be present, or librigenae fused together anteriorly. Includes the large-eyed SUPER FAMILY CYCLOPYGACEA. Thorax with six to nine segments. e.g. *Asaphus, Ogygiocaris, Nileus.*

SUBORDER 3. ILLAENINA (L. Ord. – U. Dev.): Large isopygous trilobites, smooth or tuberculate, and opisthoparian. In contrast with the Asaphina they have a large rostral plate with a pronounced rostral suture. e.g. *Illaenus, Scutellum.*

SUBORDER 4. HARPINA (U. Cam. – U. Dev.): Cephalon large with high glabella and long genal prolongations; a wide flat fringe with small irregular pits surrounds the anterolateral cephalic border: eyes very small with two to three lenses. Many thoracic segments. Short pygidium. e.g *Harpes, Loganopeltis.*

SUBORDER 5. TRINUCLEINA (L. Ord. – M. Sil.): Cephalon large without compound eyes and with marginal suture; glabella usually swollen. Thorax of six segments only. Small triangular pygidium FAMILY TRINUCLEIDAE has a pronounced fringe with regularly arranged pits. FAMILY RAPHIOPHORIDAE has no fringe but often has long anterior spines. e.g. *Trinucleus, Cryptolithus, Ampyx.*

ORDER 5. PROETIDA (Ord.–Perm.): Glabella large and vaulted, well defined, usually with genal spines, narrow and backwardly tapering rostral plate, opisthoparian, eyes holochroal and usually large, long hypostome. Thorax with eight to ten segments. Isopygous. Pygidium usually furrowed and not spiny. (See Fortey & Owens 1975.) e.g. *Proetus, Phillipsia, Bathyurus, Aulacopleura.*

ORDER 6. PHACOPIDA. (L. Ord.–U. Dev.): Large order of dominantly proparian trilobites, divided into three clearly defined suborders:

SUBORDER 1. CHEIRURINA (L. Ord. – M. Dev.): A very variable, normally proparian group. Glabella with up to four pairs of glabellar furrows and usually expanding forwards, small holochroal eyes, rostral plate present, hypostome free. Thorax with eight to nineteen segments. Pygidium usually lobed or spiny. e.g. *Cheirurus, Deiphon, Ceraurus, Sphaerexochus.*

SUBORDER 2. CALYMENINA (L. Ord. – M. Dev.): A fairly homogeneous, usually gonatoparian group characterised by an anteriorly tapering glabella with four or five pairs of lateral lobes, diminishing in size forwards. Eyes small, holochroal. Thorax with eleven to thirteen segments. Pygidium rounded or subtriangular. In FAMILY HOMALONOTIDAE the axis is very broad and the furrows are largely effaced or very shallow. e.g. *Calymene, Trimerus.*

SUBORDER 3. PHACOPINA (L. Ord. – U. Dev.): Cephalon with forwardly expanding glabella, schizochroal eyes, proparian sutures and no rostral plate. Thorax of eleven segments. SUPERFAMILY PHACOPACEA has a small pygidium and well-developed enrollment mechanisms. SUPERFAMILY DALMANITIDAE members are larger, isopygous and large-eyed and have enrollment structures poorly developed. e.g. *Phacops, Dalmanites, Acaste, Chasmops.*

ORDER 7. LICHIDA (L. Ord. – U. Dev.): Medium to very large trilobites with unmistakably distinctive cephala and pygidia. Cephalon with very broad glabella, often having fused lateral and glabellar lobes, opisthoparian sutures. Large pygidium may be bigger than the cephalon, with three pairs of blade-like pleurae. Highly tuberculate exoskeleton. Usually rare in most faunas. e.g. *Lichas, Hemiarges, Terataspis.*

ORDER 8. ODONTOPLEURIDA (U. Cam.? or L. Ord. – U. Dev.): Very spiny trilobites. Glabella with three pairs of lateral lobes, outside which on the fibrigenae lie the ocular ridges; sutures opisthoparian, rostral plate short and wide, hypostome small. Thorax with eight to ten spinose segments. Short spiny pygidium. Exoskeleton very tubercular or spiny. e.g. *Acidaspis, Leonaspis, Ceratocephala.*

Evolution in trilobites

General pattern of evolution

Almost nothing is known about the ancestors of trilobites. In common with other arthropods it may be presumed that trilobites were possibly derived from the same ancestors as the annelids, and the Ediacara genus *Spriggina*, which has a large and rather plate-like head, may have been somewhere in the ancestral line, but this is only speculation. The earliest trilobite group to appear is Suborder Olenellina which is micropygous. These, together with the secondarily blind Agnostida which appeared a little later, diversified through the Lower Cambrian and were joined by the Corynexochida and Ptychopariida, which continued through after the demise of the Olenellina at the end of the Lower Cambrian. The Upper Cambrian fauna is dominated mainly by the Ptychopariina, which shows at least superficially relatively limited variation by contrast with that exhibited by many Ordovician taxa. This could point either to a period of evolutionary stagnation in which there was both limited genetic potential and low selection pressure, or alternatively to a rather rigorous selection weeding out of all but the morphologically defined forms. The former alternative is favoured, for in the few groups in which evolution has been studied in detail (e.g. the Olenidae) there is, in spite of the retention of a standard model of organisation, a fair degree of evolutionary plasticity within the group.

At the end of the Cambrian there was a major crisis in trilobite history. Most of the comparatively unspecialised Upper Cambrian stocks died out, few reaching the Ordovician (Whittington 1966). The reasons for such extinction are unknown; perhaps the marine regressions of late Cambrian time and the rise of predatory cephalopods may be invoked as being amongst the causes. Some short-lived, rapidly evolving Upper Tremadocian groups then originated, and after their extinction came the first representatives of the important and dominant Ordovician groups: the Illaenina, Phacopida, Trinucleina and others, all highly differentiated and diverse, most of them of cryptogenetic origin, and surviving for various periods thereafter. It is an interesting feature of trilobite evolution that after this great burst of new constructional themes in the early Ordovician very few entirely new patterns of organisation arose; afterwards evolution in trilobites was very largely a matter of experimentation and development of the new genetic material that first came into being in the early Ordovician.

It seems clear that these strong and dominant new Ordovician stocks were able to colonise and exploit new environments in a way that had not been possible for their Cambrian progenitors. For the first time, for example, there are many genera colonising reef environments, some persisting for a long time, others being replaced by different genera of broadly similar morphology. These reef faunas are often difficult to correlate stratigraphically with contemporaneous faunas living in other environments. The pure limestone faunas of early Ordovician age have a preponderance of large, rather flattened and rather lightly furrowed asaphids, for instance; when these died out scutelluids (Suborder Illaenina. Family Scutelluidae) of similar morphology occupied their ecological niche. Despite the success of the Ordovician trilobites as a whole, however, many distinct groups (e.g. the Family Trinucleidae) became extinct by the end of the Ordovician. The survivors were long-ranged and presumably successful and versatile groups. Yet these 'well-tried' groups were not devoid of diversification potential, for there is evidence of several adaptive radiations in particular isolated areas of the world, where an original ancestor invading a fairly empty environment had paved the way for a burst of evolutionary diversification in its descendants.

The trilobite faunas of the Silurian and Devonian contain many of the same elements: phacopids, dalmanitids, cheirurids, calymenids, harpids, etc. Consequently a Silurian assemblage can easily be confused, at least superficially, with a Devonian one. Encrinuridae (Suborder Cheirurina) however, though so characteristic of many Silurian faunas, do not extend into the Devonian.

The extinctions of the Middle and Upper Devonian disposed of the majority of trilobite groups, yet the remaining Proetida went all the way through to the Upper Permian. Most of these, especially those living in shallow waters, were large-eyed robust genera capable of enrollment, but there were also specialised, thin-shelled, lightly constructed and often eyeless forms, especially in the deep water of the Variscan geosyncline in north-western Europe. These are found from the Upper Devonian, where reduced-eyed proetids and phacopids are present, through the Carboniferous in the Culm facies.

Extinction of the last Permian trilobites was probably related to the lowering of sea level that contemporaneously affected so many other invertebrates.

In general, some fairly clear evolutionary trends can be distinguished, such as the origin of new eye types, the improvement of superior enrollment and articulating mechanisms, the change from micropygy to isopygy, the development of extreme spinosity in certain groups and the reduction in the rostral plate. Yet trilobites as a whole remained constructed on the same archetypal plan defined in the earliest Cambrian, and, especially after the early Ordovician, changes of real significance remained surprisingly low.

Microevolution in trilobites

There have been many studies on trilobite evolution (summarised in Eldredge 1977), especially on Upper Cambrian, Ordovician, Devonian and Lower Carboniferous genera; of these only two are selected for further discussion, both illustrating allopatric speciation.

The Olenidae

Trilobites of Family Olenidae occur in vast numbers in the Upper Cambrian of Scandinavia and also in Britain, North America and Argentina. They have been studied in great detail and are of great stratigraphical value. Their evolution spans over 40 million years extending into the Ordovician. Many studies have been made of the overall pattern of evolution in olenids, and a very important contribution to our knowledge of speciation in fossils came from Kauffmann (1933). He made a statistical study of *Olenus* in a 2·5 m part of a condensed sequence of one of the lower zones of the Upper Cambrian. The pygidia were found to be particularly valuable, and he was able to elucidate changes in pygidial shape in no less than four distinct lineages. Where these trends were apparent within individual species he distinguished the different stages anteriosus, medius and posteriosus, going from conservative to progressive respectively. In all of these the character mean in the population shifts gradually, and in each case there is a general progression in pygidial shape from broad and short to long and narrow. The four groupings were:

4. (top)	*Olenus dentatus* anteriosus–medius–posteriosus	
3.	*O. attenuatus* anteriosus–medius–posteriosus	
	barren beds	
2.	*O. transversus* truncatus–wahlenbergi	
	barren beds	
1.	*O. gibbosus* anteriosus–medius–posteriosus	

Lineages 3 and 4 overlap; the others are distinct. It was postulated that the parent lineage of *Olenus* was at this time living elsewhere, and that each of the four distinctive lineages actually seen in the section began with an invasion of an evolutionary offshoot of the parent population into the area. The subsequent development of *Olenus* through time generally followed an iterative pattern of evolution, where similar trends arise in successive stocks from a more-or-less unchanging ancestor. The morphological characters of the ancestral stock are not known with certainty but can be inferred from the earliest members of lineages 1, 2 and 4. Lineage 3, which partially overlaps with 4, has somewhat different characters and seems to have come in from a related but distinct ancestor, probably living elsewhere. This suggestion of the origin of new groups from peripheral isolates of a main population is amply confirmed by other trilobite evolutionary studies, though only in lineage 3 did the observed trends lead to the development of a new species.

The *Phacops rana* complex Eldredge's (1972b) study of evolution in the Middle Devonian *Phacops rana* of North America describes a classic case of allopatric speciation. *P. rana* was widespread over North America during Middle Devonian times, having been derived probably from such a European ancestor as *P. latifrons*. To the west of the area in which *P. rana* lived *P. iowensis* resided, the two species being more-or-less mutually exclusive. Whilst *P. iowensis* is never very abundant and showed little evolutionary change throughout the Middle Devonian, there was much greater variability amongst the subspecies of *P. rana*, of which individuals are often very common. These subspecies are not very easy to distinguish, their definition being mainly based on eye morphology, and an extensive multivariate analysis was used in elucidation of this history. Eldredge's study spanned a large area of eastern North America, in which the sediments both of the shallow sea overlying the continent and of the marginal 'exogeosyncline' were represented. The region is well correlated stratigraphically, three stages being present. In the oldest (Cazenovian) stage *P. rana crassituberculata* and *P. rana milleri* were both present (conceivably they could be sexual dimorphs of a single species) in the epicontinental sea to the west, whilst *P. rana rana* lived contemporaneously in the eastern area. In the extreme west *P. iowensis* was resident. When the Cazenovian trilobites had become extinct *P. rana rana* spread from the east into the epeiric sea, and when this subspecies died out at the end of the Tioughniogan stage *P. rana norwoodensis,* a new invader from the eastern exogeosyncline, replaced it. Thus the trilobites of the epicontinental sea, though they appear to form a successive evolutionary sequence in which the eye is reduced, are actually derived from an ancestral population living in the east, as a series of invaders, originally peripheral isolates of a parental population which as in the case of the olenids changed relatively little throughout time. This study also illustrates Eldredge and Gould's model of evolution (see Bibliography following Ch. 2) as a series of 'punctuated equilibria'. The individual populations of *P. rana* once established were relatively stable and underwent little evolutionary change. It was only when a parent population became extinct that the descendants of small, rapidly evolving populations living as demes peripheral to the main ancestral population could opportunistically invade the now-vacant territory. These then became the dominant widespread species: a successful balanced population with new and stable character states.

Trilobite faunal provinces

Trilobites, as much as any other faunal group, show provincial differences throughout the Palaeozoic which can be useful indicators of the biogeography of that time. There seems to have been fairly well-marked provincialism in the Cambrian and especially in the earlier Ordovician. This provincialism decreased throughout the later Ordovician until by the late Ashgillian trilobite faunas were more or less cosmopolitan, a condition that persisted through the Silurian. In the early Devonian separate provinces began once more to differentiate, most particularly a southern Malvino-Kaffric province, but following the extinctions of late Devonian time provinciality becomes less easy to trace.

The distribution of trilobites in the Cambrian is complex (Cowie 1971). In a broad and general sense it appears that there were two Lower Cambrian main faunal realms, characterised by olenellid and redlichiid trilobites respectively.

Trilobite distribution in the Ordovician has been

much more fully explored, and attempts to work out the controls of distribution have on the whole been more successful (Whittington & Hughes 1972). It is known that the Lower Ordovician trilobites are clearly distributed in provinces, apart from a very few genera of cosmopolitan distribution. Some of these widespread trilobites (e.g. *Geragnostus, Telephina, Seleneceme*) could well have been planktonic genera. Otherwise no less than four provinces have been defined, named after characteristic groups or genera. Each province has endemic families and endemic genera of more widely distributed families. A nonmetric multidimensional scaling technique has been used to clarify the provinces. The **bathyurid** province covers most of North America, Greenland, western Norway, Ireland, Scotland and Spitzbergen, as well as the Siberian Platform (Family Bathyuridae are Illaenina). The **asaphid** province is confined to Balto-Scandia and the Urals. England, southern and eastern Europe and North Africa supported a *Selenopeltis* fauna (named after an odontopleurid), whilst a less well-known South American and Australian fauna is named after *Asaphopsis,* a peculiar asaphid. This pattern of distribution held for all the Lower Ordovician, but by the Caradocian only the *Selenopeltis* province retained its distinctiveness; the others had merged into a single '**remopleuridid** province', and even these two provinces became less distinct through the Ashgill, recognisable at the level of the genus but not the family.

Two major controls of distribution are supported by geological evidence: namely temperature and continental movement. In later Ordovician times there was a pole situated where the Sahara Desert is now. A short but widespread glaciation is known from the later Ashgill, centred on this pole. The *Selenopeltis* province lies within 30° of the pole and occurs in rock with very few carbonates; its fauna was probably adapted to cold water conditions and retained its integrity for this reason despite later continental movements. Trilobites of the bathyurid province lived on continental shelves on the opposite side of the Iapetus (or proto-Atlantic) Ocean from the other provinces. The closure of this ocean along the suture now occupied by the Scandinavian, Caledonian and Appalachian mountains is now established geologically and is well supported by the late Ordovician merging of the faunal provinces, for faunas of opposing continental shelves originally separated by a wide intervening stretch of deep water would eventually be brought into close proximity. Larvae hatching in the waters of one shelf could finally settle on the other. Though the details of plate closure are not fully clear, the hypothesis that the change in faunal distribution relates to it is an attractive one. Even so, Ordovician faunal provinces may be far more complex than is thought, for some apparent provincial differences could be accounted for by facies and community differences alone. In three benthic communities distinguished by Fortey (1975) in the Ordovician of Spitzbergen, the onshore illaenid–cheirurid community (which has high endemicity) is in part very similar faunally to the bathyurid province, the intermediate nileid community (Family Nileidae are Asaphina) can be matched with the asaphid province, and the olenid community in which endemicity is lowest has counterparts within the *Asaphopsis* province. Province boundaries, with endemicity highest in shallow waters and lowest in offshore regions, will hold good for the Ordovician also, though oceanic barriers and temperature controls were also undoubtedly important. But until the relationship between community types and faunal provinces is better understood our knowledge will be far from complete.

Silurian faunas seem to have been more-or-less cosmopolitan, whilst the differentiation of the Malvino-Kaffric province in Siluro-Devonian times (South America, Falkland Islands, southernmost Africa), which is so well elucidated in the brachiopods, appears to be substantiated in trilobites also though this is less well known. Derivatives of an acastid stock migrated into an almost empty region, and diversified exceedingly, constituting one of the most remarkably distinct of all faunal provinces. Provinciality in the later rocks is poorly known but does not seem to have been great.

In Arctic North America cosmopolitanism began to decrease after Gedinnian time, so the Devonian as a whole seems to have been a time of increasing provinciality.

Stratigraphical use of trilobites

Trilobites are of considerable stratigraphical value in the Cambrian and Ordovician but much less so in later rocks, except perhaps locally. The Cambrian system is zoned almost entirely on trilobites, other than the basal (Tommotian) zone. Cambrian trilobites obey most of the requirements of good zone fossils, being abundant and easily recognisable and often having short time ranges and wide horizontal distribution. They are limited by being distributed in the faunal provinces described above and by being rather facies controlled. Though good intraprovincial sequences are established, it is only possible to correlate between them where faunas are mixed.

The Ordovician system is zoned by graptolites (q.v.), established for the offshore sequences. But the stratigraphical subdivision of the nearshore shelly facies is based upon trilobites and brachiopods, which give a refinement of correlation of special merit in the Caradoc and Ashgill. In the British type area no stages have yet been defined in the Arenig, Llanvirn and Llandeilo, but only Upper and Lower divisions (the Llandeilo has a middle division). In the Caradocian, however, eight stages have been erected, based mainly on brachiopods though with trilobites (especially trinucleids) giving confirmatory evidence. In the Ashgillian many kinds of trilobites have been used to define several zones within the four known stages, so effectively that selected areas of lithological monotony have been mapped using the trilobite faunas as markers. There are still problems with correlation using Ordovician trilobites, mainly because of facies control and provinciality; there is still, for example, no close time correlation between the faunas of various Upper Ordovician reefs – Dalarna (Sweden), Kiesley (northern England) and Kildare (Ireland) – and those of contemporaneous bedded carbonate or mudstone facies.

In Silurian and later rocks trilobites are not of great stratigraphical value, for other fossils have generally shorter time ranges or may be otherwise used more effectively, such as the Devonian ammonoids and Carboniferous microfossils. But for local correlation they are certainly of some use, and their value as stratigraphical indicators may increase.

PHYLUM CHELICERATA

Chelicerates (Cam.–Rec.) include spiders, mites and scorpions, as well as the horseshoe crab *Limulus* and its living and fossil allies, and the extinct eurypterids. This diverse group of animals which appear so heterogeneous are united by possessing:

(a) an anterior prosoma of six segments which is equivalent to the fused head and thorax of other arthropods;

(b) a posterior opisthosoma (abdomen) with twelve or less segments;

(c) a pair of jointed pincers (the chelicerae) which give the subphylum its name and which are always present in the first (**pre-oral**) segment of the prosoma.

Two main groups within the chelicerates are of singular palaeontological and evolutionary interest:

CLASS 1. MEROSTOMATA (Cam.–Rec.): Aquatic chelicerates, often of large size. There are two subclasses: SUBCLASS XIPHOSURA which, with its ORDER XIPHOSURIDA and ORDER AGLASPIDA, includes the *Limulus* group and their relatives; and SUBCLASS EURYPTERIDA which includes the water-scorpions of Palaeozoic time.

CLASS 2. ARACHNIDA (Sil.–Rec.): Dominantly terrestrial forms, the spiders, mites, and scorpions; the latter bear a superficial resemblance to eurypterids but are not closely related, Though scorpions are known from the Silurian and mites and spiders from the Devonian, all are rare as fossils, other perhaps than the Tertiary spiders preserved in amber. Fossil scorpions are known in considerable morphological detail, but their study is beyond the scope of this book, and arachnids will not be considered further.

Class Merostomata

SUBCLASS XIPHOSURA

Limulus

The characters of xiphosures are well displayed by the modern *Limulus:* one of three closely related genera which are the only living representatives of this subclass.

Limulus polyphemus (Fig. 11.18a–d), the type species, lives in shallow waters along the north-western Atlantic shores. In dorsal view it has a vaulted semicircular prosoma, which as in all chelicerates is a fused cephalo-thorax articulating with a less vaulted plate-like abdomen, there being a narrow channel incised obliquely between the two. From the rear of the abdomen springs a strong terminal spine; the **telson.** The axial part of the prosoma (**cardiac lobe**) is defined by axial furrows so that it has a superficial similarity to a trilobite glabella. Outside the axial furrows and parallel with them are the **ophthalmic ridges** which curve anteriorly to join medially, at which point there are a pair of small **ocelli.** There are also **compound eyes** situated in the middle of the ophthalmic ridges. These eyes have many lenses, but their structure is quite unlike that of trilobites, the dioptric parts being composed of invaginations of the cuticle which form long, inwardly projecting cones. These are parabolic constructions optimised for maximum light collecting. Below these are apposition-type ommatidia, each with a marked excentric cell.

The abdomen articulates to the prosoma along a hinge which cuts across the primary segmentation. Axial ridges are present, and traces of segmentation are evident in the fringe of six movable spines. The strong styliform telson, ridged along its length, is nearly as long as the rest of the body and can turn in almost any direction.

The segmentation of the body is readily worked out from embryology but is also clear in the ventral morphology of *Limulus.* The pre-oral segment carries the short paired three-jointed chelicerae, and the five pairs of post-oral appendages are arranged radially around the mouth within the indented ring formed by the anterior doublure. These five pairs of post-oral appendages are morphologically similar to one another and function as walking legs. Each has a basal spinose coxa and several joints. The first four pairs terminate in pincer-like **chelae;** the last pair are equipped instead with numerous spines and have a peg-like outgrowth (the **flabellum**) attached to the coxa.

Within the broad ventral doublure of the abdomen are six overlapping plates which are modified appendages. The first of these is the **operculum,** morphologically the eighth segment which, as in all chelicerates, bears the **genital openings.** The seventh segment is rudimentary and is apparently represented only by a pair of reduced plates. Behind the operculum are the succeeding five gill appendages which bear the respiratory **gills** on their inner (dorsal) side, where they are protected against desiccation.

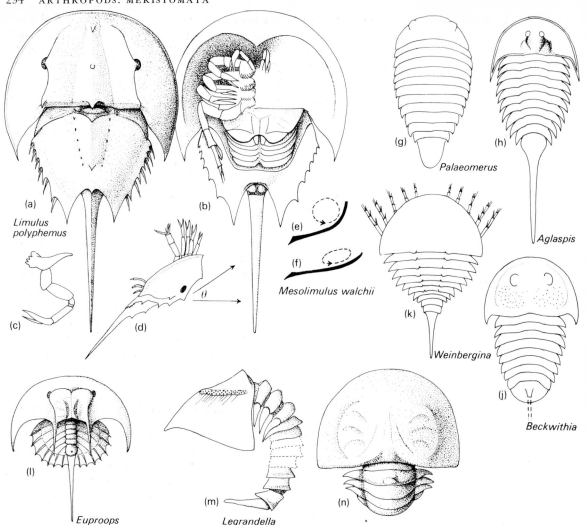

Figure 11.18 Merostomata. *Limulus polyphemus* (Rec.): (a) dorsal view; (b) ventral view; (c) fourth prosomal appendage ((a)–(c) x 1) (d) swimming upside down with the body inclined at angle θ to the horizontal; (e) section through prosoma of swimming *Limulus* showing recirculating vortex; (f) *Mesolimulus walchii* (Jur.) section through prosoma (more flattened than that of *Limulus*) showing vortex ((d)–(f) redrawn from Fisher, 1975); (g) *Palaeomerus* (L. Cam.) (redrawn from Bergstrom, 1971, *Lethaia* **4**, 393–401); (h) *Aglaspis* (reconstructed) (L. Cam.) (x 1); (j) *Beckwithia* (L. Cam.) (x 0·75) ((h)–(j) redrawn after Raasch, 1939) (k) *Weinbergina* (L. Dev) Germany (x 0·75); (l) *Euproops* (U. Cam.) (x 1) ((k)–(l) redrawn from Størmer in '*Treatise*' Part (P)); (m), (n) *Legrandella* (Dev.) in lateral and dorsal views (x 0·75) (redrawn from Eldredge, 1974).

Life habits of *Limulus*.

Limulus is a shoreline inhabitant, living in shallow waters and capable of crawling out for short distances on land. It can walk on the sea floor on the prosomal appendages, but it can also swim and right itself with the telson if overturned. It is an exceedingly versatile animal whose life habits have been used with some success in interpreting those of trilobites, to which it is the closest living relative and which have some morphological resemblances. But there must be particular functional reasons for these similarities, and the specific uses of the various organs must be fully understood before such analyses are drawn.

Swimming Limulus normally swims upside down (Fig. 11.18d), inclined at about 30° to the horizontal and moving at about 10–15 cm/sec. It is propelled by both the prosomal and opisthosomal appendages. The opisthosomal and sixth pair of prosomal appendages move in anteriorly advancing waves of metachronal rhythm, whilst the first four pairs of walking legs move in phase, extending as they push backwards, withdrawing into the prosomal cavity of vault as they recover their position. This stroking cycle of the first appendages begins directly after the sixth prosomal pair have completed their stroke.

The hydrodynamics of flow in the swimming *Limulus* have been analysed using a model *Limulus* in a flow chamber (Fisher 1975). In the normal inclined swimming attitude a strong recirculating vortex forms within the prosomal vault. This appears to break down or be shed at intervals into the wake of the animal prior to reforming. The vortex is a normal consequence of flow dynamics for an object of such a shape and inclination, but it seems to be exploited by the animal in aiding the forward movement of the anterior prosomal appendages in the recovery stroke, which would otherwise have to push against the current. At greater or lesser inclinations the vortex would be less effective or absent (Fig. 11.18e, f).

Similar models of the Jurassic *Mesolimulus walchii*, which has a much more flattened morphology, show that an equivalent vortex would operate in a similar way provided that the animal's angle of inclination was less. Hence there is a direct relationship between the shape of the prosoma and swimming ability, the different vaultings of *Mesolimulus* and *Limulus* being different answers to the same problem. But *Limulus* has more room in which to retract its legs and is more effectively adapted for burrowing.

Burrowing Limulus is active at night, but during the day it burrows in below the sediment surface for as much as twelve hours at a time (Eldredge 1970). Adult specimens have an anterior arch as do trilobites. At the start of the burial procedure the arch is first lowered so that the anterior edge of the prosoma comes in contact with the sea floor; it then digs into the sand as the prosomal legs perform normal walking movements. Sand is pushed backwards covering the prosoma and the anterior third of the opisthosoma. The telson is buried and only makes horizontal movements thereafter. The walking legs excavate a burrow, pushing the sand up between the prosoma and opisthosoma. One of these channels

remains open when the animal is completely buried. When this first stage of burial is effected, the animal is then quiescent and lies for a while oxygenating the gills. There is then a third stage, when the prosomal legs walk forwards once more whilst the opisthosoma is vigorously flapped, using the telson to some extent, and stirs up a cloud of sand which settles to cover the animal completely. The final stage is one of absolute quiescence, only a single respiratory channel being left open on one side.

It has been shown experimentally that the dorsal setae are used as mechanoreceptors which indicate to the animal when it is fully buried. The telson is covered from start to finish and is only used to a limited extent in the third stage. It functions mainly as a stabiliser or rudder in the walking or swimming individual or in righting one that is overturned. Comparative morphology of *Limulus* and trilobites in such details as the presence of dorsal sensors and the vaulted shape of the cephalon suggests that trilobites might have spent some of their time burrowed in sand, and this is amply borne out by trace fossil evidence.

Other xiphosures and their geological history

Xiphosures are rare in the fossil record and, other than *Limulus* and its modern relatives, are generally confined to non-marine sequences.

The oldest well-known xiphosures belong to the Cambrian Order Aglaspida, which are known very largely from the Upper Cambrian of south-western Wisconsin (Raasch 1939). These have a phosphatic exoskeleton in which the prosoma is typically xiphosuran though with very prominent compound eyes, in which unfortunately no structure remains. In *Aglaspis* (Fig. 11.18h) the eleven or twelve thoracic segments are free as in most aglaspids, but in *Beckwithia* (Fig. 11.18j) the posterior three or four are fused into a pygidium-like plate. In some specimens appendages are preserved. Evidently the first pair bore chelae; the rest, including the abdominal ones, were single walking legs. It would be possible to infer that evolution in the xiphosures proceeded towards fusion of the abdominal segments and modification of their appendages. But xiphosure phylogeny is obscure, and it is not certain whether aglaspids were ancestral to any later xiphosures or whether, as is more likely, they were simply an early and sterile 'experimental' group. The oldest arthropod which has been referred to Subclass Xiphosura is the Lower Cambrian *Palaeomerus* (Fig. 11.18g) but it is known from only three specimens, and could be a relative of some of the Burgess Shale arthropods ('merostomoids').

The Xiphosurida normally have the distinctive cardiac lobe and ophthalmic ridges well developed, and there are ten or fewer abdominal segments. There are two suborders:

ORDER XIPHOSURIDA (Cam.–Rec.)
    SUBORDER 1. SYNZIPHOSURINA (Ord.–Dev.): Most or all of the abdominal segments are free, and the appendages do not appear to be chelate (other than the chelicerae, which are not known with certainty).
    SUBORDER 2. LIMULINA (L. Dev.–Rec.): The abdominal segments may be fused, though not always, and the appendages are normally chelate.

Much confusion is associated with attempts to classify these animals, for the fossil record is so scanty and the preservation often poor. In such cases discovery of a new specimen, especially of a new taxon, often necessitates the taxonomic revision of the whole group. A current classification of the Synziphosurina (Eldredge 1974) includes only four genera: *Weinbergina* (Fig. 11.18k), *Legrandella* (Fig. 11.18m, n) (which has a large and very *Limulus*-like eye), *Bunodes* and *Limuloides*. The Limulina, other than the peculiar Family Pseudoniscidae, are mainly rather similar in construction to *Limulus*. The oldest, the Lower Cambrian *Eolimulus,* is known from two partial prosomas only. It was marine, as are the much better-preserved Devonian genera, but some of the Carboniferous and later groups invaded and colonised brackish water habitats. The Carboniferous *Belinurus* and *Euproops* (Fig. 11.18l) are small forms, often found in nodules deposited in coal swamps. Both these have well-developed cardiac lobes and opthalmic ridges. *Belinurus* like the Devonian *Neobelinuropsis* has some free abdominal segments, but in *Euproops* and *Limulus* the whole of the abdomen is fused, though surface traces of segmentation are very pronounced and incidentally very trilobite-like. It is of some interest that the first free larva of *Limulus*, when hatched from the egg, has a similar pronounced

segmentation in the fused abdomen to the adult *Euproops*. It is colloquially known as a 'trilobite larva' without any implication of a close relationship. The prosomal appendages of all Limulina are very similar to those of *Limulus*.

From the known fossil Limulina it may be inferred that there was a general evolutionary trend towards a fused short abdomen and larger size. The Permian *Palaeolimulus*, the Triassic *Limulitella* and the Jurassic *Mesolimulus* differ from modern genera mainly in their smaller dimensions and their prosomal lobation. They were all marine forms. Amongst fossil limulids one specimen is unique: an Upper Jurassic *Mesolimulus* from the Sölnhofen Limestone preserved at the end of the trail in which the imprints of its appendages can be clearly seen.

SUBCLASS EURYPTERIDA

The eurypterids are relatively rare but always spectacular fossils, normally confined to brackish or fresh-water sediments of Ordovician to Permian age. Most representatives are less than 20 cm long, but there are several giant forms including the Devonian *Pterygotus (Erettopterus)*, the largest arthropod of all time which reached nearly 2 m in length.

Eurypterid morphology is well displayed by the Silurian *Baltoeurypterus tetragonophthalmus* (Fig. 11.19a–g), formerly known as *Eurypterus fischeri*. The morphology of this species was one of the first to be really well understood, largely from the work of Holm (1898) who isolated fragments of the exoskeleton from rock, as he did with graptolites, so that they could be studied as transparencies. In dorsal view the **prosoma** is large and trapezoidal, with prominent compound eyes between which are small paired organs, probably ocelli. Behind this is the **opisthosoma,** consisting of a broad flattened **preabdomen** of seven segments, and a narrower and more cylindrical **post-abdomen** of only five segments, terminating in a stout pointed telson. Each of the abdominal segments is composed of a dorsal tergite and a ventral sternite. The ventral appendages can be seen from the dorsal surface projecting well beyond the body. The ventral morphology is complex. The prosoma has a broad inflected doublure with a marginal suture, used during ecdysis, and often with paired connective sutures isolating a median doublural plate (the **epistoma**). Within the

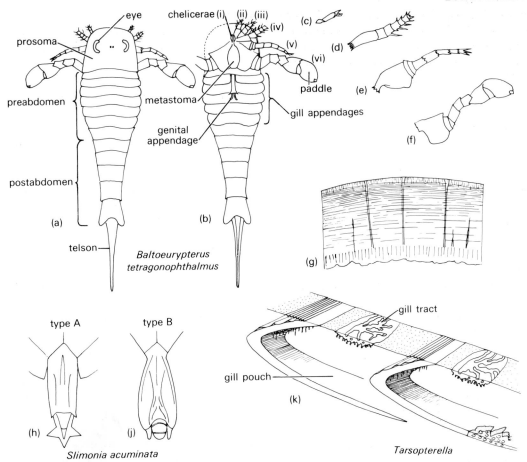

Figure 11.19 Eurypterida. *Baltoeurypterus tetragonoph-thalmus* (Sil.): (a) dorsal view; (b) ventral view; (c) chelicera (i); (d) walking leg (prosomal appendages (ii)–(iv)); (e) fifth prosomal appendage; (f) sixth, paddle-like appendage (all x 0·35) ((a)–(e) redrawn from Størmer in '*Treatise*' Part (P)). (g) Section (diagrammatic) through cuticle of *B. tetragonophthalmus*, showing inner non-laminar layer with fine vertical canals; middle (laminar) layer with laminae more closely spaced externally; and outer layer with indications of vertical elements (based on Dalingwater, 1975). (h) *Slimonia acuminata* (Sil.), type A (?male) genital appendage; (j) type B (?female) genital appendage (redrawn from Waterston, 1960, *Palaeontology* **3**, 245–59); (k) *Tarsopterella* (Dev.) simplified reconstruction of a gill chamber; ((l)–(k) redrawn from Waterston, 1975).

doublure is a softer integument surrounding the mouth, to which the appendages are attached. The first (pre-oral) pair of appendages are small cheli-cerae. Behind these post-orally are four pairs of stout jointed walking legs, increasing in size towards the rear legs, which are cylindrical and spiny. The sixth pair of prosomal appendages are very large and have the terminal parts flattened like paddles. These last appendages were capable of being protracted and retracted and were probably used as in swimming, in a breast-stroke manner. The coxa of each 'leg' is large and armoured internally with spines projecting as gnathobases. A small U-shaped plate (the **en-dostoma**) borders the mouth; this is normally covered by a much larger plate (the **metastoma**) which is actually part of the abdomen and is all that remains of the reduced seventh segment. It also covers the proximal parts of the coxa.

The pre- and post-abdomen of seven and five segments respectively have been defined on the positon of the waist, but the abdomen can also be separated into a **mesosoma** and **metasoma** of six segments each. The former bears appendages; the latter does not. These appendages look like sternites but are in fact true appendages. The seventh segment is reduced as the metastoma; the eighth as in all merostomes has the genital aperture. This segment is known as the operculum and, together with the four succeeding appendages, is plate-like and attached along the anterior border so that they all overlap. In the centre of the operculum and directed posteriorly is the **genital appendage,** an elongated and elaborately sculptured rod of which two kinds may be present in a single eurypterid population, suggesting sexual dimorphism. In *Baltoeurypterus* one kind (type A) is elongated, and the other kind (type B) though tubular is much shorter; other genera have genital appendages that are of approximately equal length though differently shaped (Fig. 11.19h, j). In *Pterygotus* and some other genera type A is consistently associated with 'clasping organs' on the prosomal appendages and seems to have been a male copulatory organ; the other type is best interpreted as female ovipositor. Traces of internal ducts have been distinguished in both kinds.

The operculum and the succeeding four appendages of the pre-abdomen covered chambers in which the gills were situated. The structure of these is known in *Slimonia* and better still in *Tarsopterella*, where the different patterns of internal and external surface sculpture have allowed a full interpretation of morphology to be made, even in flattened specimens (Waterston 1975). Here the gills are specialised vascular tracts of the ventral body wall protected by the appendages (Fig. 11.19k), which would perhaps have allowed the gills to remain moist and so enabled the animal to move about for short periods on the land.

Eurypterid cuticle

The cuticle of eurypterids is very thin, and specimens are usually crushed. It may bear different kinds of external sculpture, notably terrace lines round the border and sometimes on the lower surfaces, and a characteristic scale-like ornament from which fragmentary remains in sedimentary rock can be immediately distinguished as being eurypterid in origin.

Recent ultramicrographic work (Dalingwater 1973a, 1975) has shown that eurypterid cuticles are preserved as silica, though undetermined organic

material was originally present too. The outer cuticular zone is thick and finely laminated; below this lies a thinner non-laminated zone with a proximal thin laminated zone (Fig. 11.19g). The laminations resemble those of *Limulus* in nature, which could be of phylogenetic importance.

Range in form, evolution and ecology of eurypterids (Fig. 11.20)

Eurypterids are relatively rare fossils, and though the morphology of some genera is well known others are represented only by very poorly preserved material, and there are acute taxonomic problems in classifying such isolated remains. Rather than describing the taxonomic problems in detail, the following discussion is confined to a few selected types which show notable and presumably adaptive differences in size, prosomal shape, the location of the eyes, and the morphology of the body, the appendages and the telson. A current classification (Størmer 1970–1974) divides eurypterids into three suborders, each with associated families and superfamilies.

SUBCLASS EURYPTERIDA
  ORDER EURYPTERIDA
    SUBORDER 1. EURYPTERINA (Ord.–Perm.): Genera with small untoothed chelicerae. e.g. *Drepanopterus*, *Baltoeurypterus*, *Stylonurus*, *Slimonia*, *Mixopterus*, *Hughmilleria*.
    SUBORDER 2. PTERYGOTINA (Ord.–Dev.): Eurypterids with enormous denticulate chelicerae. e.g. *Pterygotus*, *Jaekelopterus*.
    SUBORDER 3. HIBBERTOPTERINA (Carb.): Genera in which the posterior prosomal legs have a basal extension e.g. *Hibbertopterus*, *Campylocephalus*.

The Eurypterina, by far the largest suborder, have the greatest diversity. *Hughmilleria* is relatively small and unmodified. *Slimonia* has a quadrate prosoma, antero-lateral eyes and a laterally expanded telson. *Carcinosoma* has a large discoidal pre-abdomen with a marked waist and a cylindrical post-abdomen and, like the fiercely armoured *Mixopterus*, it has a telson, like that of a scorpion, apparently modified as a poison spine. The prosomal appendages of *Dolichopterus* are of typical form though enormously extended, whereas in *Stylonurus* and its relatives all the

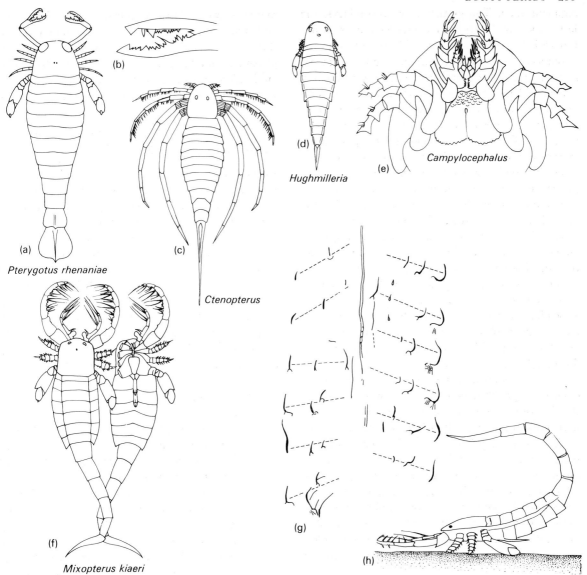

Figure 11.20 Eurypterida. *Pterygotus rhenaniae* (Dev.): (a) dorsal view; (b) detail of chelicera (x 0·1); (c) *Ctenopterus* (Sil.), a stylonurid (x 0·5); (d) *Hughmilleria* (Sil.) (x 0·35); (e) *Campylocephalus* (Carb.) – ventral view of prosoma showing large metastoma and basal extensions on posterior legs typical of Hibbertopteridae (x 0·2) (redrawn from Waterston, 1957, *Trans. Roy. Soc. Edin.* **63,** 265–88); (f) *Mixopterus kiaeri* (Sil.), dorsal and ventral view (x 0·1); (g) a trail in Silurian redbeds at Ringerike, Norway, believed to have been made by *M. kiaeri* (see text) whilst walking in the position illustrated in (h) (x 0·25 approx.) ((a)–(d), (f) Redrawn from Størmer in '*Treatise*' Part (P), (g) from Hanken and Størmer, 1975)

legs are slender and elongate, even the sixth prosomal appendage, and presumably modified for lightly walking over muddy surfaces.

Though inference of the modes of life of eurypterids remains somewhat speculative, more concrete evi-dence is provided by a large trail in the Silurian of Ringerike, Norway, believed to have been made by a large *Mixopterus* (Hanken and Størmer 1975) (Fig. 11.20f–h). The trail is 520 mm long and about 160 mm broad and consists of three sets of paired parallel tracks made up of separate successive imprints and decreasing in size inwards. The outermost 'A-tracks' are hook-shaped and would fit with the imprint of the paddle-shaped end of the sixth prosomal appendage. The intermediate 'B-tracks' and the innermost 'C-tracks' seem, however, to have been made by appendages with flattened spinose ends. A long median groove seems to have been made by the tip of the genital appendage scraping along the ground. These various facts fit with the concept of the trail being made by a *Mixopterus* walking on the fourth, fifth and sixth pairs of appendages, whilst keeping the grasping second and third pairs of appendages held

out in front of it as a kind of cage. *Mixopterus* was contemporaneous and was the right size to have made such a trail. Presumably the eurypterid could also swim, using the retractable sixth appendage as a paddle.

The mode of life of other eurypterids was probably broadly similar. They must have been active benthos, sometimes swimming. And as indicated by the modified chelicerae of *Pterygotus* and by the spinose second and third prosomal appendages of *Mixopterus*, they were undoubted predators, their prey being other eurypterids and armoured fishes.

Bibliography

Books, treatises, symposia

Bergstrom, J. 1973. Organisation, life and systematics of trilobites. *Fossils and Strata.* **2,** (Large scale work; presents a different classification from that in the '*Treatise*')

Clarke, K. U. 1973. *The biology of arthropods.* London: Arnold. (Short simple text)

Levi-Setti, R. 1975. *Trilobites: a photographic atlas.* Chicago: Univ. Chicago Press. (Large numbers of full page plates, with discussion of eyes)

Moore, R. C. (ed.) 1969. '*Treatise*' Part (R) Vols. 1 & 2. Arthropods. (Crustaceans)

Moore, R. C. (ed.) 1961. '*Treatise*' Part (Q). Arthropoda 3. (Ostracods)

Moore, R. C. (ed.) 1959. '*Treatise*' Part (O). Arthropoda 1. (Trilobites and related groups)

Moore, R. C. (ed.) 1955. '*Treatise*' Part (P). Arthropoda 2. (Chelicerates, pycnogonids and spiders)

Manton, S. M. 1977. *The Arthropods: habits, functional morphology and evolution.* Oxford: O.U.P. (Definitive work on arthropod morphology and evolution)

Individual references

Campbell, K. S. W. 1975a. The functional anatomy of phacopid trilobites: musculature and eyes. *J. Proc. Roy. Soc. New South Wales* **108,** 168–188. (Lens structure)

Campbell, K. S. W. 1975b. The functional morphology of *Cryptolithus. Fossils and Strata* **4,** 65–86. (Trinucleid fringe function, and ichnology)

Cisné, S. 1975. Anatomy of *Triarthrus* and the relationships of the trilobites. *Fossils and Strata* **4,** 45–64. (X-radiography, with limb morphology)

Clarkson, E. N. K. 1969. A functional study of the Silurian trilobite *Leonaspis deflexa* (Lake). *Lethaia* **2,** 329–44. (Lateral view and functional evolution of odontopleurids)

Clarkson, E. N. K. 1975. The evolution of the eye in trilobites. *Fossils and Strata* **4,** 7–31. (Summary and bibliography on eye function)

Clarkson, E. N. K. and J. L. Henry 1973. Structures coaptatives et enroulement chez quelques trilobites ordoviciens et siluriens. *Lethaia* **6,** 105–32. (Enrollment and interlocking mechanisms)

Clarkson, E. N. K. and R. Levi-Setti 1975. Trilobite eyes and the optics of Des Cartes and Huygens. *Nature* **254,** 663–7. (Lens structure and function)

Cowie, J. W. 1971. Lower Cambrian faunal provinces. In *Faunal provinces in space and time,* 21–46. Geol. Journ. spec. issue, no. 4. (Recent study of trilobite provinces)

Dalingwater, J. E. 1973a. The cuticle of a eurypterid. *Lethaia* **6,** 179–86. (Morphology)

Dalingwater, J. E. 1973b. Trilobite cuticle microstructure and composition. *Palaeontology* **16,** 827–39. (Fine structure)

Dalingwater, J. E. 1975. Further observations of eurypterid cuticles. *Fossils and Strata* **4,** 271–80. (Eurypterid cuticles were siliceous)

Eldredge, N. 1970. Observations on burrowing behaviour in *Limulus polyphemus* (Chelicerata, Merostomata) with implications on the functional anatomy of trilobites. *Novit. Amer. Mus. Nat. Hist.* **2436,** 1–17. (Valuable study of living animals)

Eldredge, N. 1972a. Patterns of cephalic musculature in the Phacopina (Trilobita) and their phylogenetic significance. *J. Paleont.* **45,** 52–67. (Reconstruction of internal organs)

Eldredge, N. 1972b. Systematics and evolution of *Phacops rana* (Green 1832), and *Phacops iowensis* (Delo 1935) (Trilobita) from the Middle Devonian of North America, *Bull. Amer. Mus. Nat. Hist.* **147,** 49–113. (Allopatric evolution)

Eldredge, N. 1974. Revision of the suborder Synziphosurina (Chelicerata, Merostomata) with remarks on merostome phylogeny. *Novit. Amer. Mus. Nat. Hist.* **2543,** 1–41. (Taxonomy and morphology)

Eldredge, N. 1977. Trilobites and evolutionary patterns. In *Patterns of evolution*, A. Hallam (ed.), 305–32. Amsterdam: Elsevier. (What do trilobites have to tell us about evolutionary processes?)

Fisher, D. 1975. Swimming and burrowing in *Limulus* and *Mesolimulus*. *Fossils and Strata* **4,** 281–90. (Important functional study)

Fortey, R. A. 1975. Early Ordovician trilobite communities. *Fossils and Strata* **4,** 331–52. (Three onshore–offshore benthic and one planktonic assemblage)

Fortey, R. A. and R. M. Owens 1975. Proetida: a new order of trilobites. *Fossils and Strata* **4,** 227–40. (Taxonomy; used in the classification presented here)

Hanken, N. M. and L. Størmer 1975. The trail of a large Silurian eurypterid. *Fossils and Strata* **4,** 255–70. (How eurypterids used their legs)

Henningsmoen, G. 1957. The trilobite family Olenidae. *Norsk. Vid. Akad. Mat-Natur. Kl. Skr.* **1,** 1–303. (Morphology, taxonomy and evolution)

Henningsmoen, G. 1975. Moulting in trilobites. *Fossils and Strata* **4,** 179–200. (Exuvial assemblages and various ways in which trilobites moulted)

Henry, J.-L. and E.N.K. Clarkson 1975. Enrollment and coaptations in some species of the Ordovician trilobite genus *Placoparia*. *Fossils and Strata* **4,** 87–96. (Interlocking mechanisms and their evolution)

Hughes, C. P., J. K. Ingham and R. Addison 1975. The morphology, classification and evolution of the *Trinucleidae* (Trilobita). *Phil. Trans Roy. Soc. Lond. B.* **272,** 537–607. (Definitive work with descriptions of all trinucleid genera and evolutionary charts)

Kaufmann, R. 1933. Variations statistiche Untersuchungen über die Artabwandlung und Artumbildung an der oberkambrischen Trilobitengattung *Olenus* Dalm. *Abh. Geol. Pal. Inst. Univ. Griefswald* **10,** 1–54. (Classic study of evolution in *Olenus*)

Lindström, G. 1901. Researches on the visual organs of the trilobites. *K. Svensk. Vetensk. Akad. Handl.* **34,** 1–85. (First work on trilobite eyes, hypostome may have ventral eyes)

Manton, S. M. 1973. Arthropod phylogeny – a modern synthesis. *J. Zool. Soc. Lond.* **171,** 111–30. (Classification of arthropods)

Miller, J. 1975. Structure and function of trilobite terrace lines. *Fossils and Strata* **4,** 155–78. (Environmental monitoring)

Miller, J. 1976. The sensory fields and life mode of *Phacops rana* (Green 1832) (Trilobita). *Trans Roy. Soc. Edin.* **69,** 337–67. (Functions of cuticular sensory organs)

Öpik. A. 1960. Alimentary caeca of agnostid and other trilobites. *Palaeontology* **3,** 410–38. (Surface sculpture and its interpretation)

Osgood, R. G. 1970. Trace fossils of the Cincinnati area. *Palaeontographica Americana* **41,** 281–444. (Trilobite trails, burrows and methods of locomotion)

Orlowski, S., A. Radwanski and P. Roniewicz 1970. The trilobite ichnocoenoses in the Cambrian sequence of the Holy Cross Mountains Poland. In *Trace Fossils*, T. P. Crimes and J. C. Harper (eds) 345–60. Liverpool: Liverpool Geol. Soc. Seel House Press. (Description of *Rusophycus* and *Cruziana*)

Palmer, A. R. 1957. Ontogenetic development of two olenellid trilobites. *J. Paleont.* **31,** 105–28. (Changes in proportion during ontogeny)

Raasch, G. O. 1939. *Cambrian Merostomata*. (Spec. Pap.) Geol. Soc. Amer. **19,** 1–146. (Aglaspids)

Raymond, P. E. 1920. The appendages, anatomy and relationships of trilobites. *Connecticut Acad. Arts. Sci. Mem.* **7,** 1–169. (Descriptions of trilobite appendages with fine photographs)

Robison, R. A. 1967. Ontogeny of *Bathyuriscus fimbriatus* and its bearing on affinities of corynexochid trilobites. *J. Paleont.* **41,** 213–21. (Use of ontogeny in phylogenetic affinities)

Rolfe, W. D. I. 1962. Grosser morphology of the Scottish Silurian phyllocarid crustacean *Ceratiocaris papilis* Salter in Murchison. *J. Paleont.* **36,** 912–32. (Description)

Seilacher, A. 1955. Spüren and Lebensweise der Trilobiten; Spüren und Fazies im Unterkambrium. In *Beitrage zur Kenntnis des Kambriums in der Salt Range* (Pakistan), O. H. Schindewolf and A. Seilacher (eds). *Akad. Wiss. Lit. Mainz. Math-nar. Kl. Abh.* 86–143. (First report of sideways-crawling trilobites)

Seilacher, A. 1962. Form and Funktion des Trilobiten – Daktylus. *Palaont. Zeitschr.* (Herta Schmidt Festband) 218–27. (Matching of trilobite trails with appendage-structure)

Seilacher, A. 1964. Biogenic sedimentary structures. In *Approaches to Palaeoecology*, J. Imbrie and N. Newell (eds) 296–316. New York: John Wiley. (Includes discussion of trilobite tracks)

Stockton, W. L. and R. Cowen 1976. Stereoscopic vision in one eye – palaeophysiology of the schizochial eye of trilobites. *Palaeobiology* **2,** 304–15. (New interpretation of eye function)

Størmer, L. 1930. Scandinavian Trinucleidae. *Norske. Vid. Akad. Skr. Mat.-Natur.-Kl. Skr.*, **4,** 1–111. (Detailed morphology of fringe and sensory organs)

Størmer, L. 1939. Studies on trilobite morphology Part 1. The thoracic appendages and their phylogenetic significance. *Norsk. Geol. Tidsskr.* **19,** 143–273.(*Ceraurus* appendages; a very detailed study)

Størmer, L. 1970–74. Arthropods from the Lower Devonian (Lower Emsrain) of Alken an der Mosel Germany. 1970: Part 1, Arachnida; 1972: Part 2, Xiphosura; 1973: Part 3, Eurypterida Hughmilleridae; 1974: Part 4, Eurypterida Drepanopteridae. *Senckenbergiana lethaea* **51,** 335–69; **53,** 1–29; **54,** 119–295, 359–451. (Classification of eurypterids)

Stubblefield, J. 1936. Cephalic sutures and their bearing on current classification of trilobites. *Biol. Rev.* **11,** 407–40. (Classic paper pointing out limitations of sutures as taxonomic criteria)

Stuermer, W. and J. Bergstrom 1973. New discoveries on

trilobites by X-rays. *Palaont.* **47,** 104–41. (Appendage morphology)

Teigler, D. J. and K. M. Towe 1975. Microstructure and composition of the trilobite exoskeleton. *Fossils and Strata* **4,** 137–49. (Cuticle of several species described)

Waterston, C. D. 1975. Gill structures in the Lower Devonian eurypterid *Tarsopterella scotica. Fossil and Strata* **4,** 241–54. (Detailed morphology of gills)

Westergård, A. 1922. Sveriges Olenidskiffer. *Sver. geol untersokn.* **18,** 1–205. (Classic taxonomic study of olenid-bearing shales of Scandinavia)

Whittington, H. B. 1956. Silicified Middle Ordovician trilobites; the Odontopleuridae (Trilobita). *J. Paleont.* **30,** 304–520. (Detailed morphology and functional interpretation)

Whittington, H. B. 1957. The ontogeny of trilobites. *Biol. Rev.* **32,** 421–69. (Standard reference)

Whittington, H. B. 1966. Phylogeny and distribution of Ordovician trilobites. *J. Paleont.* **40,** 696–737. (Evolutionary charts and discussion of faunal provinces)

Whittington, H. B. 1975. Trilobites with appendages from the Middle Cambrian, Burgess Shale, British Columbia. *Fossils and Strata* **4,** 97–136. (Description and functional morphology of *Olenoides* appendages)

Whittington, H. B. and C. P. Hughes 1972. Ordovician geography and faunal provinces deduced from trilobite distribution. *Proc. Roy. Soc. Lond. B.* **263,** 235–78. (With detailed maps; updates Whittington (1966))

12   Exceptional faunas

In normal circumstances only the hard-shelled animals of any living community are likely to be preserved as fossils. Thus the preserved assemblage at any horizon is no more than a narrow band of the whole biotic spectrum. But rarely, as a result of exceptional conditions, a much wider range of the living community may be preserved, including some or all of the soft-bodied or thin-shelled forms. Such 'windows', through which a larger part of the biota living at one time can be examined, alter the whole perspective of palaeontology.

A prime example of such exceptional preservation is the Ediacara fauna (Ch. 3) which, in the few localities where it has been found, gives our only insight into the nature of late Precambrian metazoan life. Three such horizons in the Phanerozoic will be considered here.

BURGESS SHALE FAUNA

In 1909 the American geologist Charles D. Walcott was engaged in a reconnaissance survey of the Cambrian geology of the Mount Field area in British Columbia, Canada. In the process he found a single dislodged slab of rock containing a remarkable assemblage of previously unknown fossils, and he went back the next year to find the horizon from which it came. When the bed in the Middle Cambrian Burgess Shale had been successfully located, he excavated and intensively quarried it, discovering not only trilobites with their appendages preserved but also a very diverse suite of other arthropods, numerous worms, a curious creature like the modern *Peripatus*, echinoderms, brachiopods, early molluscs, as well as sponges and algae. Walcott's quarrying activities between 1910 and 1917 produced several tens of thousands of specimens from what he called the 'phyllopod bed' and resulted in diverse publications by himself and other authors. In 1967–68 under the direction of H. B. Whittington, many thousands of new specimens, including parts and counterparts, were collected from Walcott's phyllopod bed and from other productive horizons which were found, and these have been intensively studied since.

All the soft-bodied and thin-shelled fossils are flattened, preserved as a dark film which wholly or partly reflects light; pyrite may be associated with the film (Whittington 1971b). The majority of specimens (e.g. the arthropod *Marrella*) lie flat upon the bedding planes, but many others lie at an oblique angle to the bedding with their spines and appendages at different levels (Fig. 12.1). In an oblique preservation different parts of the body separate on part and counterparts, so both are needed for a full interpretation of structure. Since Walcott did not make use of both faces for study, some of his interpretations have been found to be in error.

The Burgess Shale was apparently deposited in relatively deep water off a submarine limestone escarpment on or near a deep sea fan. The fauna lived on the mud surface below the bank or swam just above it, and it was overwhelmed by a turbidity current of fine suspended sediment which flowed down the slope and transported the animals to an anaerobic environment, not far away, where they were preserved. This explains the attitudes of the animals in the sediment and the completeness of most of the specimens.

Burgess Shale arthropods

Several genera, which have been redescribed recently, will be treated in more detail than others, which are only mentioned in passing. It is probable that continuing research will yield equal detail on genera that have not been redescribed since Walcott's time or whose interpretation is still controversial. The arthropods of the Burgess Shale were until recently

Figure 12.1 (a) *Marrella splendens* (M. Cam.), Burgess Shale, British Columbia. Specimen from Geol. Surv. Canada, with most of body preserved though posterior cephalic horns are damaged (x 4). (b) *Opabinia regalis* (M. Cam.), Burgess Shale. Complete specimen in lateral view (x 1 approx.). (U.S. National Museum no. 57683)

classified together as Class Trilobitoidea, since their appendages seemed to be generally biramous with leg and gill branches like those of trilobites. However, recent work has shown that this supposed resemblance is evident only in a few genera and even in these is not especially close. The possession of biramous limbs in any case may be no more than a heritage from an original arthropod ancestor with serially uniform limbs.

Marrella splendens (Figs 12.1a, 12.2b), perhaps the most elegant of the arthropod fauna, has a wedge-shaped cephalic shield with four long, backwardly directed horns, the posterior pair of which have crenulated margins. On the ventral surface is a two-spined labrum near which are attached the two pairs of antennae; the first pair are long flexible multijointed rods, the second pair six-segmented with setose distal joints. Behind the head is the cylindrical segmented body which (contrary to original belief) is devoid of pleura. Each of the twenty-five somites has a pair of biramous appendages which possess a jointed walking leg and a feathery gill branch above it. The latter was probably capable of rotation backwards

and forwards about a horizontal axis, and the combined rotary effects of all gill branches could have enabled *Marrella* to swim. Rare specimens are preserved with the intestine squashed out and isolated from the body. *Marrella* is still classified as a trilobitoid though its affinities with other arthropods are unclear (Whittington 1974). The resemblance between the Cambrian *Marrella* and the Devonian *Mimetaster* (q.v.) is close enough to point to a real affinity and indicates the survival of this stock until much later.

Opabinia regalis (Figs 12.1b, 12.2d) has an elongated segmented body, in which the head possesses five [*sic*] mushroom-like eyes (see Whittington 1975). From the front of the head extends a long flexible process terminating in two groups of spines facing each other in a pincer-like fashion. Behind the head is an elongated cylindrical body of fifteen segments with a tailpiece having upwardly turned lobes. All the segments bear a pair of appendages, each being a folded gill-blade overlying a flat paddle-like lobe. These were fixed and quite rigid, though possibly capable of movement in an up-and-down plane. *Opabinia* was probably benthonic, moving slowly over the sea floor pushing with its paddles, or perhaps feebly swimming above it. The flexible frontal process could reach round to the mouth which was located ventrally and behind the head, and it was probably used to explore for and convey food to the mouth. This animal does not seem to bear close affinity to either annelids or arthropods but is probably a descendant of the segmented ancestral stock of both.

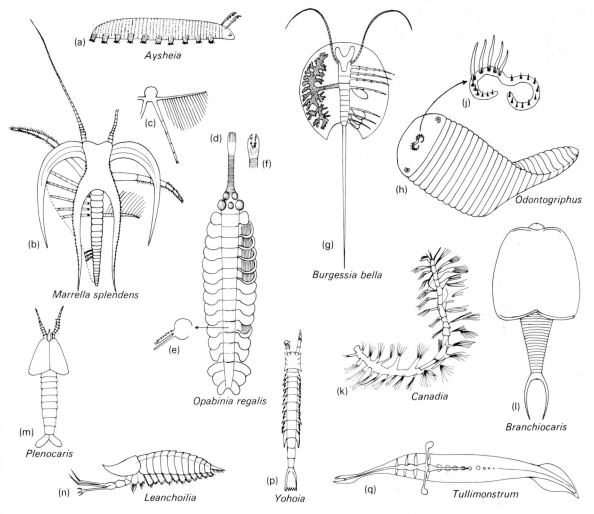

Figure 12.2 Elements of the Burgess Shale fauna (except (q)): (a) *Aysheia*, an onychophoran (x 1); (b) *Marrella splendens*, dorsal view (x 2 approx.); with (c) transverse section showing inferred disposition of appendages; (d) *Opabinia regalis* with (e) section through posterior end and (f) side view of frontal organ (x 1); (g) *Burgessia bella* (x 3); (h) *Odontogriphus*, and (j) reconstruction of its lophophore with 'teeth' and inferred tentacles (x 0·75); (k) *Canadia*, a polychaete worm (x 1); (l) *Branchiocaris* (x 1); (m) *Plenocaris* (x 1); (n) *Leanchoilia* (x 0·5); (p) *Yohoia* (x 2·5); (q) *Tullimonstrum*, from the Mazon Creek fauna (x 0·4) (Pennsylvanian). ((a) Redrawn from Walcott in '*Treatise*' Part (O); (b), (c) Whittington, 1976; (d–f) Whittington, 1975; (g) Hughes, 1975; (h)–(j) Conway Morris, 1976; (k) Walcott, in Piveteau (ed.) *Traité de Palaeontologie*, Part III; (l) Briggs, 1976; (m), (p) Whittington, 1974; (n) Walcott in '*Treatise*' Part (O); (q) Johnson and Richardson (1969)).

Burgessia bella (Fig. 12.2g), redescribed by Hughes (1975), has a large convex carapace covering the whole body except for the terminal tail spine, which emerges under a posterior indentation and is nearly twice as long as the body. Below this is the rod-like body, whose cephalic region is followed by nine segments and a telson in front of the spine. A pair of very large crenulated kidney-shaped organs occupy the lateral parts of the carapace and are interpreted as digestive diverticulae of the gut. There is a pair of long multijointed uniramous antennae projecting in front of the carapace and projecting outwards. The cephalic region bears three pairs of jointed legs, each with a coxa and six joints, identical with the seven pairs of legs on the trunk. Three other pairs of cephalic appendages have the form of whip-like flagellae. The walking legs of the trunk bear small lateral leaf-like plates which were probably gills, seemingly attached to the coxa.

Yohoia (Fig. 12.2p) is a rare form in which the cephalic shield is subrectangular and the cylindrical trunk has fourteen segments; the first ten segments have lateral pleura and the remainder taper to a flat telson. The head bears a pair of stout jointed appendages with terminal spines, behind which are three pairs of walking legs, whilst the trunk has lobate setally fringed appendages quite unlike those of trilobites.

Plenocaris (Fig. 12.2m) may be a primitive crustacean (Whittington 1974). Its bivalved carapace covers the anterior part of the body, which bears one pair of antennae and probably three pairs of indeterminate appendages. Behind this the twelve-segmented body terminates in a telson and a pair of uropods, forming a tail fan. Though little is known of the limb morphology of *Plenocaris*, in other respects its construction approximates that of phyllocarids, the oldest known group of higher crustaceans, and it is referred tentatively to this group.

Branchiocaris (Fig. 12.2l) (Briggs 1976) also has a large bivalved carapace, but the trunk it covers has as many as forty-six somites. The head end carried a stumpy pair of antennae, followed by a pair of stout appendages which appear to have been pincer-like. The serially repeated appendages of the trunk segments are flat and lamellar and go all the way down to the telson, which terminates in a pair of spines. Whilst many features resemble those of recent notostracan brachiopods it is not believed that *Branchiocaris* was actually ancestral to this group, and it cannot readily be placed in an extant taxon.

One single immense limb, spiney and segmented, could belong to a great millipede-like creature which, judging by the size of the appendage, must have been in excess of 2 m long.

Various other arthropods are known only from older descriptions or from controversial works. *Waptia* is another shrimp-like form with a carapace and long antennae. *Leanchoilia* (Fig. 12.2n) has a long trilobed body, with a well-defined cephalic part and ten thoracic segments with pleura. The head appendages are long and powerful, either grasping or tactile, whilst those of the trunk are gill-like. *Emeraldella* has a likewise trilobed body and possibly trilobite-like appendages. *Naraoia* has a large head-shield and a somewhat larger tail, round which a fan of appendages is clearly visible. *Sidneyia* and *Helmetia*, relatively large for the fauna, seem to have merostome affinities. Other arthropods are also present in the Burgess Shale fauna, which, like those discussed above, show remarkable diversity but also exhibit puzzling combinations of characters which make it very difficult to understand their evolutionary relationships. But further studies of these remarkable creatures may give more information on the nature of the early arthropod stock and hopefully lead to a tenable classification.

Other Burgess Shale invertebrates

Many remarkable species of annelids have been described from the Burgess Shale, e.g. the polychaete worm *Canadia* (Fig. 12.2k). A single specimen of a hitherto unknown animal was found in association with another polychaete, *Eldonia*. This organism, *Odontogriphus* (Fig. 12.2h, j), is about 6 cm long, and its body is flat and annulated with a poorly defined head of semicircular form. On the head are a pair of lateral 'palps' (sensory organs) of rather indistinct morphology and also a bilaterally symmetrical median structure forming a pair of loops. This apparatus lies at the front end of tubular gut and bears some twenty-five thorn-like 'teeth'. These were probably not biting or rasping teeth; they have been interpreted as the supports for a food-gathering apparatus having the form of a 'tentacular lophophore' (Conway-Morris 1976). Such tentacular lophophores are found in modern brachiopods, tube-dwelling phoronid 'worms' and bryozoans, which are all commonly linked together in the Superphylum Lophophorata. In these the lophophore, at least in its initial stages of development, is of remarkably constant form and is, incidentally, bilaterally looped like that of *Odontogriphus*. There seems to be a good case for aligning *Odontogriphus* with the Lophophorata as an early derivative of the same superphylum. But does *Odontogriphus* have any other relatives in the fossil record, or living today? Here the teeth occurring with the lophophore could prove to be enormously significant, for they are very similar to conodonts of the same age, which are probably the supports of some kind of food-gathering mechanism in the extinct animals that bore them.

If *Odontogriphus* is really a conodont-bearing animal, it is the only known conodontophorid in which the conodonts, supporting the lophophore, are indisputably preserved in situ. Nevertheless the chemistry of the teeth of *Odontogriphus* is unknown, for in the sole specimen they have been leached away and so a vital bit of evidence is lacking.

One of the invertebrates originally described by Walcott is *Aysheia* (Fig. 12.2a): an annulated caterpillar-shaped creature with serially paired but unjointed legs. It is apparently similar to the modern *Peripatus* (Class Onychophora): a terrestrial uniramian which lives in damp soil in the jungles of Minas Gerais and elsewhere in Brazil. The characters of *Peripatus* are in many ways intermediate between those of annelids and arthropods, preserving, for example, the paired segmental excretory organs of the segmented worms, together with an annelid-like eye and a muscular body wall covered by a thin cuticle. Yet the coelom is arthropodan, and the presence of tracheae – ramifying capillaries connecting to the outside and bringing in air to the tissue like those of insects – links *Peripatus* with arthropods. The legs of *Peripatus* move in metachronal rhythm but are unjointed. Their rigidity depends upon the antagonistic operation of muscles against coelomic fluid, so that they can lengthen and be rigid whilst pressing backwards, thereafter shortening before the next forward stroke. *Peripatus* is highly adapted for terrestrial life and well defended, being capable of entangling predators with a sticky secretion from glands below the eye. Nevertheless, though *Aysheia* was evidently marine, it is also classified as an onychophoran and is the earliest uniramian so far known.

Only some of the Burgess Shale animals have been described here, but they are enough to show the great diversity that soft-bodied and thin-shelled animals had attained so early in the Phanerozoic.

HÜNSRÜCKSCHIEFER FAUNA

The famous localities in the Lower and Middle Devonian of Bundenbach, Wissenbach and Gemunden in the Hünsrück Shales of the German Rhineland yield exquisitely preserved starfish, trilobites, other arthropods and also cephalopods, which are largely preserved in pyrite. The appli-

cation of soft X-radiography to these fossils, initiated in the 1930s and recently resumed by Stuermer (1970) and Stuermer and Bergstrom (1973, 1976) has provided a rather surprising fund of information, not only on the hard parts but also in many cases on the soft tissue of the animals. It is clear from the X-ray photographs that some of the straight-shelled cephalopods were ectocochlear, with the shell outside the body, whilst others were endocochlear with the outside of the shell invested with soft tissue. Tentacles and other external organs appear in some radiographs, both of the orthocone *Lobobactrites* and in *Goniatites*, and embryonic cephalopods have also appeared unexpectedly on the photographic plates.

In the Gemunden fauna specimens of the enigmatic conical-shelled tentaculitids are abundant. On X-ray plates their tentacles show up, as does the gut, which tends to confirm the view of some workers that

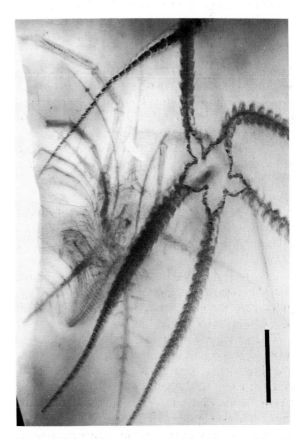

Figure 12.3 *Mimetaster*, associated with an ophiuroid (Dev.), Hünsrückschiefer, Germany. Length of bar is 10 mm. (X-radiograph by courtesy of Prof. W. Stuermer)

these septate shells with their long body chambers are actually the remains of cephalopods.

Phacops, *Asteropyge* and other trilobite genera have revealed details of limb morphology (q.v.). The *Marrella*-like arthropod *Mimetaster*, which in about half the known specimens is found associated with an ophiuroid, was known in gross morphology prior to Stuermer and Bergstrom's work, as was *Vasconisia*: a large arthropod with a bivalved carapace. *Mimetaster*, through the X-ray photographs (Fig. 12.3), is now known to have paired stalked eyes, stout jointed appendages on the head with pyritised strands of muscle, and about thirty pairs of bifid appendages down the body. *Vasconisia* has turned out to have a surprisingly similar body under the large carapace and is probably related. These two genera could be survivors of the *Marrella* stock, persisting to the Devonian and known only from these exceptional faunas.

MAZON CREEK FAUNA

The Pennsylvanian beds of Illinois have long been strip-mined for coal. Over wide areas the nodular shale overlying the coal has been removed to spoil tips, from which the weathered ironstone nodules have proved the source of an outstandingly diverse fauna. These Mazon Creek beds were all deltaic, and the fauna has two components: a terrestrial (Braidwood) assemblage and a marginal marine (Essex) assemblage deposited along the delta front. Both faunal assemblages occur in several bands of ironstone nodules throughout the Mazon Creek beds.

The Braidwood assemblage consists mainly of plants and non-marine arthropods and bivalves and is thus typical of a standard coal-swamp assemblage. The Essex fauna on the other hand is a mixed assemblage with both terrestrial and marine elements. Insects and millipedes represent the terrestrial elements, whilst the marine fauna includes holothurians, annelids, malacostracan crustaceans, other arthropods of unknown affinities, organic-plated barnacles, jellyfish and other molluscs (such as cephalopods with arms and hooks attached), as well as *Tullimonstrum* which is an animal of unknown affinities. Fragments of small and large fish are found, also coprolites and even young lampreys still with their yolk sac in place. Such a mixed fauna probably resulted from a storm surge which moved the marine faunas forwards over the delta, stranding the animals on its surface. Increased rain during a storm is thought to have swept down masses of sediment to bury the stranded animals almost immediately. Concretions formed round the specimens very rapidly before anaerobic decay set in, though the conditions of formation of such concretions is unknown.

Specimens of *Tullimonstrum* (Fig. 12.2q) ('Tully monsters', named after their discoverer) are unique to the Essex fauna and have no living relatives. *Tullimonstrum* was a soft-bodied, bilaterally symmetrical animal some 8 cm long. Its head tapers to an elongated proboscis with a pair of pincer-like jaws at the end, their inner face armed with minute stylets. Where the head grades into the trunk a pair of transverse bars project laterally from the body, each terminating in a hollow globular bar-organ; these have been variously interpreted as eyes or stabilisers. The trunk is segmented, often showing a median impression, probably the gut. At the rear a spatulate tail is marked with paired fins.

Tullimonstrum was probably a pelagic carnivore, catching its prey with the proboscis. Since the transverse bar and the proboscis are unique it is hard to assign it to any known higher taxon, and it is probably a member of an entirely extinct phylum. It was Johnson and Richardson (1969) who first described it and pointed out its value as 'a reminder that our conception of the diversity of the organic world is based on a small sample consisting almost entirely of animals with preservable hard parts'. The same applies to all of the three unique faunas here reviewed.

BIBLIOGRAPHY

Books, treatises, symposia

Moore, R. C. (ed.) 1959. '*Treatise*' Part (O). Trilobitomorpha (Burgess Shale Fauna).

Individual references

The following are mainly recent descriptions of the Burgess Shale fauna.

Briggs, D. E. G. 1976. The arthropod *Branchiocaris* n. gen., Middle Cambrian, Burgess Shale, British Columbia. *Geol. Surv. Canada Bull.* **264**, 1–29.

Briggs, D. E. G. 1977. Bivalved arthropods from the Cambrian Burgess Shale of British Columbia. *Palaeontology* **20**, 595–622.

Conway-Morris, S. 1976. A new Cambrian lophophorate from the Burgess Shale of British Columbia. *Palaeontology* **19**, 199–222.

Conway-Morris, S. 1977. A new metazoan from the Cambrian Burgess Shale of British Columbia. *Palaeontology* **20**, 623–40.

Conway-Morris, S. 1977. A new entoproct-like organism from the Burgess Shale of British Columbia. *Palaeontology* **20**, 833–45.

Conway-Morris, S. 1977. *Fossil priapulid worms*. Spec. pap. in palaeont., no. 20, 1–155.

Hughes, C. P. 1975. Redescription of *Burgessia bella* from the Burgess Shale, British Columbia. *Fossils and Strata* **4**, 415–36.

Johnson, R. G. and E. S. Richardson 1969. Pennsylvanian invertebrates of the Mazon Creek area, Illinois: the morphology and affinities of *Tullimonstrum*. *Fieldiana. Geol.* **12**, 119–49. (Mazon Creek Fauna)

Stuermer, W. 1970. Soft parts of cephalopods and trilobites: some surprising results of X-ray examination of Devonian slates. *Science* **1170**, 1300–2. (First work by Stuermer on X-ray applications; illustrated)

Stuermer, W. and J. Bergstrom 1973. New discoveries on trilobites by X-rays. *Paläont. Z.*, **47**, 104–41.

Stuermer, W. and J. Bergstrom 1976. The arthropods *Mimetaster* and *Vachonisia* from the Devonian Hünsrück Shale. *Paläont. Z.* **50**, 78–111.

Whittington, H. B. 1971a. The Burgess Shale: history of research and preservation of fossils. *Proc. North. Am. Paleont. Convention, Chicago 1969*, (I), 1170–1201.

Whittington, H. B. 1971b. Redescription of *Marrella splendens* (Trilobitoidea) from the Burgess Shale, Middle Cambrian, British Columbia. *Geol. Surv. Canada Bull.* **209**, 1–24.

Whittington, H. B. 1974. *Yohoia* Walcott and *Plenocaris* n. gen. arthropods from the Burgess Shale, Middle Cambrian, British Columbia. *Geol. Surv. Canada Bull.* **231**, 1–24.

Whittington, H. B. 1975. The enigmatic animal *Opabinia regalis*, Middle Cambrian, Burgess Shale, British Columbia. *Phil Trans Roy. Soc. Lond.* B **271**, 1–43.

Whittington, H. B. 1977. The Middle Cambrian trilobite **Naraoia**, Burgess Shale, British Columbia. *Phil Trans Roy. Soc. Lond.* B **280**, 409–43.

Systematic Index

General Index